KATYDIDS AND BUSH-CRICKETS

A VOLUME IN THE CORNELL SERIES IN ARTHROPOD BIOLOGY

EDITED BY *John Alcock*

The Tent Caterpillars
by Terrence D. Fitzgerald

Army Ants: The Biology of Social Predation
by William H. Gotwald Jr.

Solitary Wasps: Behavior and Natural History
by Kevin M. O'Neill

The Wild Silk Moths of North America: A Natural History of the Saturniidae of the United States and Canada
by Paul M. Tuskes, James P. Tuttle, and Michael M. Collins

KATYDIDS AND BUSH-CRICKETS

REPRODUCTIVE BEHAVIOR AND EVOLUTION OF THE TETTIGONIIDAE

Darryl T. Gwynne

University of Toronto

Comstock Publishing Associates A DIVISION OF

Cornell University Press | ITHACA AND LONDON

Copyright © 2001 by Cornell University

All rights reserved. Except for brief quotations in a review, this book, or parts thereof, must not be reproduced in any form without permission in writing from the publisher. For information, address Cornell University Press, Sage House, 512 East State Street, Ithaca, New York 14850.

First published 2001 by Cornell University Press

Printed in the United States of America

Library of Congress Cataloging-in-Publication Data
Gwynne, Darryl T.
  Katydids and bush-crickets: reproductive behavior and evolution of the Tettigoniidae/Darryl T. Gwynne.
    p. cm.—(Cornell series in arthropod biology)
  Includes bibliographical references and index.
  ISBN 0-8014-3655-9 (cloth: alk. paper)
  1. Tettigoniidae—Behavior. 2. Tettigoniidae—Reproduction. 3. Sexual selection in animals. I. Title. II. Series.
  QL508.T4 G89 2001
  595.7'26—dc21   00-012065

Cornell University Press strives to use environmentally responsible suppliers and materials to the fullest extent possible in the publishing of its books. Such materials include vegetable-based, low-VOC inks and acid-free papers that are recycled, totally chlorine-free, or partly composed of nonwood fibers. Books that bear the logo of the FSC (Forest Stewardship Council) use paper taken from forests that have been inspected and certified as meeting the highest standards for environmental and social responsibility. For further information, visit our website at www.cornellpress.cornell.edu.

Cloth printing   10 9 8 7 6 5 4 3 2 1

To four of my teachers: Jim Potter, Glenn K. Morris, Howard E. Evans, and Randy Thornhill

Contents

Preface	ix
1. Mormon Crickets and Mating Meals	1
1.1. Insects on the March 1	
1.2. Variable Mating Behavior 6	
1.3. Questions and Topics 14	
2. Katydids, Longhorns, and Bush-Crickets: Diversity and Evolution of the Tettigoniidae	18
2.1. Katy and Katydid 18	
2.2. Relationships with Other Orthoptera 19	
2.3. Diversity 28	
2.4. Fossil History and Diversification 43	
3. What Katy Did: Habits and Life Cycles of Tettigoniids	49
3.1. Oviposition 49	
3.2. Eggs and Development 53	
3.3. Growth 57	
3.4. Food Habits 59	
3.5. Economic Importance and Gregarious Behavior 63	
4. Survival Strategies: Natural Enemies, Counteradaptations, and Population Regulation	68
4.1. Natural Enemies 68	
4.2. Population Regulation 78	
4.3. Defenses 79	
5. Entomological Choristers: Song and Mate Attraction	89
5.1. The Choristers of Summer 89	
5.2. Producing and Hearing Songs 90	

5.3. Evolution and Adaptive Significance of Katydid Song   114
5.4. Nuptials and an "Extraordinary Wallet"   119

6. A Nuptial Banquet and Seminal Sac: Evolution of the Spermatophylax Meal   122
6.1. A Mating Mystery   122
6.2. The Courtship Gift Menu   130
6.3. Why Katydids Feed Their Mates   133
6.4. Discussion and Conclusions   158

7. The Nature of Sexual Selection: Females Choosing and Males Competing   161
7.1. The Struggle for Reproduction in a Michigan Marsh   161
7.2. Mate Choice   166
7.3. Competition for Mates   191

8. The Hazards and Costs of the Mating Game   199
8.1. Eavesdropping Flies   199
8.2. Eavesdropping Bats   207
8.3. Factors That Affect Mating Risk   210
8.4. The Survivorship of the Sexes   217
8.5. Conclusions and Future Research   219

9. Can Katydids Tell Us Why the Sexes Are Different?   222
9.1. Role Reversals in Courtship   222
9.2. Variable Courtship Behavior in Flower Lovers   226
9.3. Courtship Role Reversal in Other Katydids   232
9.4. What Cues Do Katydids Use to Alter Their Behavior?   234
9.5. A General Test of Sexual Differences Theory   235
9.6. Reversed Sexual Selection and Darwinian Devices in Females   251

Appendix   261

References Cited   270

Index   309

Preface

A current roster of insect groups (in CSIRO 1991) includes over 800 families in 28 different orders. This book is about the Tettigoniidae, one of the 28 families listed in the order Orthoptera. Tettigoniids do not usually spring to mind when most of us think of insects. We are much more likely to consider everyday animals such as house flies (order Diptera: family Muscidae), hornets (Hymenoptera: Vespidae), or honey bees (Hymenoptera: Apidae). And among the Orthoptera, grasshoppers (Acrididae) in the meadow or crickets "on the hearth" (Gryllidae) are more familiar to most of us than are tettigoniids. This lack of familiarity may be why the Tettigoniidae has yet to receive a universal common name. In a book on British Orthoptera (1936), Burr[1] suggested the common name "bush-cricket," which remains in use in England and Europe.[2] In North America, Australia and New Zealand tettigoniids have been called "katydids," at least since C. V. Riley's (1874) report on the family, and the name probably hails back to John Bartram's (1751) reference to singing "catedidists" in his *Travels in Pensilvania and Canada* (see *The Oxford English Dictionary*). In other parts of the world, there are a number of different common names for different groups within the family, including "long-horned grasshopper."

So why do we need a book about Tettigoniidae? One reason is simply that given their diversity and importance, katydids deserve to be better known; there are over 6,000 described species (Naskrecki and Otte 1999),

---

[1] Burr did not use "bush-cricket" in his earlier (1897) book on British Orthoptera.
[2] Common names for tettigoniids in other European languages are *sauterelles* (France), *esperanças* (Portugal), *grillos* (Spain), and *Laubheuschrecken* (Germany) (Nickle and Naskrecki 1997).

a diversity that is second only to the grasshoppers (almost 12,000 species) and more than double that of crickets (some 3,000 species). Moreover, species of katydids are dispersed across all continents (except Antarctica) and can be very common in certain field, forest, and bushland communities. Despite their abundance in these habitats, most katydid species are not often seen, as they hide in vegetation during the day. But they are frequently heard (Swan and Papp 1972). The raspy mating songs of males dominate the nocturnal summertime choruses of animal sounds in locations as disparate as Europe and Australia. In fact, this acoustic harbinger of midsummer has been immortalized in poem and prose.

Although the tettigoniids are unfamiliar to many biologists, they have been very well studied by some. Biologists interested in katydids have always included the systematists who lay important groundwork in species descriptions and taxonomic revisions. But more recently these insects have become important subjects for animal behaviorists in two separate but connected research areas. First, along with their fellow orthopterans, the crickets and grasshoppers, katydids have been examined for their sexual communication systems, the calls of males and the attraction of sexually responsive females. The second area includes my own main research interest: sexual selection. Katydids have turned out to be an ideal system to understand sexual selection (Whitten 1991), a process first suggested by Darwin (1859, 1874) to explain traits he saw as threatening to survival. Darwin's many illustrated examples of these traits included not only peacock's tails but also the singing apparatus of male katydids. My interest in katydids and sexual selection began after reading Randy Thornhill's (1976b) paper in which the observations of naturalists such as Rene-Guy Busnel (1956) supported the inclusion of katydids on the list of animals in which males feed females. It was common knowledge then that male insects such as dance flies (Diptera: Empididae) and scorpionflies (Mecoptera) feed prey items to mates, but not so well known was that katydid males also provided meals—nuptial offerings that are forged from the secretions of their abdominal glands. Thornhill's paper pointed out how nurturant contributions by male insects could have important consequences for male reproductive strategies and the degree of choosiness shown by the sexes when mating (see also Thornhill 1986).

I began to examine "courtship feeding" in katydids in a collaboration with Randy Thornhill in 1980 (Thornhill and Gwynne 1986). This, my subsequent work, and the studies of other researchers form the main focus of this book, a focus that I hope will point to productive directions for future research. I begin in Chapter 1 by describing my first observations of mate-feeding and mating behavior in the Mormon "cricket," one of the best-known katydids because unlike most other species, it can form dense bands that devastate crop fields. As with most initial studies, this work raised many more questions than it answered. I return to these questions in Chapters 5 to 9. The last part of Chapter 1 provides a detailed breakdown of the questions and topics that are covered in subsequent chapters.

This book is a natural history but a natural history in a modern scientific sense, that is, where observations of animals in nature lead to hypotheses whose expectations and predictions are tested with further comparisons and observations or with some sort of manipulative experiments. The hypotheses are derived from "individual selectionist thinking" (Darwin 1859; Williams 1966) applied to insect mating systems (Alcock and Gwynne 1991). This approach is particularly evident with the studies I cover in Chapters 4 to 7, in which observations on the reproductive behavior of katydids led to much insight into the control and consequences of Darwinian sexual selection. I hope that anyone interested in sexual selection and the evolution of sexual differences, as well as naturalists seeking an introduction to a fascinating group of animals, will find my book useful.

Many people have provided critical comments and other assistance to me during the course of this project. The series editor, the late George Eickwort, read and provided valuable comments on early versions of Chapters 1 to 4, as did Winston J. Bailey, Françoise Berlandier, and Ian Dadour. Several people endured the whole manuscript. These include Luc Bussière, Howard E. Evans, Dan Otte and Glenn K. Morris. Special thanks to the editor John Alcock, for wading through several drafts and providing important comments within a general lesson on how to write a book! For comments on individual chapters, I thank Mormon cricket expert Chuck MacVean (Chapter 1); Sigfrid Ingrisch, who tackled Chapters 2 and 3 and assisted me greatly with an understanding of diapause and with his knowledge of European and Asian tettigoniids; James Fullard (Chapter 5); Karim Vahed (Chapter 6); Marianne Feaver, Carl Gerhardt, and Leo Shapiro (Chapter 7); Geoff Allen and Klaus Gerhart Heller (Chapter 8); and Patrick Lorch and Leigh Simmons (Chapter 9). For comments or discussions on various aspects of Orthoptera biology, I thank pioneer Mormon cricket researcher, the late Frank Cowan (for his theory of "phase" polymorphism in Mormon crickets), Win Bailey, Glenn Morris, Fernando Montealegre, Dan Otte, and David Rentz. I thank Irina Raskolnikova, for translations of parts of Boldyrev's (1915) Russian and Gottskalk Jensson for the translation of the Latin of Brunellius (1791). Julie Clark, David Nykamp, Angela Lange, and Dianne Scott provided lessons in using the software necessary to produce the line art for this book. Alison Dias printed photographs, provided a lot of advice concerning the artwork, and helped clean up and redraw some of my artwork. Cynthia Thomas also helped with some of this work. For allowing me to reproduce their photographs or drawings, I am indebted to Howard E. Evans, Brigitte Helfert, Reinhard Lakes-Harlan, David Funk, Klaus-Gerhart Heller, Dan Otte, Bob Pemberton, Heather Proctor, David Rentz, Heiner Römer, Karl Saenger, and Delia Scott. I am especially indebted to Glenn Morris, Dita Klimas, Karim Vahed, Fernando Vargas, and Fernando Montealegre for allowing me to pick freely from their slide collections of katydids. Glenn Morris also provided some of the original line art.

Finally, to my family. Thanks to my wife Sarah for reading the manuscript and especially to her and our children, Rhiannon, Rhys, and Willy, for putting up with the long hours of my distractions with pen, pencil, and computer keyboard. And to my parents, Trevor and Joan, for their encouragement of my work over many years; and for good company at our Bob's Lake cottage, where many of the later chapters were written.

KATYDIDS AND BUSH-CRICKETS

1

Mormon Crickets and Mating Meals

> ... in the latter part of May, when the fields had put on their brightest green, there appeared a visitation in the form of vast swarms of crickets, black and baleful as the locust of the Dead Sea. In their track they left nothing behind them, not a blade or leaf, the appearance of the country which they traversed in countless and desolating myriads being that of lands scorched by fire.
>
> H. H. Bancroft and A. Bates (1889), *History of Utah*

1.1. Insects on the March

Each spring on high mesas in the western United States, small groups of cricket-like insects, newly hatched from the soil, gather in sheltered locations. The nymphs thrive and grow, even under the cool conditions of the Great Basin just west of the Rockies. After the first two growth stages, or *instars*, small gatherings coalesce into much larger groups called *bands* and begin daily marches over longer distances (MacVean 1987; Anonymous 1991). Marching occurs only during warm daylight hours and continues throughout the life cycle into the flightless adult stage. The mass movements come to a standstill only during rain, extreme heat, and the cooler hours after dark when band members cluster in roosts beneath vegetation. With the return of warmer conditions they descend to bask in sun patches in preparation for the next march.

These migrations were first described by explorer and U.S. cavalry officer Captain John Feilner (1864). He reported that the insects "start in the morning, and, from their precision of movement, they appear like a vast army on parade. The course once marked out, they never deviate from it on account of any obstacle, but move straight forward over houses and all else" (Fig. 1.1, Plate 1). These highly directed movements have been confirmed in subsequent reports. There are even observations of different groups, when first meeting, at first merging and then literally flowing through each other as they continue on their individual headings (MacVean 1987). The great hordes of dark-colored adults present an awesome sight; dozens of the hefty individuals per square meter have been reported to swarm in bands up to 16 km long and 2 km wide (MacVean 1987). Such a mass of insects can become a menace to traffic, owing to the oily residues from their crushed bodies on roads (Fig. 1.2). Surviving crickets also can be a real threat to rangeland crops when they wander

2 Mormon Crickets and Mating Meals

Figure 1.1. Mormon crickets migrating across a sand dune near St. Anthony, Idaho.

Figure 1.2. Obvious hazards from a Mormon cricket crossing (a road sign from the 1930s in Nevada). From Schweis et al. (1939).

into fields in their endless search for food (Cowan and McCampbell 1929; Wakeland 1959; MacVean 1987). An early case of spectacular crop damage gave the insects their common name, the Mormon cricket (*Anabrus simplex*). In 1848 bands descended on wheat fields planted by the first Mormon settlers in the Great Salt Lake area, as H. H. Bancroft and A. Bates (1889) vividly described in the passage that opens this chapter.

Since this first record of a cricket "visitation," outbreaks of Mormon

Figure 1.3. Control of Mormon crickets in the 1930s. A worker involved in the killing of these flightless katydids poses next to a mound of his victims caught in a trap after the insects were corralled by metal barriers. From Schweis et al. (1939).

crickets have continued to occur, prompting newspaper headlines such as "War on Six Legs" (Anonymous 1990) and "Crickets Prepare to Darken the Land" (Coates 1990). This species affected widespread areas in the 1930s, stimulating drastic control measures that included the use of fire, loud noises (to drive the insects away), toxic arsenic-based pesticides[1] that could cause painful burns to the sprayers (Cowan 1990), and elaborate systems of sheet metal fences. These barriers were used to protect gardens (Feilner 1864) or to corral crickets into oil- and water-filled ditches where they could be drowned or burned (Cowan and McCampbell 1929) (Fig. 1.3). These past control campaigns could well have had costs that exceeded the value of the crops saved, but as MacVean (1987) pointed out, swarming crickets have impacts on humans that cannot be underestimated, including the "psychological [one] of hordes of large black crickets not

---

[1] In their precautions for using these pesticides to control Mormon crickets, Mills and Hitchcock (1938) offered advice that is both practical ("learn to take advantage of the wind") and mysterious ("underwear should be worn at all times while workers are handling arsenite dust").

only traversing the range but also invading houses, barns . . . , destroying vegetable gardens and contaminating well water with thousands of dead bodies. . . ." Cowan (1929) recalled an invasion of a dance pavilion in a Wyoming town in which "crushed bodies [of Mormon crickets] soon made the floor so slippery and nauseating that the dancers had to stop and leave the place." His later report on Mormon crickets in Colorado (Cowan 1932) included a poignant photograph of a schoolhouse that was abandoned halfway through construction as a result of residents being "driven from the country" by Mormon cricket invasions. This event occurred during the longest Mormon cricket infestation on record (1918–1938), which resulted in the number of farms in northwestern Colorado being reduced by about 50 percent (Cowan and McCampbell 1929; Wakeland 1959; MacVean 1987).

Although serious damage can occur when a phalanx of marching crickets engulfs a field of alfalfa or other crops, recent research showed that they are not the "predators" of cattle forage that they were once thought to be (Raffelson 1989); grasses are quite low on their preferred-food list. For the vegetation in their diets Mormon crickets actually favor broad-leafed succulent plants (MacVean 1987, 1989, 1990; Redak et al. 1992). For their diet as a whole, like most katydids, Mormon crickets tend to seek more nutritious items, in part because of the importance of food to reproduction, as we will see later. Especially favored are seeds, carrion, and occasionally even live prey (Criddle 1926; Cowan 1929; MacVean 1987). In fact, band members are not beyond an opportunistic snack on a sick or injured conspecific. This is one of the reasons why so many band members congregate at cricket "road kill" sites. Cannibalism in the closely related coulee cricket (*Peranabrus scabricollis*) was described graphically by well-known insect anatomist R. E. Snodgrass (1905). He reported seeing "many gruesome sights" among the crickets, including a female that "pulled off a leg [of a living band-mate] and sucked out the contents through the open end with great relish"!

As might be expected, the large moving mass of animal matter in a band attracts predators and other natural enemies. A description of a particularly effective predator, one revered in Utah where a monument to it can be seen (Fig. 1.4), continued Bancroft and Bates's (1889) account of the Mormon crop invasion: " . . . behold, from over the lake appeared myriads of snow-white gulls. . . . Settling upon all the fields and every part of them, they pounced upon the crickets, seizing and swallowing them. They gorged themselves. Even after their stomachs were filled they still devoured them. On Sunday, the people, full of thankfulness, left the fields to the birds, and on the morrow found on the edges of the ditches great piles of dead crickets that had been swallowed and thrown up by the greedy gulls. Verily, the Lord had not forgotten to be gracious!"

Dozens of lesser known natural enemies of Mormon crickets have also

Figure 1.4. Monument outside of a church in Salt Lake City, Utah, dedicated to a predator of katydids. California gulls were reported to have saved the crops of early Mormon settlers by consuming large numbers of Mormon crickets. Photo by H. E. Evans.

been recorded. The list includes pathogens such as a new species of microsporidian protozoan (*Variomorpha cowani*) (Lange et al. 1995) that shows considerable promise as a control measure (MacVean and Capinera 1991, 1992). More obvious enemies include digger wasps (*Palmodes laeviventris*, Sphecidae) that are often seen flying around the band. A female digger wasp stings its prey into paralysis and seals it in a burrow to provide fresh meat for a developing larva (Evans 1966). La Rivers (1945) used quadrat samples of wasp nests to estimate that some 30,000 wasps had dispatched about half a million Mormon crickets! Vertebrate predators include lizards, rodents, badgers, over two-dozen bird species (Cowan 1929; Wakeland 1959; MacVean 1987), and fishes such as salmonids and even endangered native minnows (Cyprinidae) that eat migrating crickets rafting across rivers (Tyus and Minckley 1988).

The list of natural enemies also includes *Homo sapiens*. Evidence of a long history of humans snacking on Mormon crickets comes from cave deposits in Wyoming over 4,000 years old that contain cooked remains of the insects (Anonymous 1991). Meal preparation by aboriginal North Americans began with the herding of Mormon crickets into sagebrush and greasewood corrals that were set alight. The roasted captives were "stripped of their legs and heads" before consumption (Lorenzo Young 1847, cited in Bancroft and Bates 1889; Riley et al. 1880). Raw or cooked crickets also were made into a soup, or dried and ground with grass seeds to make a paste that was baked into "cricket cakes" (Wakeland 1959; Evans 1985; MacVean 1987). These orthopteran meals were also noted by John Wesley Powell among the Ute and Southern Paiute while he explored the southwestern United States in the 1870s. Prepared crickets have a substantial fat content and all the amino acids required for human diets (Madsen 1989). As a culinary experience I prefer the deep-fried over the uncooked version. When prepared this way, Mormon crickets are surprisingly tasty!

Mormon crickets can benefit humans in other ways, as I discovered in the summer of 1980 when I encountered scientists from the University of Wisconsin collecting and freezing sack loads of crickets from bands in western Colorado. Subsequent research with the frozen harvest led these researchers to estimate that Mormon crickets, at average adult densities in a 1-km^2 band, would yield 11 to 22 tons of harvestable protein, a nutritious dietary supplement in the production of broiler chickens (DeFoliart et al. 1982).

1.2. Variable Mating Behavior

My own interest in Mormon crickets centers on their sexual habits. The mating behavior, although not as well known as the infamous feeding behavior of this pioneer pest, is nevertheless every bit as fascinating. My work began in July 1980 when I headed off on a field trip to observe migrating Mormon crickets. My research was triggered by the published reports of Captain John Feilner in 1864, and also of Colorado state entomologist C. P. Gillette (1904), on a "peculiarity" in Mormon cricket mating that appeared to me could be useful in understanding sexual differences in animal courtship.

My expedition took me to northwestern Colorado where Gillette had studied the katydids over 70 years previously. My primary site was a sagebrush flat near the town of Greystone where local residents had seen a large number of Mormon crickets. The exact location of the site was easy to spot because of the scattered cricket corpses from the previous day's road kill. The morning air was still cool but the sky clear as I arrived to find most of the insects clustered like bunches of brown grapes in the branches of sagebrush and some small lodgepole pines. Some individuals had already descended from these overnight roosts as they readied for the day's march. Others were basking in sun patches to elevate their body temperatures to the sweltering preferred temperatures of 35°C and above (Turnbow and O'Neill manuscript) (Fig. 1.5). Occasionally a distant burst of male song broke the morning's silence, an indication that reproductive activities were already underway. Research with close relatives of the Mormon cricket confirmed that the "shrill, whistling sound" of the male does indeed "call his mate," as Feilner (1864) had surmised.

A mature male singing insect will call whenever conditions are appropriate, although he usually stops advertising once he makes tactile contact with a female. But the sagebrush around me was not filled with the calling cacophony I expected from the thousands, if not millions, of adult males within earshot, each equipped with fully functional stridulatory equipment. Even when a male did sing, he did so for only a minute or so. When I focused my observations on these occasional singers, I discovered that the brevity of male signaling was not a result of any waning sexual interest in males but instead was due to the sexual eagerness of females who accosted the males very quickly. In fact, these ardent females frustrated my first few attempts to watch singing males because the females located each male before I could! Finally, my persistence paid off, and subsequent observations of calling and mating finally revealed the behavior of Mormon crickets to be quite different from that reported for most other species, not just other insects, but for animals in general (Gwynne 1981, 1983a, 1984c; Gwynne and Dodson 1983).

First, unlike most male animals, which tend to engage in perilous reproductive activities (Burk 1982), it was the female Mormon crickets that appeared to take the risks while hopping across open patches of ground toward the callers. We found that digger wasps (*P. laeviventris*), a species that hunts at the same time the Mormon crickets mate, preyed on females more than they did males (Gwynne and Dodson 1983). Second, because males were in greater demand as mates than females, behavior during courtship and mating was quite the reverse of what is typical for most animals (Darwin 1874; Trivers 1972); females, not males, competed for

Figure 1.5. Mormon crickets basking in the early morning sun.

mates while males proved to be coy and choosy about their mating partners. I often saw more than one female moving toward a caller. If females made contact en route, they grappled and fought in disputes more typical of male-male interactions in other species. Eventually one of the females subdued her rivals and mounted the male by crawling onto his back (Fig. 1.6). She quickly coupled with him by lowering her ovipositor and allowing him to grasp her with his abdominal claspers (cerci: Fig. 2.11). These

Figures 1.6 to 1.9. Mating in Mormon crickets.
Figure 1.6. A female mounts a male. It is during this part of mating that males appear to weigh their partner.

winners of fights did not always secure a mating, however. In more than half of the pairings, copulation ended with the male promptly uncoupling from his mounted suitor and pulling forward to leave her motionless with her ovipositor still lowered beneath her (Fig. 1.7). Rejection by males appeared to follow some sort of weight assessment during the mounting phase; I found that acceptable partners were almost 20 percent heavier than rejected individuals. For the large successful females (Fig. 1.8), all the risk-taking and persistence paid off with a reward that is also relatively uncommon among animals: male Mormon crickets feed their mates.

The Mormon cricket's nuptial meal is an enormous wad of material accounting for an extraordinary 20 to 30 percent of the male's body weight that is squeezed from enlarged abdominal glands and passed out of his genital pore in a process that looks more like giving birth than copulation. The massive meal is attached to the male's seminal sac just before the mating pair breaks apart. Afterward, the female holds the tip of her abdomen elevated (so that the sac and attached meal are held above the ground) (Fig. 1.9) and at intervals bends to take mouthfuls of her prize (Fig. 1.10). She takes several hours to eat her way through her meal, as she pulls off and consumes long strands that remind me of pulling strings of warm cheese from a pizza!

Zoologists refer to a seminal sac such as that produced by the male Mormon cricket as a *spermatophore* (Mann 1984). In this species, however,

Figure 1.7. A male (on left behind grass stems) rejects a female by pulling away from her.

Figure 1.8. A male accepts a female as a mate by curling beneath her.

Figure 1.9. A newly mated female (as viewed from the top and side) "carrying the hominy," the spermatophylax meal.

the spermatophore is more than just a package of sperm because it has a meal attached. After transferring the sperm sac (in these insects it is termed the *ampulla*, literally "small bottle") to the female, the male surrounds it with a sperm-free mass, the cheeselike, edible material called a *spermatophylax* (Boldyrev 1915). Once the gift is transferred, the pair breaks up. The male then ejaculates *in absentia* as sperm from the ampulla move into a tube inside the female and finally into her sperm-storage organ, the *spermatheca*. Females at this stage of the reproductive process are easy to see,

Figure 1.10. A female pulling a long strand of material as she begins her spermatophylax meal.

as the large, white, glistening spermatophylax is quite conspicuous, even in dense vegetation. The sight of spermatophore-carrying females was well known before any scientists came on the scene. It had even acquired its own terminology; in 1904 Gillette reported hearing ranchmen refer to females carrying "white sacs" or "blubber," and to the first workers attempting to control Mormon crickets, newly mated females were described as "carrying the hominy" (Cowan 1990).

The final course in the female's nuptial meal is the empty sperm ampulla, which she eats quite quickly. By this point the sperm have been transferred, and the sated female rejoins the band, with the life cycle being completed in the days ahead when she oviposits in the late afternoon sunshine. Then, a "bayonet-shaped" ovipositor injects into the soil eggs that resemble "in form, pieces of vermicelli of one-forth of an inch in length" (Feilner 1864) (Fig. 1.11). As in other insects, the eggs are fertilized from stored sperm as they enter the genital chamber. Eggs can withstand blistering soil temperatures (of 50°C and above: Cowan 1990) before they hatch in subsequent springs (see Chapter 3), ensuring lots of band members for future seasons.

I completed my first season's field studies of Mormon crickets by visiting another study site, a canyon several hundred kilometers east of Greystone. In these eastern locations crickets have never been a nuisance as population densities are far lower than in the west and the insects do not migrate in large moving bands. With their lower densities and more seden-

Figure 1.11. A female Mormon cricket has left her band-mates to oviposit.

tary habits, eastern populations are similar to other species of singing insects in that they are heard more often than they are seen. The males are much more vocal than their western relatives, and calls are a common summer morning sound in fields and canyon roadsides.

My eastern site, "Indian Meadows," was located in the Poudre Canyon, just west of Fort Collins, Colorado. The site was an area of low prairie cloaked with the long seedheads of needle-and-thread grass (*Stipa comata*) and the occasional large patch of sweet clover (*Melilotus officinalis*). Many males were perched in the sweet clover, and the population was much more noisy than at my western Greystone site. Virtually every adult male was calling from a perch in vegetation or while walking over the ground. Beneath the canopy of plants and chorusing males, I spotted the occasional female moving over the ground. A few females responded to callers but without the urgency I had seen in the phonotactic movements of the western females. Moreover, approaching females never fought for access to callers nor were they rejected by males after mounting a potential partner. Instead females regularly rejected suitors by passing up certain calling males before finally settling on one as a mate.

Poudre Canyon males also behaved quite differently from those in dense western populations, as they engaged in territorial border disputes in which they appeared to use song to intimidate rivals. These interactions often escalated to physical contact when males grappled and fought in a manner very similar to females in the west. Despite the differences in premating behavior, copulation itself in eastern Mormon crickets was very similar to that in the western counterparts, including the transfer of a spermatophylax meal every bit as large as those provided by the Greystone band members.

The behavior of the sexes in eastern populations (Gwynne 1984c) is not much different from that of other animals in which males call. In many singing insects and frogs, males call for long periods from "acoustic territories" where females can come and inspect several males before mating with one (Gerhardt 1994b). Moreover, the premating behavior of eastern Mormon crickets is typical of most animals in that the males compete for mates whereas noncompetitive females appear to choose among potential partners (Darwin 1874; Trivers 1972; Andersson 1994). In contrast, the mating behavior in band-forming western populations involves mating roles that are completely reversed.

1.3. Questions and Topics

My first observations of Mormon crickets in the summer of 1980 at Greystone and in Poudre Canyon raised many questions. Answers to these questions, based on later work with Mormon crickets and their relatives, comprise the focus of much of this book. Before I get to these main questions and chapter contents, however, I should say something about the taxonomy of Mormon crickets and these relatives: just where do these insects fit into the systematic groupings of orthopterans, the crickets, grasshoppers, and their relatives? The common name would suggest that *A. simplex* is one of the "true crickets," that is, a member of the family Gryllidae, the field crickets and their allies. However, the term *cricket* is a misnomer for *A. simplex*. The species resembles many gryllids, with a dark coloration, robust body, and short wings (Fig. 2.11). But in fact it is a katydid (Tettigoniidae), a family that along with acridids, the grasshoppers and locusts, heads the list of the most speciose families in the order Orthoptera (see Chapter 2). Except for their looks (many katydids are green, cryptic species) and gregarious habits (quite rare for Tettigoniidae), Mormon crickets are typical tettigoniids. The biology of the family is presented in Chapter 3, followed by a discussion of predators, parasites, and defenses against these natural enemies in Chapter 4.

Chapter 2 begins with a review of the phylogenetic history and systematics of Tettigoniidae and related groups. A reconstruction of the phylogenetic tree of a group provides not only an understanding of evolutionary relationships but also an important template for tracing the evolution of the habits of living members of the group. Contemplating the ancestral habits of katydids, nineteenth-century naturalist and pioneer katydid behaviorist J. H. Fabre (1917) pointed out that "It is unfortunate for our curiosity that the fossil remains are silent on this magnificent subject," but Fabre recognized that "Luckily we have one resource left, that of consulting the successors of the prehistoric insects. There is reason to believe that the Locustidae [= Tettigoniidae] of our own period have retained an echo of the ancient customs and can tell us something of the manners of olden time." Examples of Fabre's "echo" can now be examined using modern comparative methods (Brooks and McLennan 1991; Harvey and Pagel 1991). I use these methods throughout the book to answer questions such as, Did the ancestral male sing using forewing (tegmina) stridulation? and Is this behavior present in both katydids and crickets because these groups both had a common ancestor that sang or because male song originated independently in each group? Questions such as these can be addressed by mapping (tracing) the presence and absence of the traits onto a phylogeny of the group and determining ancestral states. The answers to these "sound" questions come in Chapter 5, one devoted to acoustical behavior, the most familiar aspect of katydid biology both to naturalists and to biologists who have used it as a model for the study of animal communication.

The origins and evolution of another typical tettigoniid trait, the spermatophylax food gift, is addressed in the first of the four chapters (Chapters 6 and 9) that deal with my main conceptual theme—the evolution of katydid mating. Why did Mormon crickets and other katydids evolve to feed their mates? This question is of general interest in the study of animal reproductive behavior because mate-feeding represents an evolutionary departure from the typical mating strategy of males, which is to invest little in offspring production beyond the costs of insemination (Trivers 1972; Clutton-Brock 1991). "Why" questions such as this address how the trait increases an individual animal's reproductive success. A male that feeds his mate might gain by distracting her with the meal and so increase his success in insemination (and thus fertilization). Alternatively, males may be nurturant individuals that benefit from siring offspring that receive a prezygotic boost from nutrients in the courtship meal. I address these and other hypotheses with regard to how a spermatophylax meal might have evolved in the first place (again using comparative phylogenetic methods) and how mate-feeding in present-day populations might benefit the investing male.

The remaining chapters further examine the behavior of katydids in light of Darwin's (1859, 1874) theory of sexual selection. Observations of Mormon crickets showed plenty of evidence of both mate choice (recall, for example, western males rejecting lightweight females) and fighting for access to mates. A theoretical understanding of these aspects of mating behavior began with Darwin (1859, 1874). He noted that for most species the males exhibited traits such as striking color patterns, elaborate weaponry, and aggressive behavior, adaptations that seem designed to enhance their success in obtaining mates, even though they appeared to reduce the chance of survival. Darwin argued that these male-typical traits evolved either because they provided an advantage in competition for mates or because they were attributes preferred by females. In theory, both these processes impose *sexual selection* on males, causing the spread of any trait that promotes success in increasing the fertilization of eggs. However, attracting or otherwise obtaining mates is only the first step in obtaining fertilization. Sexual selection also should promote the spread of devices that enhance fertilization at later stages when, for example, competition may occur among different ejaculates or females may bias the fertilization success of certain males over others (Lloyd 1979a; Eberhard 1985, 1991).

I explore the nature and evolution of sexual competition and mate choice in Chapter 7. Katydids are a model system for such studies for several reasons. First, they present an exceptional opportunity to study not only female choice and male competition but also the sex reversal in these roles in populations such as the Greystone Mormon crickets. Second, the male calling song is just the sort of behavioral display that females might use in discriminating among males (Searcy and Andersson 1986). There is an obvious potential direct benefit to the choosy female—good males may supply a better spermatophylax meal. There is also the possibility that a female might obtain indirect (purely genetic) benefits by selecting a mate of high quality, thus producing genetically superior offspring (Kirkpatrick and Ryan 1991; Andersson 1994).

Female discrimination of calling song characters may reflect not only within-species processes but also a history of interactions between species. The fitness of the sexually receptive female may be compromised if she responds to the signals of other katydid species that are often within earshot. Increased discrimination by females for conspecific males could then impose sexual selection among these males so that calling songs evolve to emphasize species-specific characters (Ryan and Rand 1993).

Chapter 7 also examines competition for mates. As described for Mormon crickets, fights between individuals competing for mates occur in both male and female katydids, depending on the site. In males, sexual competition can be mediated via acoustical interactions; it is the presumed escalation of such interactions that leads to fights. Finally, a powerful form of sexual competition among males is the competition between sperm in the ejaculates of different males mated to the same female (Parker 1970; Birkhead and Møller 1998).

Sexual competition comes with costs. In western populations of Mormon crickets, the apparent eagerness of females to locate calling males appears to expose them more often to predation, as noted earlier in this chapter. In most mating systems, however, it is males that incur more risks (Burk 1982; Sakaluk 1990). The length of time that males sing is probably one reflection of the level of risk because there are natural enemies that detect these sounds (Alexander and Borgia 1979). There is evidence that the calls of katydids attract acoustically orienting killers such as parasitoid flies (Burk 1982) and predatory bats (Belwood and Morris 1987). In the more eastern populations of Mormon crickets, competing males produce long bouts of uninterrupted calls. However, in populations that show female competition for mates, males seem to be more reticent to call. Since producing calling songs is a risky endeavour, males in the populations showing mating-role reversal may be exposed to less risk than females. Chapter 8 examines the premise that sexual competition often carries with it a greater risk of predation. It reviews recent studies of predation on katydids during mating, as revealed by the survivorship of male and female katydids. One prediction is that competing females are exposed to a greater risk of predation in populations showing courtship-role reversal.

A final question about sexual selection concerns the variable courtship behavior of Mormon crickets, to me the most interesting problem raised by my first summer's observations. The observations clearly showed that insect behavior is not always inflexible and species-specific. I discuss the causes of the plasticity in katydid behavior in relation to theories about the factors thought to determine variation in mating roles (Williams 1966; Trivers 1972; Emlen and Oring 1977). In 1980 I was aware that there had been very little in the way of empirical tests of the theories proposed. Perhaps these ideas could be useful in understanding the behavior of Mormon crickets, or to turn it around the other way, perhaps Mormon crickets could serve as a study system to test hypotheses about the factors controlling the mating roles. The empirical examination of this sexual-differences theory showing how Mormon crickets and other katydids have been valuable test subjects is the topic of my final chapter.

2

Katydids, Longhorns, and Bush-Crickets: Diversity and Evolution of the Tettigoniidae

> ... long even before birds had been fashioned to pour forth their vocal melody, there is good palaeontological evidence that [long-horned] grasshoppers, not greatly different from present forms, fiddled away among the carboniferous ferns, and enlivened the dense atmosphere of those preadamic times.
>
> C. V. Riley (1874), *Katydids*

2.1. Katy and Katydid

A characteristic sound of summer nights in the eastern United States is the loud multiple-phrase call of the male *Pterophylla camellifolia*. In his discourse on sexual differences (1874), Charles Darwin described the song of this "true katydid" (Fig. 2.1) with a quote from T. W. Harris (1842): the "groves resound with the call of Katy-did-she-did the live long night." Later, Cornell naturalist J. H. Comstock (1899), in a pioneering entomology text, reported a more mixed message from the same species: "Katy did, Katy did; she did she didn't"![1] Thus, it is from the call of a single species that we have a common name for a family of over 6,000 described species (Otte 1997; Naskrecki and Otte 1999) found on all continents except Antarctica and in an assortment of habitats including the mountains of California "far above the last outposts of trees" (Tinkham 1944), Canadian peat bogs (Vickery and Kevan 1985), and tropical rain forests.

Katydid has remained the common name for the family in the New World and Australia and New Zealand (where one also hears the name "long-horned grasshopper") (Riley 1874; Rentz 1985), but in Britain and Europe, tettigoniids are called bush-crickets (Burr 1936) (see Preface). The "katy" part of the common name may have onomatopoeic connections with the insect's noisy nature as well as a woman's equivocal reputation. Sue Hubbell (1993) brings up this dual origin, noting that *The Oxford English Dictionary* entry for "katy" includes "wanton" and that "cate-" has

---

[1] Sue Hubbell's (1993) delightful account of what Katy may actually have done includes a tale from North Carolina about newlyweds found lifeless, poisoned in the nuptial bed after the handsome groom had spurned his wife's sister, Katy. Following the crime, the "bugs ... debate as to whether Katy was the responsible party." Such accounts probably inspired O. W. Holmes's verse that opens Chapter 3.

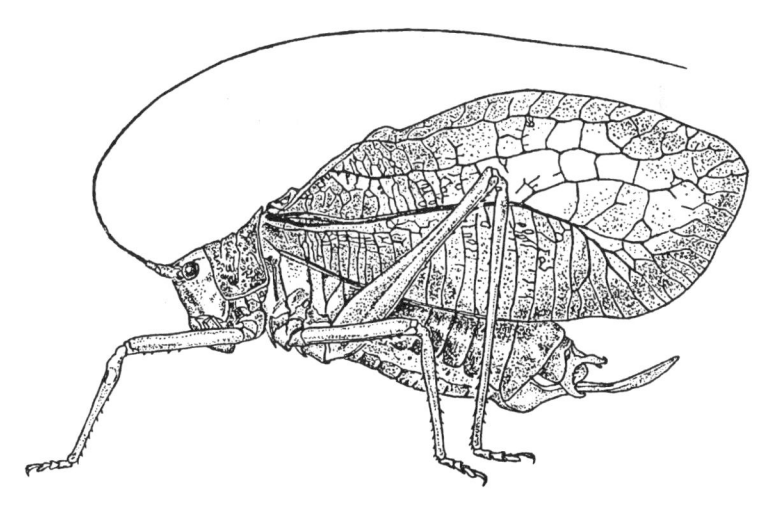

Figure 2.1. A male true katydid, *Pterophylla camellifolia*. From Vickery and Kevan (1983, 1985).

Greek origins for "resound, to din in one's ears" (according to *The Oxford English Dictionary*, John Bartram's [1751] "catedidist" had become "katy did" by the time of John Smyth's [1784] description of *Pterophylla* sounds). The scientific name for the family, Tettigoniidae, almost certainly has onomatopoeic origins in ancient Greek, with its root "*tettix*" (τέττιξ) meaning (and reflecting the call of) the insect songster "grasshopper" or "cicada" (Myers 1929; Jaeger 1955; Brown 1956).

But what are katydids? I answer the question in this chapter by listing the characters that define the family, discussing traits that link the group with other orthopteran insects, and outlining the diversity and evolutionary history of the Tettigoniidae.

2.2. Relationships with Other Orthoptera

Katydids are placed in the suborder Ensifera. The Ensifera and Caelifera (grasshoppers and allies) comprise the order Orthoptera (Kevan 1982; Rentz 1991; Flook and Rowell 1997). The ordinal name means "straight-winged" and refers to the leathery forewings or tegmina. A similar forewing appearance is shared by three related orders—Blattodea (roaches), Mantodea (mantids), and Phasmatodea, or Phasmida to some authors (stick insects)—that together with the Orthoptera are generally referred to as "orthopteroid insects" (CSIRO 1991).

Ever since Aristotle (Kirby and Spence 1826), the "saltatorial orthopteroids," or the Orthoptera, have been regarded as a natural group. But

Kevan (1986) suggested that Ensifera and Caelifera are taxonomically joined in large part only because of jumping hind legs and sound production (but using quite different mechanisms). On this basis Kevan separated the two groups into separate orders, Grylloptera and Orthoptera, respectively. However, most other orthopterists have not supported a two-order scheme, mainly because there are a number of other taxonomic characters that unite living species of Ensifera and Caelifera as sister groups within a clade or "natural" group, one that includes all descendants from a common ancestor (Kamp 1973; Flook and Rowell 1998, Flook et al. 1999).

One set of characters, from DNA sequences coding for ribosomal RNA, that supports Orthoptera as a "natural" group also shows that (1) Phasmatodea is the sister group to Orthoptera; (2) ice-crawlers (order Grylloblattodea) and a mantid-roach clade are the basal groups of orthopteroids (Flook et al. 1999; Maekawa et al. 1999) (Fig. 2.2); and (3) earwigs (order Dermaptera) are the most closely related order to the orthopteroids (Maekawa et al. 1999).

The hypothesis that Ensifera is a natural group becomes problematic when extinct groups are considered; some extinct ensiferans may be closer to species in the suborder Caelifera than they are to living Ensifera species (Sharov 1968). Should this contention be supported by a full character analysis of fossil and living species, a splitting of the suborder Ensifera may be warranted. However, analyses of characters of living ensiferan families support the monophyly of this suborder (Gwynne 1995; Flook et al. 1999) (Figs. 2.2 and 2.3).

In addition to katydids, the suborder Ensifera includes eight other families (Fig. 2.3). The elevation of some families, including Tettigoniidae, to superfamily level by some orthopterists (e.g., Kevan 1982; Flook et al. 1999) reflects the disagreement and uncertainties that have typified systematic studies of orthopteroid insects (Kevan 1977). As most classification schemes use a family-level classification for the major ensiferan subgroups (e.g., Ragge 1955; Rentz 1991; Otte 1997), I will too.

The 10 families of living Ensifera (Table 2.1) have a number of distinctive characters shared by descent from a common ancestor (Gwynne 1995). The subordinal name itself is derived from the distinctive external ovipositor, which is "ensiform"or swordlike in most families, although it is needle-like in most gryllids. Another key character is the pair of "long horns"; ensiferans possess very long, threadlike antennae that are used in a graceful "fencing" routine during courtship. Many ensiferans also share a shieldlike pronotum that conceals the thoracic pleura. There are also internal anatomical characters common to the group such as a globular-shaped foregut and the structure of the spines and teeth that line this part of the alimentary tract (Judd 1947; Elzinga 1996). Finally, ensiferans

2.2 Relationships with Other Orthoptera

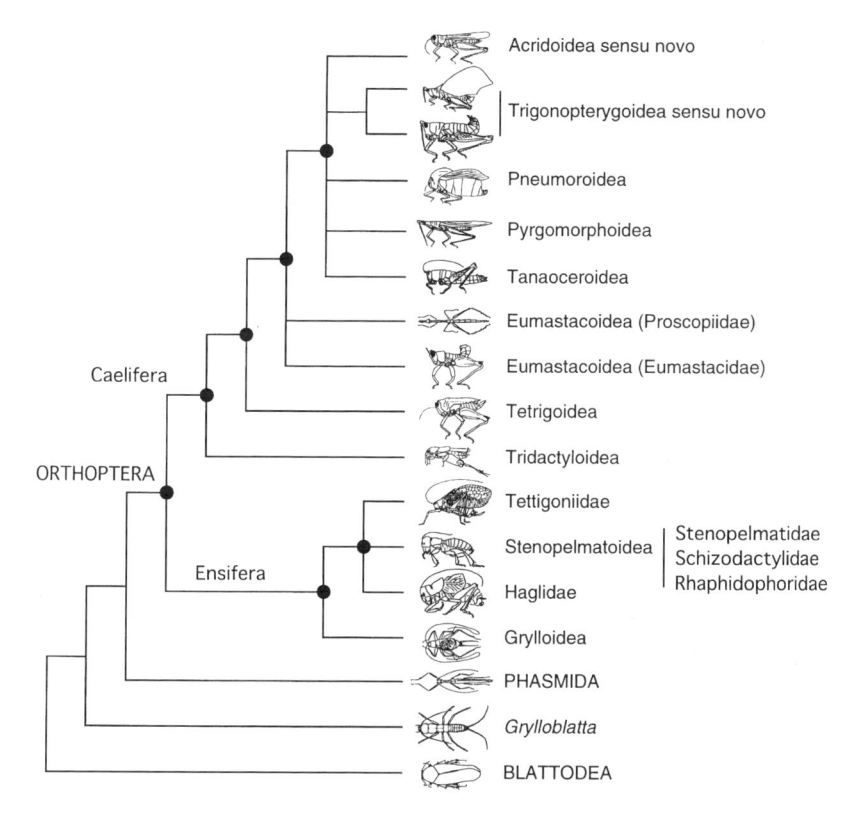

Figure 2.2. Relationships among the two suborders of Orthoptera and other orthopteroids. This figure summarizes the phylogenetic conclusions of Flook et al. (1999), which elevated Haglidae and Tettigoniidae to superfamily status. This tree is based on both a maximum likelihood and a maximum parsimony analysis of the characters. Dots indicate nodes in the tree that were clearly resolved (using statistical bootstrap or likelihood techniques) in analyses of molecular characters (from 12s, 16s [mitochondrial] and 18s [nuclear] DNA sequences coding for ribosomal RNA). The analyses included two tettigoniids, *Tettigonia viridissima* and *Ruspolia nitidula*; one haglid, *Cyphoderris monstrosa*; a schizodactylid, *Comicus campestris*; a stenopelmatid, *Hemideina crassidens*; a rhaphidophorid, *Ceuthophilus carlsbadensis*; and two gryllids, *Acheta domesticus* and *Gryllus campestris*. The Tettigoniidae-Haglidae-Stenopelmatoidea grouping is shown as an unresolved clade because different analyses gave conflicting results: a "maximum evolution" analysis showed haglids and tettigoniids as sister groups (congruent with Fig. 2.3), whereas a "maximum-likelihood" analysis showed tettigoniids and stenopelmatoids as being more closely related. Adapted from Fig. 8 of Flook et al. (1999).

share certain unique DNA sequences (Flook et al. 1999) (see legend to Fig. 2.2).

There is no consensus on the evolutionary (phylogenetic) relationships among ensiferan families; virtually every proposed scheme is different

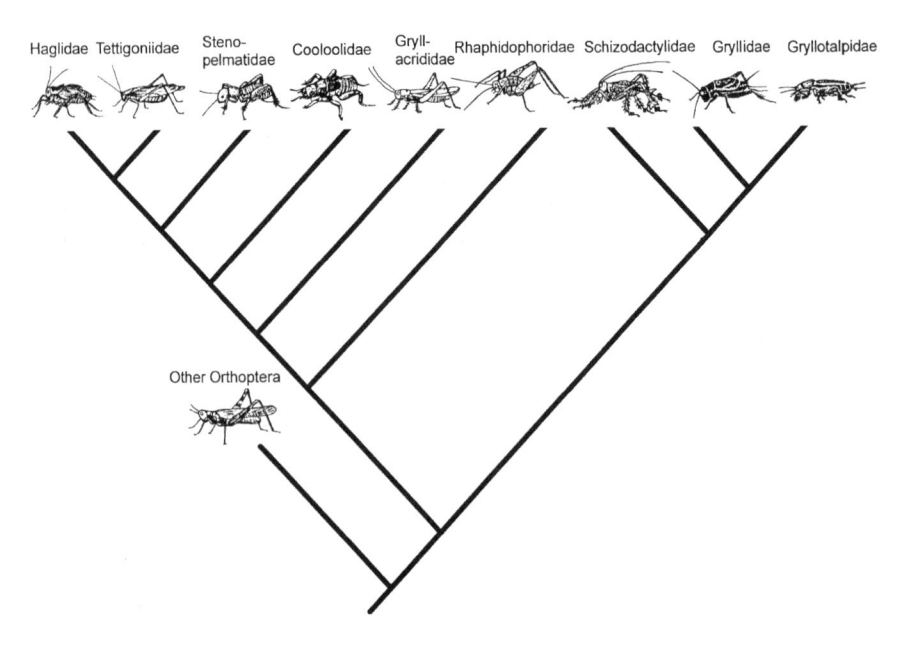

Figure 2.3. A phylogeny of katydids and their kin: the nine subgroups of Ensifera. Redrawn from Gwynne (1997b), which was redrawn from Gwynne (1995). Artwork of insects by R. Osti.

Table 2.1. Classification of the Orthoptera suborder Ensifera showing the two superfamilies representing two ensiferan clades (Ander 1939, Fig. 2.3)[a]

Tettigonioidea	Grylloidea
Haglidae (ambidextrous crickets)	Gryllidae (true crickets)
Tettigoniidae (katydids)	Gryllotalpidae (mole crickets)
Stenopelmatidae[b] (Jerusalem and king crickets and most weta)	Mogoplistidae (scaly crickets)
Cooloolidae (cooloola monsters)	Schizodactylidae (splay-footed crickets)
Gryllacrididae (wood crickets, etc.)	
Rhaphidophoridae (camel crickets, cave weta)	

[a]Some authors retain only Tettigoniidae and Haglidae in Tettigonioidea, placing the other tettigonioid families in a third superfamily, Gryllacridoidea (e.g., Rentz 1991; Otte 1997). Other authors separate haglids and tettigoniids and recognize four superfamilies (Gorochov 1995; Flook et al. 1999; see legend to Fig. 2.2).
[b]Includes Anostostomatidae of Johns (1997).

(Ander 1939; Zeuner 1939; Judd 1947; Ragge 1955; Sharov 1968; Gwynne 1995). A tree of living ensiferan groups that I have suggested (Gwynne 1995) is used here, as it is based on a cladistic (parsimony) analysis of character distributions in a large sample of genera from all families (mainly

Figure 2.4. A male of the decorated cricket, *Gryllodes sigillatus*, a cosmopolitan species (Gryllidae).

from Ander 1939). Parsimony is one of a number of methods used to analyze characters; a cladistic analysis determines the most plausible evolutionary tree by seeking the most parsimonious arrangement, that is, the tree with the fewest evolutionary character changes. The single shortest tree that I found (Gwynne 1995) (Fig. 2.3) has a topology very similar to that of Ander (1939) and Judd (1948) in which there are two main monophyletic groups (clades): the superfamily Tettigonioidea, containing six families (in Table 2.1), and the Grylloidea (Fig. 2.4), containing the true crickets (Gryllidae and Mogoplistidae) and two related families (Table 2.1, Figs. 2.2 and 2.3). The phylogenetic separation of Grylloidea (Gryllidae and Gryllotalpidae) and Tettigonioidea is supported in a preliminary analysis of ribosomal RNA gene sequences (Flook et al. 1999) (Fig. 2.2).

As its family name indicates, katydids are in the Tettigonioidea, a group that also includes five other families. The sister group to Tettigoniidae are the Haglidae (Fig. 2.5), the hump-winged and ambidextrous (including the sagebrush) crickets, so named because a male can switch the overlap of his tegmina during singing (Spooner 1973). The four other families are Stenopelmatidae—the Jerusalem crickets, king crickets, and New Zealand weta[2] (Figs. 2.6 and 2.7); Cooloolidae—the cooloola monsters; Gryllacrididae—the raspy and leaf-rolling crickets (Fig. 2.8); and Rhaphidophoridae—the camel crickets and cave weta (Fig. 2.9, Table 2.1).

---

[2] *Weta* is a maori word; the singular and plural forms are the same. Johns (1997) puts weta and king crickets into the family Anostostomatidae.

Figure 2.5. A mating of ambidextrous crickets, *Cyphoderris strepitans* (Haglidae: United States). **a.** The female is feeding on the male's fleshy hind wings (see Chapter 6). Photo by D. Klimas, reproduced from Gwynne 1997a with permission of Cambridge University Press. **b.** The pair is about to break up. Note the sexual dimorphism in forewings. The male (bottom) has long forewings for singing, whereas the female's (top) are very small and barely visible at the back of her pronotum. Photo by G. K. Morris.

Key characters shared by all tettigonioid families are a swordlike ovipositor, shape of the foregut teeth, and the ability to produce the large complex and edible spermatophore that we saw for Mormon crickets in Chapter 1. The analysis of DNA sequences by Flook et al. (1999) supports the group-

2.2 Relationships with Other Orthoptera   25

Figure 2.6. A Jerusalem cricket, *Stenopelmatus* species (Stenopelmatidae: United States).

Figure 2.7. A tree weta, *Hemideina maori* (Stenopelmatidae: New Zealand).

ing in a single clade of Tettigoniidae (two species [subfamilies] sampled), Haglidae, Rhaphidophoridae, and Stenopelmatidae. (But, this clade also includes schizodactylids) (Fig. 2.2).

The traits that katydids are best known for—male calling using tegmi-

Figure 2.8. A female gryllacridid (Australia).

Figure 2.9. A male sand-treader camel cricket, *Ammobaenetes* species (Rhaphidophoridae: United States).

nal stridulation and ears just below the "elbows" of each front tibia (Fig. 2.10)—are not unique to Tettigoniidae, as they are found in four other families as well. Two of these families, Gryllidae and Gryllotalpidae, may have evolved the characters independently of the Tettigonioidea (the conclusion, at least, when the characters are traced onto the phylogenies of Ander

2.2 Relationships with Other Orthoptera 27

Figure 2.10. A giant weta, *Deinacrida* species (Stenopelmatidae: New Zealand) showing the left tibial tympanum (on far right of the figure) typical of these ensiferans, haglids, many gryllids, and virtually all katydids.

[1939] or Gwynne [1995]; details are provided in Chapter 5). Within the Tettigonioidea, the Tettigoniidae, Haglidae, and Stenopelmatidae share tibial ears, but tegminal stridulation occurs only in the tettigoniid-haglid clade (Fig. 2.3). Stenopelmatids make sounds using a different mechanism altogether: the hind femur is rubbed on the abdomen (Field 1993).

The characters that come to mind as typical of katydids (Fig. 2.11), such as male tegminal stridulation, a swordlike ovipositor, and a spermatophylax mating gift, are also found in other ensiferan families. However, some characters appear to be unique to the katydids. These include a left-over-right overlap of the singing wings (tegmina) in males; a fully functional stridulatory file on the underside of the left tegmen, with the right wing's file being vestigial; four tarsal segments (also found, however, in the apparently extinct haglid, *Prophalangopsis obscura*); and certain DNA sequences (Flook et al. 1999). Finally, I can point to the katydid life history as diagnostic of the group: most tettigoniid species use vegetation as a microhabitat, particularly during inactive periods when the insects retreat into (or onto) locations such as leaves and crevices in plants during daylight hours. These habits are most obvious in species that resemble parts of plants, but even ground-dwelling flightless species such as the Mormon cricket usually retreat into bushes during periods of inactivity (see Chapter 1). The few species that use tree holes as retreats (e.g., pseudophyllines such as *Nastonotus*: Montealegre 1997) are almost certainly secondarily derived from vegetation-inhabiting ancestors (see Gorochov 1988). In con-

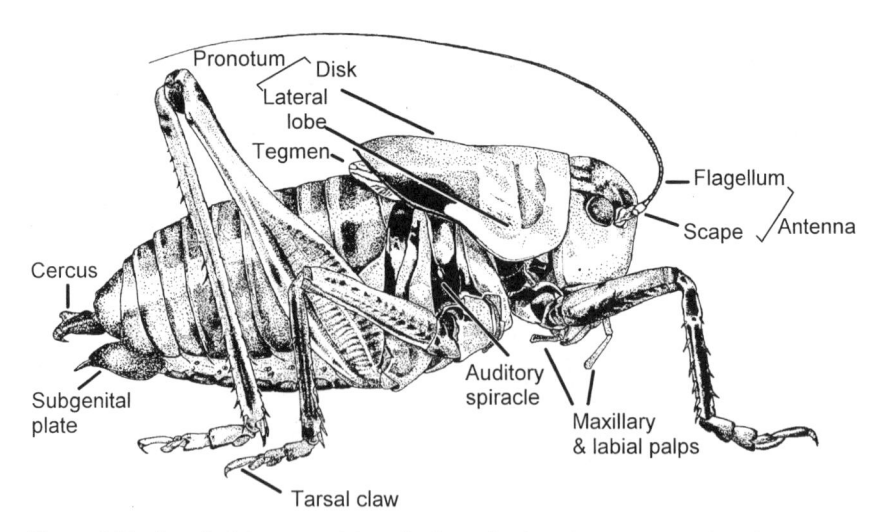

Figure 2.11. A male Mormon cricket, *Anabrus simplex*, showing some key features of the structure of katydids. From Vickery and Kevan (1983, 1985).

trast to most Tettigoniidae, most other ensiferans retreat to tree holes or burrows in soil (Kevan 1989). Like katydids, some crickets (some Mogoplistidae, Eneopterinae, Oecanthinae, Podoscirtinae, and Trigonidiinae) also live mainly in vegetation, but a comparative analysis suggests that as in katydids, this habit was derived from a burrow- or crevice-using ancestral ensiferan (Gwynne 1995). Differences in the use of a "retreat" by ensiferans appear to be reflected in structure; ensiferan taxa that burrow or hide in crevices tend to have forewings that wrap around the body, whereas the tegmina of katydids, unconstrained by burrow use, can be held away from the body as "rooflike" structures and modified for acoustical or leaf-mimicry purposes (references in Gwynne 1995).

2.3. Diversity

Systematists have arranged katydids into a number of different subfamilies (Table 2.2). Some subfamilies originally recognized by Zeuner (1939) and Ander (1939) have been changed or added to in more recent works (Rentz 1979; Kevan 1982; Gorochov 1988; Otte 1997) (Kevan [1982], Gorochov [1988], and Ingrisch [1995] raised several katydid subgroups to family rank). The subgroup arrangement probably will continue to change as knowledge of the world's fauna improves (e.g., see the revisionary studies of Rentz 1979, 1985, 1993; Naskrecki 1994, 1996; and Ingrisch 1995), and phylogenetic methods of character analysis are increasingly used to

Table 2.2. Number of described species (Otte 1997) in each major katydid group (Rentz 1979; Otte 1997; Naskrecki and Otte 1999—www site: http://viceroy.eeb.uconn.edu/Orthoptera)

Subfamily and Tribe	Common name	No. of species
Austrosaginae	Flightless predatory katydids	33
Bradyporinae	Ground katydids	56
Conocephalinae		962
Conocephalini	Meadow katydids	
Copiphorini	Cone-headed and snout-nosed katydids	
Agraeciini	Spine-headed forest katydids	
Coniungopterini	Gondwanan katydids	
Hetrodinae	Armored ground katydids	77
Lipotactinae		25
Listroscelidinae	Spiny predatory katydids	50
Meconematinae	Diurnal predatory katydids	445
Mecopodinae	Long-legged katydids	159
Microtettigoniinae	Micro katydids	2
Phaneropterinae	Leaf and broad-winged katydids	2,013
Phasmodinae	Stick katydids	3
Phyllophorinae	Giant leaf katydids	69
Pseudophyllinae	False leaf katydids	1,051
Saginae	Predatory katydids	50
Tettigoniinae	Shield-backs and others	868
Tympanophorinae	Balloon-winged katydids	7
Zaprochilinae	Pollen and nectar katydids	18

determine whether subfamily taxa are monophyletic (natural) groups (Nickle and Naskrecki 1997).

I have arranged the list of subfamilies into a tentative phylogeny adapted from the only tree published recently (Gorochov 1988), with additional details from Rentz (1979, Rentz and Colless (1990) and Ingrisch (1995) (Fig. 2.12). Note the assumption in this arrangement, as in any phylogeny, that each of the terminal taxa is a natural group. This assumption is likely to be violated by some of the currently recognized tettigoniid subfamilies (Nickle and Naskrecki 1997).

The katydid phylogeny has one group that includes four subfamilies of bush- and tree-inhabiting species that represent the image of a katydid familiar to most people. Most of the species in this clade are tropical and share (apparently by common descent; Fig. 2.12) broad forewings (tegmina) that bear a striking resemblance to leaves (Wallace 1891; Kevan 1982; Belwood 1990) (see Chapter 4). One subfamily in this clade, the Phaneropterinae (Plates 2 and 3; Figs. 2.13, 2.14, and 6.15), includes leaf katydids and broad-winged katydids, comprising a large group of some 2,000 species distributed worldwide, including a species, *Ducetia japonica*,

```
                    ┌──────── HAGLIDAE          🍃
                    │    ┌─── Phaneropterinae
              🍃    │    ├─── Phyllophorinae
                    │    │┌── Acridoxeninae
                    │    ││··· Lipotactinae
                    │    │└── Mecopodinae
                    │    └─── Pseudophyllinae
                    │   ┌──── Listroscelidinae
                    │   │···· Austrosaginae
                    │   │ ┌── Tympanophorinae  🍃
                    │   │ ├── Saginae
                    │   │ │··· Microtettigoniinae 🍃
                    │   │ │└── Conocephalinae  🍃
                    │   │ ├── Hetrodinae
                    │   │ │┌── Tettigoniinae   🍃
                    │   │ │└── Bradyporinae
                    │   │  ┌── Phasmodinae
                    │   │  │··· Zaprochilinae
                    │   │  └── Meconematinae   🍃
```

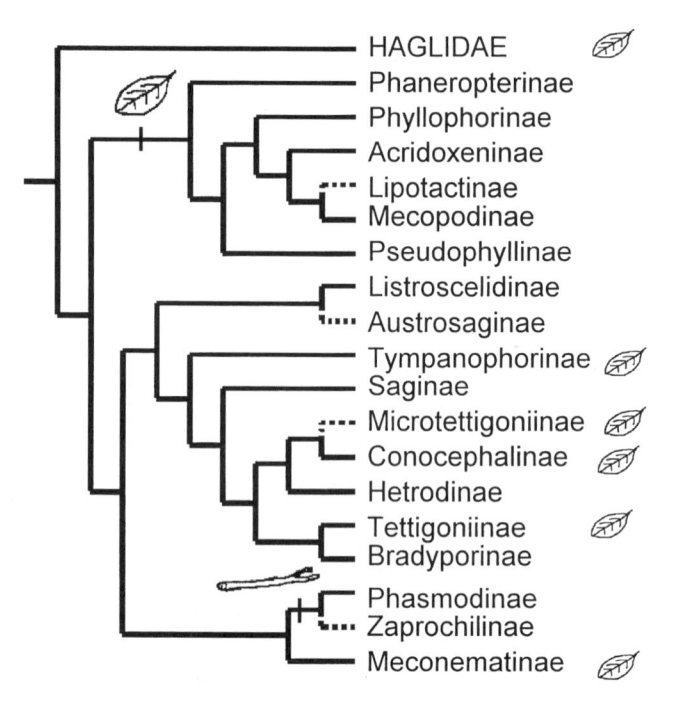

Figure 2.12. A phylogeny of katydid subfamilies following the scheme of Gorochov (1988) and including katydid subgroups of recent treatments (Otte 1997; Naskrecki and Otte 1999). Suggested relationships of recently described (or recently separated) subfamilies are shown as dotted lines. These include Microtettigoniinae, Austrosaginae and Zaprochilinae from Rentz (1979) and Rentz and Colless (in Rentz 1993) and Lipotactinae from Ingrisch (1995). Traced onto the phylogeny (bars and associated symbols) is a hypothesis about the independent origins of mimicry to leaves or sticks (the most parsimonious hypothesis when these characters are traced onto the phylogeny). The leaf symbol next to the taxon name indicates a group in which some members have (apparently independently evolved) tegmina that resemble leaves or grass blades.

kept as a singing pet in China (Jin 1994). Some phaneropterines ("barbistines") no longer mimic leaves, as they have very short wings and can be quite active on the ground surface during the day (Kevan 1982). A second subfamily, Pseudophyllinae, contains just over 1,000 known species of false leaf katydids and the true katydid, *P. camellifolia*. This family is extremely diverse in the tropics where a large spectrum of different morphological forms are found, including species with reduced wings. Also included are species showing remarkable leaf, bark, and other "background" mimicry (Plates 4, 10 to 13, and 24; Figs. 2.1 and 2.15) (Belwood 1990). Pseudophyllines are found mainly in the Americas and Old World tropics. The Phyllophorinae, the giant leaf katydids, are another group of leaf-eaters (Fig. 2.16) and are the largest tettigoniids, with wingspans reaching 25 cm

Figure 2.13. A long-winged phaneropterine, *Elephantodeta nobilis* (from Australia), with her spermatophore. From Gwynne (1997a).

Figure 2.14. A short-winged phaneropterine, *Leptophyes punctatissima*, with a spermatophore. Photo by K. Vahed.

(Rentz 1996). They possess a heavily sclerotized, wide pronotum, and the males have lost tegminal stridulation, a characteristic of virtually all other Tettigoniidae (see Chapter 5). There are 69 described species, all from Australasian tropical forests.

Figure 2.15. *Eubliastes chlorodictyon* female (Pseudophyllinae) eating the elaborate spermatophore. Photo by D. Klimas.

Figure 2.16. A phyllophorine nymph (*Phyllophora*). Photo by D. Klimas.

The subfamily Mecopodinae, long-legged katydids, includes about 160 species of large insects that are related to phyllophorines and also found in the Australian region (as well as Oceania and the Old World tropics) (Naskrecki 1994). Species of *Segestes* and *Segestidea* are well known in New

Figure 2.17. Fang Zhi Niang, the weaving lady, *Mecopoda elongata* (Mecopodinae). This species is kept as a singing pet in China. The female has a small spermatophore. This species does not have a spermatophylax (see Chapter 6). Photo by K. Vahed.

Guinea as defoliators of coconut palms (Room et al. 1984; Young 1985). Katydids such as *Mecopoda elongata*, the "weaving lady" (Fig. 2.17), have a long history as pets both in China (Hsu 1928–29) and in Japan. Katydid pets were woven into wheat-stem cages without doors or housed in elaborate bamboo or brass cages (Chou 1960; Kevan 1982; Jin 1994) (Figs. 2.32, 2.33). They were originally kept in China as symbols of "thriving prosperity," and after A.D. 618 (Tang dynasty), also for the pleasure of their songs. In contrast, male crickets (Gryllidae) are raised for both singing and fighting prowess (Hsu 1928–29; Meng 1993). The different uses of the two ensiferans affect their market prices; crickets can sell for over $100 (U.S.), whereas a "cultured" katydid songster will fetch at most about $16 (Jin 1993).

Related to Mecopodinae is a new subfamily, Lipotactinae (Ingrisch 1995) (Fig. 2.18). The predatory lipotactines rove around the vegetation, apparently spotting prey with their large eyes. Males have long tegminae and females can be wingless.

The other main subgroup (clade) of Tettigoniidae revealed in Figure 2.12 includes 12 subfamilies. The Australian endemic subfamilies Phasmodinae (Fig. 2.19), the stick katydids (3 species), and Zaprochilinae (see Chapter 9), the pollen- and nectar-feeding katydids (26 species), bear a striking resemblance to stick insects (order Phasmatodea), owing to their mimicry of twigs and grass stems when at rest during the day (Figs. 4.8, 9.2, and 9.6). As the name suggests, the "phasmatoid" appearance is most striking

Figure 2.18. A newly mated *Lipotactes* female. She is eating the spermatophylax. The rest of the spermatophore is visible at the base of her ovipositor. Photo by S. Ingrisch.

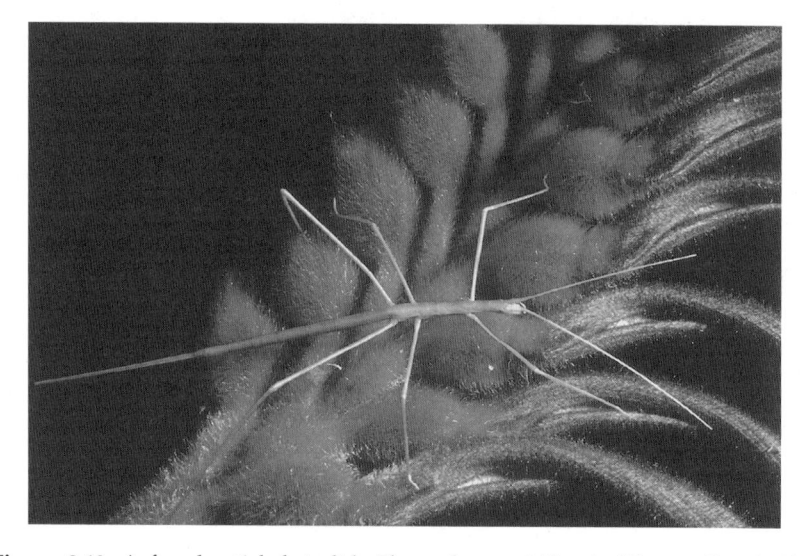

Figure 2.19. A female stick katydid, *Phasmodes ranatriformis* (Phasmodinae). The obvious cue that this is not a stick insect (Phasmida) is the long ovipositor.

in the Phasmodinae (Rentz 1993; Bailey 1998b), the subfamily with katydids that have lost both male calling behavior and the tympanal hearing organ (see Chapter 5). Members of both subfamilies eat nectar, pollen, and flower parts (Phasmodinae).

The subfamily Meconematinae includes the small-bodied, diurnal

Figure 2.20. An undescribed genus and species of Australian listroscelidine. This male produces an ultrasonic call (Gwynne et al. 1988).

predatory katydids. There are 445 described species of this Palearctic and African subfamily (Naskrecki 1996). Another group of hunters, the spiny predatory katydids of the subfamily Listroscelidinae, comprise some 50 species, mainly in the Neotropics and Australasian regions (Plate 5; Fig. 2.20). They range from short-winged species in the genus *Requena* (Fig. 6.4), common in gardens in southwestern Australia (and a key subject of mating behavior research: see Chapter 6), to long-winged species such as the singing pet *Hexacentrus unicolor* ("small weaving lady" in China). Most listroscelidines have long spines on the front tibiae that are used in capturing arthropod prey (Heller 1986; Rentz 1995). Austrosaginae is a recently described subfamily of 33 species of flightless predatory katydids endemic to Australia, particularly the southwest region of the country (Rentz 1993) (Figs. 2.21 and 2.22). All species inhabit heathland vegetation.

Two other subfamilies of predatory species are the Saginae (Fig. 4.7), a flightless group (Kaltenbach 1990), and the Tympanophorinae, the balloon-winged katydids (Riek 1976). The 50 species of sagines are distributed widely from Asia through the Mediterranean region into Africa. One species, *Saga pedo* (European but adventive in North America), lacks males and is one of the few parthenogenetic tettigoniids (Cantrall 1972; see Warchalowska-Sliwa 1998). The Tympanophorinae (7 species from Australia) is similar to the Lipotactinae in eye shape, predatory habits, and sexual dimorphism in wing size. Tympanophorines have prominent eyes

Figure 2.21. *Pachysaga australis* male (Austrosaginae: Australia).

Figure 2.22. *Hemisaga denticulata* male (Austrosaginae: Australia).

that appear to be important for their mobile lifestyle, in which they rove around seeking prey (Rentz 1996). The sexes of *Tympanophora* species are strikingly dimorphic, as the males have long tegmina and the females are apterous and possess a very long ovipositor (Figs. 2.23, 2.24).

The Conocephalinae (literally "cone-headed") is an important group of katydids distributed worldwide and containing about 1,000 species, including the Chinese grass katydids used as singing pets, *Conocephalus maculatus* and *Ruspolia lineosus*. Of the four tribes, the most dominant is the Conocephalini, the meadow katydids (Plates 21 and 22; Fig. 5.1), a group that resembles the short-horned grasshoppers (Acrididae) with their narrow tegmina and diurnal habits. *Conocephalus* is a particularly speciose genus. The Copiphorini includes the cone-heads or snout-nosed katydids (Figs. 2.25 and 2.34). The remarkable pointed head appears to facilitate the dartlike "nose dive" that these animals employ when disturbed (see Faure and Hoy 2000). The insects remain motionless where they land (the resemblance of the narrow tegmina of the winged species to grass leaves no doubt aids in avoiding detection by the predator). Most species in both tribes inhabit low vegetation (grasses and reeds) in fields and swamps and are common in northern regions of North America and Europe. Some South American and Australian cone-heads live in woodlands and forests. Coniungopterini (Gondwanan katydids) and Agraeciini (spiny-headed katydids (Plates 6, 14 to 17; Fig. 2.26) also inhabit trees (Rentz and Gurney

Figure 2.23. *Tympanophora similis* male (Tympanophorinae: Australia).

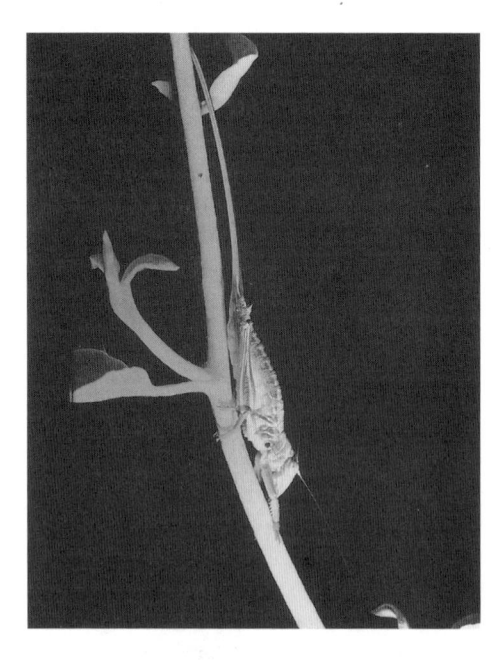

Figure 2.24. *Tympanophora similis* female (Tympanophorinae: Australia). Note the striking difference between this female and the conspecific male (see Fig. 2.23).

Figure 2.25. Newly mated *Ruspolia nitidula* female. The tiny spermatophore of this European species is barely visible at the base of her ovipositor. Photo by K. Vahed.

Figure 2.26. The agraeciine (Conocephalinae) *Veria colorata* is found in northeastern Australia.

1985). Other agraeciines inhabit understory plants. This group is especially common in the tropics (Ingrisch 1998a). Apparent relatives of Conocephalinae species are the two described species of heath-inhabiting micro katydids, the Australian endemic Microtettigoniinae (Rentz 1979). As the name suggests, these are the world's smallest katydids, with body lengths of less than a centimeter.

The subfamilies Hetrodinae (Figs. 2.27 and 4.10) (77 species) and Bradyporinae (Fig. 6.14) (156 species) comprise large-bodied, flightless katydids that live on the ground, usually in dry habitats. These katydids range from the Palearctic into Africa. Hetrodines include the "armored ground crickets" (or "koringkrieks"), endemic to arid areas in southern Africa (Grzeschik 1969; Skaife et. al. 1979; Schmidt 1990; Glenn 1991; Irish 1992; Mbata 1992a, 1992b). The armoring consists of a strong spined plate on the pronotum. Bradyporines include the European *Ephippiger* species whose acoustic behavior has been studied extensively (Busnel et al. 1956), and *Bradyporus dasypus*, an enormous "corpulent, sluggish" and flightless species reported to be kept as a caged pet in Macedonia (Burr et al. 1923).

The subfamily Tettigoniinae (Figs. 2.28–2.31) is a cosmopolitan group that includes about 900 species of some well-known katydids such as shield-backs and wart biters (Duncan 1843).[3] At one time shield-backs,

---

[3] This common name (Caudell 1908b) originated "from a belief said to have once prevailed among Swedish peasantry, that its bite and the black liquid which it disgorges into the wound were useful in removing warts" (Duncan 1843).

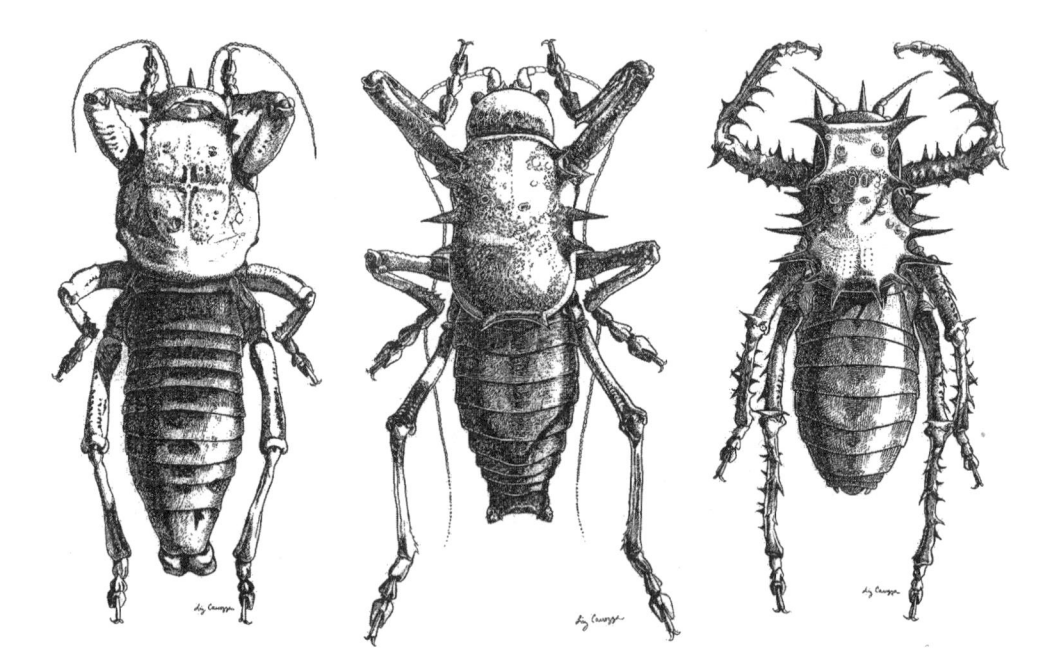

Figure 2.27. Males of three species of "armored ground crickets," African Hetrodinae, showing the large variation in spination that is probably a defense against predators (see Chapter 4). Drawing by Liz Carossi, commissioned by D. Otte.

named for the prominent pronotal shield (Fig. 2.11), were assigned to the tribe Decticini (Caudell 1908b) and other tettigoniines to the Tettigoniini. However, knowledge of additional fauna and phylogenetic trees of the genera no longer justify this split (Rentz 1985; Rentz and Colless 1990). Shield-backs are typically flightless species, some of which are ground dwellers, such as Mormon (Figs. 1.1–1.11, 2.11, and 3.5) and coulee crickets (Fig. 3.8). Two tettigoniines, the white-faced decticus (*Decticus albifrons*) (Figs. 6.13 and 7.1) and the great-green grasshopper (*Tettigonia viridissima*) (Figs. 2.29 and 2.30), were subjects of the first detailed observations on katydids carried out by nineteenth-century naturalist J. H. Fabre (1917) in one of a series of books featuring a large cast of insect and spider characters (*Souvenirs Entomologiques*).

Tettigoniines are found in a diversity of habitats. *Metrioptera sphagnorum* (Fig. 2.31) lives in Canadian peat bogs (Vickery and Kevan 1985) and *Acrodectes philophagus* in rocky alpine habitats (Tinkham 1944). They are an especially diverse group in hot, dry habitats such as the deserts of America (Tinkham 1948; Rentz 1972b; Rentz and Gurney 1985) and Australian heathland (Rentz 1985).

Desert tettigoniines such as *Eremopedes* species were feared in the Navajo

Figure 2.28. An Australian tettigoniine (male), *Rhachidorus*.

Figure 2.29. A great-green grasshopper female (*Tettigonia viridissima*: Tettigoniinae), with her large spermatophore. Photo by K. Vahed.

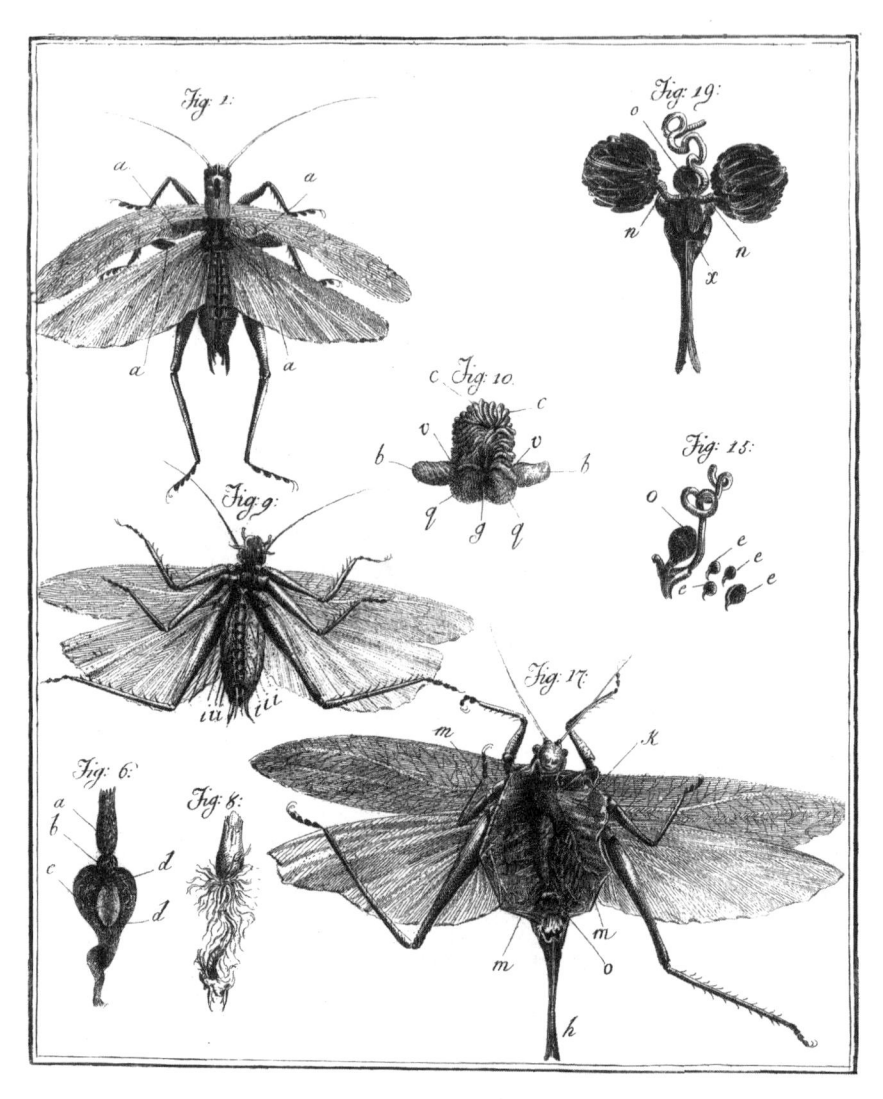

Figure 2.30. The anatomy and structure of the great-green grasshopper, *Tettigonia viridissima*, reproduced here in reduced form from the original plate of 'The anatomy of the "locust"' by Gabriel Brunellius (1791). A dorsal view of the male is shown in "Fig: 1" on this plate and a ventral view in "Fig: 9." "Fig: 10" of the male reproductive anatomy shows the testes (b), smooth accessory glands (q), and rough accessory glands (c) (both glands producing components of the spermatophore). "Fig: 17" is a ventral view of a dissected female with her reproductive parts. "Fig: 19" shows ovaries (o), oviducts (n), spermatheca (n), and the genital chamber (x). "Fig: 15" shows the oviducal gland (long structure) spermatheca (o), which contained four spermatodoses (e). Finally, "Fig: 6" and "Fig: 8" show the digestive system.

Figure 2.31. The bog katydid, *Metrioptera sphagnorum* (Tettigoniinae; male), is found only in Canada. Photo by G. K. Morris and D. Klimas.

Indian culture, as they were believed to be associated with spirits and were said to be attracted to corpses (Kevan 1979). In addition to the species already mentioned, shield-backs, such as *Uvarovites inflatus* ("singing sister"), *Gampsocleis gratiosa* ("singing brother") (China: Jin 1994), and *G. burgeri* (Japan), have a long history as caged singers (Pemberton 1990) (Figs. 2.32 and 2.33).

2.4. Fossil History and Diversification

C. V. Riley's (1874) quote at the opening of this chapter remains valid today. The first known katydids are from late Permian fossils, some 250 million years old (Sharov 1968), whereas the first avian songsters did not appear for another 100 million years. Katydids probably evolved from haglids (Zeuner 1939), the closest-related family to the Tettigoniidae (Fig.

Figure 2.32. Katydids as pets in China. Caged callers (*Gampsocleis gratiosa*) on sale in a market. From Pemberton (1990). Photo by R. W. Pemberton.

2.3). Haglidae were formerly very diverse (Sharov 1968) but are represented today by only a few species, mostly in the genus *Cyphoderris*, the ambidextrous crickets (Fig. 2.4). These insects are adapted to the extreme conditions of high altitudes in North America's Rocky Mountains. Males emerge as adults in spring and can be heard calling at night when snow is still present, even when ambient temperatures are below freezing (Morris and Gwynne 1978).

The fossil record of Tettigoniidae shows that they were present during the Tertiary and Jurassic periods; a katydid calling song as the first noise on the soundtrack of the movie *Jurassic Park* was historically accurate—tettigoniids did indeed sing during the era of dinosaurs. Tettigoniid diversity increased during the same period that the great adaptive radiations of flowering plants, birds, and mammals occurred: from the Tertiary and Quaternary to the present (the last 50–60 million years) (Sharov 1968). The "stem" of the katydid radiation probably sprouted and grew in the tropics (D. Rentz, pers. comm. 1994). Support for a tropical center of origin for the family comes from Gorochov's (1988) contention that the mainly tropical

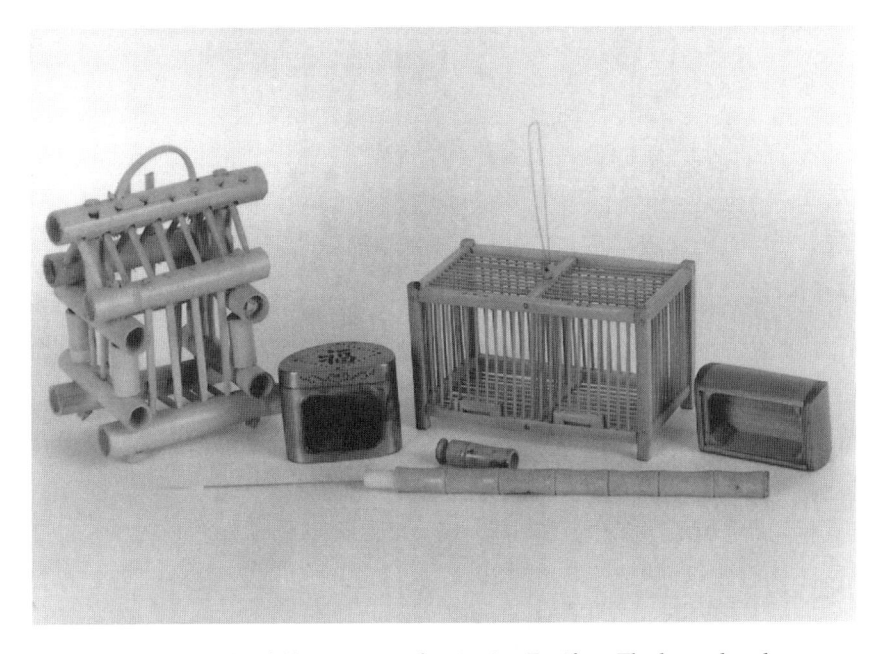

Figure 2.33. A sample of Chinese cages for singing Ensifera. The larger bamboo cages are used for katydids (the fine hair brush in the foreground is used to tickle ["antennate"] crickets [Gryllidae] prior to a cricket fight). Photo by Steve Jaunzems.

clade containing Pseudophyllinae, Mecopodinae, and Phyllophorinae (Fig. 2.12) retains ancestral tettigoniid characters, mainly comprising adaptations to living in plants, such as cryptic coloration and a diet containing a large number of leaves.

Diversification of a taxon such as Tettigoniidae is fueled by population differentiation and subsequent speciation. We can investigate the history of diversification and even speciation mechanisms by examining the relationships and geographical distributions of living species. The evidence for katydids suggests that speciation occurs as a result of restricted gene flow when populations are isolated by geographical barriers such as the ocean (Barendse 1986, 1990) or mountain ranges (Shapiro 1998). However, speciation caused by such barriers is probably not the only mechanism at work. For example, the great diversity of tropical faunas may indicate past speciation without geographical barriers.

A clear example of allopatric speciation comes from work by Willliam Barendse on *Mygalopsis*, a robust, slow-moving, and flightless genus of Australian cone-heads (Figs. 2.34 and 4.7). Four morphologically distinct *Mygalopsis* species are restricted to a narrow coastal strip around Australia's southwestern corner (Fig. 2.35). There is also a fifth species, an undescribed taxon ("*marki 2*" in Fig. 2.35) found just south of the range

Figure 2.34. A male of *Mygalopsis sandowi* (Conocephalinae: Copiphorini), one of several *Mygalopsis* species that have speciated in Australia's southwest.

of "*marki 1.*" The two species are almost morphologically identical but genetically separate groups that form a hybrid zone (Dadour and Johnson 1983).

Barendse concluded that speciation in the genus was due to geographical separation that isolated parts of the ancestral *Mygalopsis* population (Fig. 2.35); the present ranges of closely related ("sister") species contact each other (Fig. 2.35). The four morphologically distinct species diverged between 1 and 10 million years ago, corresponding with the timing of Pliocene incursions of the sea that would have isolated ancestral populations on islands near the ranges of the current species. A subsequent decline in sea level presumably allowed ancestors of these four species to invade the new coastal areas. Another inundation of coastal land in the Pleistocene some 100,000 years ago split the coastal plain and thus could explain the differentiation of *M. marki* from its cryptic sister species to the south. Finally, the typically small population size and low gene flow in the slow-moving *Mygalopsis* probably aided geographical isolation and consequent speciation in these conocephaline katydids.

Although *Mygalopsis* species move slowly compared to most other katydids, they are similar to other members of the family in that they grow

2.4 Fossil History and Diversification 47

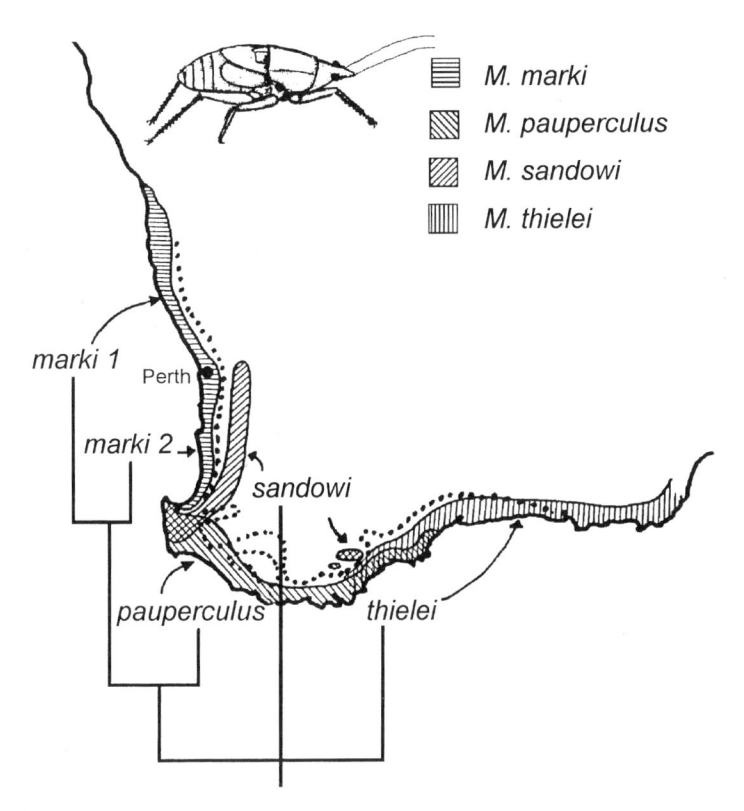

Figure 2.35. A scenario for speciation in *Mygalopsis* katydids in southwestern Australia. The phylogenetic analysis (bottom) was based on electrophoretic characters and shows an unresolved trichotomy of ancestral taxa because there were two shortest trees of equal length; either *M. sandowi* (see Fig. 2.34) or *M. thielei* could be ancestral (i.e., the most distant from the *pauperculus-marki* group). The dotted line represents the location of the Pliocene coastline. The high Pliocene sea level would have divided the coastline at the present *marki-pauperculus* and *thielei-sandowi* borders as well as create an island in the center of the current *M. pauperculus* distribution. Distribution map redrawn from Sandow (1980). Other information was derived from Barendse (1986, 1990).

from egg to adult in just a few months (Lymbery 1987). This turnover time is rapid compared to related families such as Haglidae (Morris and Gwynne 1978) and Stenopelmatidae (Ramsey 1955; Wahid 1978; Gwynne and Jamieson 1998), which take 2 years or more to grow to adulthood. The relatively reduced generation time in the life histories of katydids is expected to increase the rate of evolutionary change and thus might explain the high speciation rate the family has experienced compared to the other ensiferans. Supporting this argument is that the two other diverse families of orthopterans, Gryllidae and Acrididae, also have rapid generation times. An alternative hypothesis for the impressive diversity in the three families is sexual selection. In theory, sexual selection can rapidly produce differences in male calling songs in isolated populations, with the incidental effect of speciation via behavioral isolation (Lande 1981; West-Eberhard 1983).

It is to topics such as life history and male calling that we now turn, beginning with life histories and habits in Chapter 3.

3

What Katy Did: Habits and Life Cycles of Tettigoniids

> *O tell me where did Katy live,*
> *And what did Katy do?*
> *And was she fair and young,*
> *And yet so wicked too?*
> *Did Katy love a naughty man,*
> *Or kiss more cheeks than one?*
> *I warrant Katy did no more*
> *Than many a Kate has done.*
>
> Oliver Wendell Holmes, "To an Insect"

3.1. Oviposition

In late summer in southern Canada, females of the black-sided meadow katydid, *Conocephalus nigropleurum*, leave the riparian sedge and grass clumps. The insects move up onto the branch tips of willow bushes in search of buds that have swollen into galls resembling small pinecones (Fig. 3.1). The galls are created by cecidomyiid gall gnats that oviposit in willow buds. Gaps between gall bracts serve as oviposition sites for the katydids, as first noted by C. V. Riley (1874) and W. M. Wheeler (1890), the latter describing how a female "slowly and sedately... thrust[s] her sword-like ovipositor down between the leaves" of a gall. A single 2-cm long gall may contain up to 150 of the large ($\frac{1}{2}$cm long) eggs (Blatchley 1920).

An individual katydid female can lay hundreds of eggs[1] if she survives long enough. Eggs are laid through an ovipositor consisting of three pairs of valves that slide along each other but remain connected, owing to an intriguing tongue-and-groove mechanism (Fig. 3.2). Eggs are not cemented in a foamy matrix, as are the egg pods of short-horned grasshoppers (Caelifera). However, *Pseudosubria* species (Agraeciini) are known to secrete a soft envelope around each egg (Ingrisch 1998a).

Eggs are laid in a number of different substrates. Many species oviposit in soil (Fig. 1.11) where eggs of some species can withstand very high temperatures (see Chapter 1). Soil oviposition is seen in species in the

---

[1] This fecundity elevated katydids to a symbol of "thriving prosperity" in China so that people "blessed their friends to have as many children as the katydids" (Meng 1993).

Figure 3.1. A female *Conocephalus nigropleurum* on a willow gall. She oviposits between the bracts of the gall.

subfamilies Saginae (Burr et al. 1923; Cantrall 1972; Kaltenbach 1990), Austrosaginae (D. Rentz, pers. comm. 1998), Bradyporinae (Fabre 1917; Boldyrev 1928; Hartley and Dean 1974), Listroscelidinae (*Requena*), many Tettigoniinae (Fabre 1917; Rentz 1985; MacVean 1987), Tympanophorinae (D. Rentz, pers. comm. 1998), and Mecopodinae (*Segestes*: Young 1985) and in a zaprochiline (*Kawanaphila nartee*) (D. T. Gwynne and L. W. Simmons, unpublished observations [1990]). Oviposition in soil is probably an ancestral trait for the family, as it also occurs in the close relatives of katydids, such as the Stenopelmatidae (e.g., Richards 1973) and Gryllacrididae (G. Allen, pers. comm. 1998). The oviposition habits in the sister group to katydids (Haglidae) are unknown, but the eggs are probably laid in a burrow, as the short ovipositor exhibited by *Cyphoderris* is typical of Ensifera, in which females tend eggs or hatched larvae in a subterranean chamber (Gwynne 1995).

Many other katydids deposit their eggs in plant material. This habit may have been derived independently several times in the family. Various conocephalines lay eggs in leaf sheaths or into the center of a plant stem (Table 3.1). The *Conocephalus discolor* female spears the stem with her ovipositor. If she is unsuccessful in one location, she swivels her body around the stem to another spot "with the sharp movements of a well-trained gymnast" (Boldyrev 1915). In another meadow katydid, *Orchelimum erythrocephalum*

	Leptophyes punctatissima	*Conocephalus dorsalis*	*Tettigonia viridissima*
Ovipositor shape:			
Egg shape:			
Oviposition site:	Crevices in bark	In stems	In soil

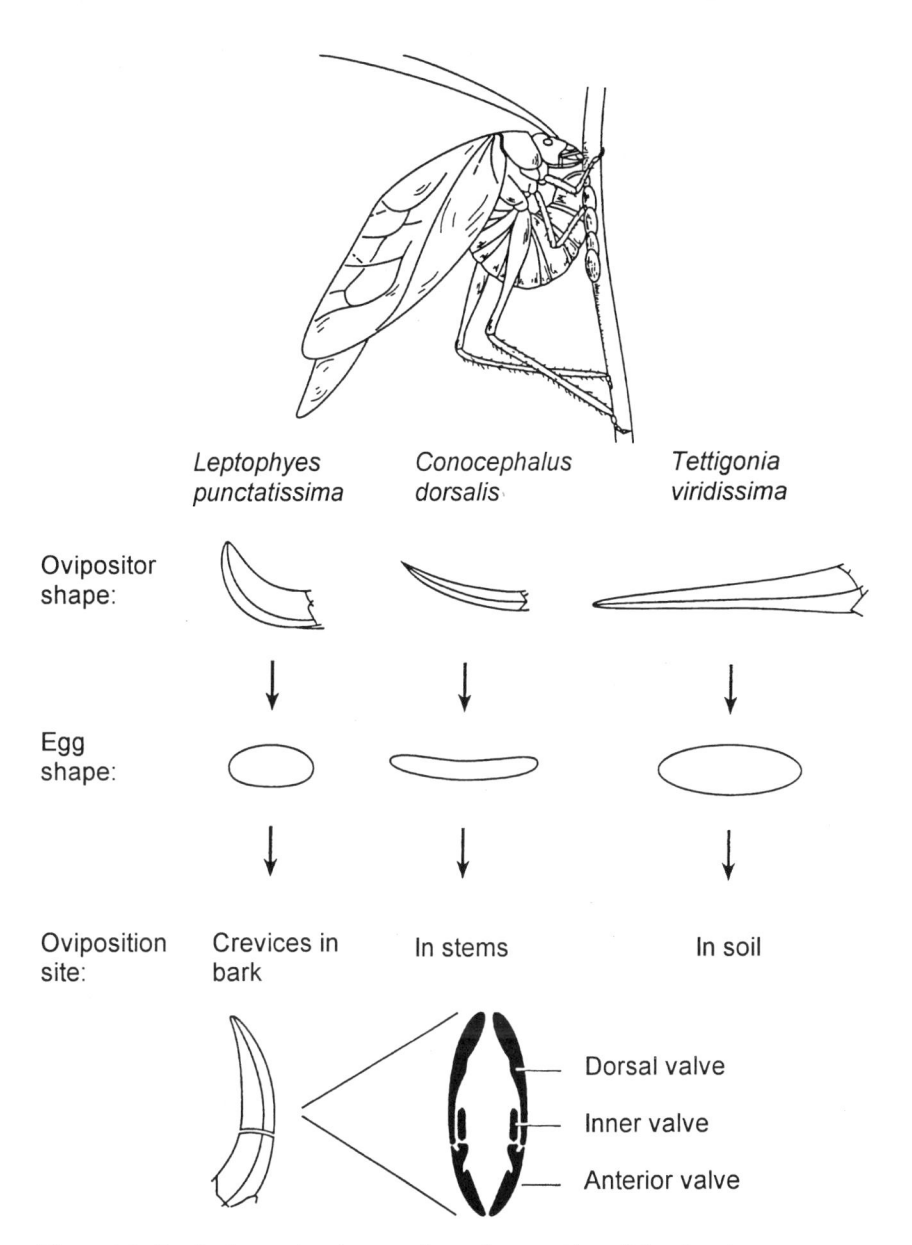

Dorsal valve
Inner valve
Anterior valve

Figure 3.2. Egg-laying, ovipositors, and egg shapes of katydids. The phaneropterine *Microcentrum* (from Riley 1874) at the top places her flat eggs on a stem like overlapping roof shingles. Soil and plant usage produces divergent ovipositor shapes, as seen in *Tettigonia*, *Leptophyes*, and *Conocephalus* species. The lower figure is a cross section of an ovipositor showing the tongue-and-groove mechanism of valve attachment. From Brown (1983) and Marshall and Haes (1988). Redrawn by Alison Dias.

Table 3.1. Oviposition substrates used by some Tettigoniidae in Central Europe (see Table 11 in Ingrisch and Köhler 1998, and references therein)

Species	Leaf tissue	Leaf sheaths	In pith of stems	Dry plants	Bark crevices	Soil
Phaneroptera falcata	X					
P. nana	X					
Leptophyes punctatissima				X	X	
L. albovittata		X	X		X	
Isophya kraussi						X
Barbitistes serricauda					X	
Meconema thalassinum					X	
Conocephalus discolor		X	X			
C. dorsalis		X	X		X	
Ruspolia nitidula		X				X
Tettigonia viridissima						X
T. cantans						X
T. caudata						X
Decticus verrucivorus						X
Platycleis albopunctata		?	X			
P. tessellata		X	X			
Metrioptera brachyptera		X				X
M. saussuriana		?				
M. roeselii		X	X	X		
M. bicolor		?	?			
Pholidoptera griseoaptera		X	X			X
Saga pedo						X

(Hancock 1904), and the tettigoniine *Metrioptera roeselii* (Kevan et al. 1962), females make the job considerably easier by biting a hole in the stem before inserting the ovipositor. Pseudophyllines also oviposit in plant stems, twigs, and wood (Belwood 1988), with the true katydid's oviposition into bark (Riley 1874), according to Jaeger (cited in Riley), being "announced" by conspecific males "in loud tones that katy-did-it." Also known to oviposit in bark are the phaneropterine *Stictophaula armata* (S. Ingrisch, pers. comm. 1999) and some agraeciine conocephalines (Rentz and Gurney 1985), some *Conocephalus* species, certain phaneropterines (Duncan 1960), the zaprochiline *Anthophiloptera* (Rentz and Clyne 1983), and some Meconematinae species. Other meconematines lay in galls made by cynipid wasps (Kevan 1982) and in mosses (Hartley 1990) (*Vetralla* species). Table 3.1 shows the diversity of oviposition substrates for just the species found in Central Europe.

Ovipositor and egg shapes tend to reflect oviposition habits (Cappe de Baillon 1922; Hartley 1964, 1990; Leroy 1969; Brown 1983; Ingrisch 1998a) (Fig. 3.2): for example, the sausage-shaped eggs of many *Conocephalus* species are inserted between leaf sheaths or bracts using a slightly curved, narrow ovipositor; shield-backed katydids such as Mormon crickets use a straight ovipositor to place a more ovoid egg into the soil (Fig. 1.11).

Ovipositor shape can vary greatly within groups. Examples are agraeciines from Southeast Asia (Ingrisch 1998a) and particularly the Phaneropterinae. The ovipositor of *Phaneroptera* (Grasse 1924, Table 3.1), described by Fabre (1917) as a "short yataghan [a dagger-like sabre] bent into a reaping hook," is used to lay eggs between the epidermal layers of leaf tissue. Similar habits have been reported for many other species including *Scudderia curvicauda* (Riley 1874) (Plate 8) and *Euthyrrachis* (Leroy 1969). *Leptophyes* species lay eggs into bark crevices (Duncan 1960), *Microcentrum* katydids cement flat eggs onto twigs like overlapping roof shingles (Riley 1874) (Fig. 3.2), and *Zabalius apicalis* injects eggs into the pith of twigs (Eluwa 1975a).

Some phaneropterines have evolved secondarily to use soil as an oviposition substrate (Kevan 1982; Hartley 1990). Several species have more elaborate habits and will use soil and other substances to construct a mud nest. For example, *Poecilimon* species use water to coagulate the soil surrounding the eggs (Heller et al. 1998), and *Arethaea grallator* has a simple "nest" where the tips of the eggs are barely placed in soil and the exposed portions are covered with chewed plant and mud fragments (Isely 1941). A similar sort of nest was noted by Power (1958) for the hetrodine *Acanthoplus discoidalis* in which several eggs are enclosed in a clay pellet placed in a shallow depression in the soil. Nests are more complex in some antipodean phaneropterines such as *Caedicia simplex*, the kikihipounami (from the maori word *kikihi*, "to make a faint sound" and *pounami*, "green"), one of only four katydid species found in New Zealand (Hudson 1972). The female kikihipounami plasters her eggs onto stones or twigs using mud (Lysaght 1925, 1931). A still more elaborate version is seen in a west Australian *Polichne* species, for which a substantial mud nest attached to vegetation surrounds the eggs (T. Houston, pers. comm. 1985). Given the amount of material used, nest construction in these two species must involve a number of return trips from the ground to the nest site.

3.2. Eggs and Development

Harvard entomologist William Morton Wheeler is best known for his work on the taxonomy and social life of ants, but his research career began with a treatise on katydid reproduction (Evans and Evans 1970). As we have seen, Wheeler found females of the North American *Conocephalus nigropleurum* laying eggs in an unusual substrate—willow galls. He used this easy-to-locate source of katydid eggs to study the development of a representative exopterygote insect (i.e., one showing simple metamorphosis). In his thesis on this topic, Wheeler (1893) coined the term *diapause* to describe the temporary cessation of development he observed in the eggs (Hartley and Ando 1988; Hartley 1990). Diapause functions to prevent development during favorable conditions so that vulnerable later developmental stages are not exposed to severe conditions such as winter cold. The eggs of *C. nigropleurum* enter diapause and survive the winter cold before hatching the following spring. This single generation per year, with the egg as the overwintering stage, is common for species that experience distinct seasons, as found in the colder parts of northern North America and Europe (Ingrisch 1990). It is also the standard life cycle for most katydids in warmer, but still seasonal, areas such as central Texas (23 of the 24 species) (Isely 1941) and southwestern Australia (14 of the 16 species inhabiting *Eucalyptus* bushland, where a single generation per year is revealed by a single peak of male singing in spring or summer: Gwynne et al. 1988).

A series of detailed experiments by Ingrisch (1984, 1986a, 1986b, 1986c, 1990) revealed a number of different diapause strategies among the species that overwinter as eggs in northern climes (Fig. 3.3). Ingrisch found the annual life cycle that we have already discussed for *C. nigropleurum* (Wheeler 1893) in European *Conocephalus* and *Ruspolia* (Conocephalinae) species, a tettigoniine (*Platycleis*), and a phaneropterine (*Phaneroptera*) (Fig. 3.3A). In these species, postwinter hatching occurs after rapid development for 4 weeks at about 24°C. The longer development (8–12 weeks) at similar temperatures for the eggs of some other tettigoniines, certain species of *Decticus*, *Metrioptera* and the phaneropterine *Leptophyes*, means that eggs laid late in summer cannot complete development before the first winter frosts. Consequently these species show a second type of life cycle

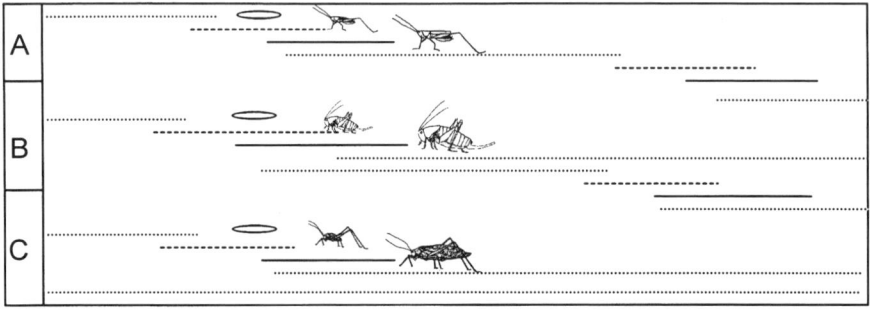

Figure 3.3. Three different development and diapause strategies of katydids from Europe. The dotted line represents the egg stage; the dashed line, the larval stage; and the continuous line, the adult stage. Redrawn from Ingrisch and Köhler (1998).

(Fig. 3.3B) in which there are two patterns of egg development, depending on the season in which the eggs are laid, cued by day length. Eggs of species such as *M. roeselii*, if laid by midsummer, have time for full development and so exhibit a typical annual life cycle by entering the obligate late-embryonic winter diapause before hatching in the following spring. In contrast, eggs laid in late summer and autumn take two seasons to develop; they first enter a prewinter dormancy, a facultative diapause, at an early stage of embryonic development. These eggs then resume development during the following spring and summer, and enter the obligate late-embryonic diapause and end it during the second winter, to hatch in their second spring (Ingrisch 1984). Thus, several bouts of egg-laying by a long-lived female *M. roeselii* may result in the eventual coexistence of adult offspring and grand-offspring of similar ages since the adult molt.

Ingrisch (1986b) identified another type of developmental flexibility in the facultative life cycle of the widely distributed *Decticus verrucivorus*, the wart biter. Populations of this species from different latitudes differ in the critical day length necessary to trigger the double-diapause, 2-year cycle.

A third life cycle identified by Ingrisch (Fig. 3.3C) was anticipated by the earlier work of Hancock (1916), Dumortier (1967), and Dean and Hartley (1977a, 1977b), who showed that the eggs of a phaneropterine *Amblycorypha* and species of *Ephippiger* and *Decticus* required up to three winters to hatch. Ingrisch found a multiple-winter diapause in the eggs of the tettigoniines *Tettigonia*, as well as *M. saussurina*, and three sagines (*Saga* species), which entered an initial (early-embryonic) obligate diapause (i.e., a diapause not influenced by day length as in species such as *M. roeselii*). The initial diapause can be extended so that at least two, and up to seven, winters may be necessary for all the eggs to hatch. By not placing all of their eggs in a single year's "basket," females of these species may have adapted to a variable, harsh environment with a "bet-hedging" strategy that ensures some eggs will hatch in a suitable year (see Dean and Hartley 1977a). Some years may be much more dry than others. Ingrisch (1986b) found that the number of eggs maintaining the initial diapause through successive years increased with increasing dryness of the soil. The eggs of some katydids tend to be drought resistant and lose little of the water that is so necessary for their long-term survival (Ingrisch 1988).

Bet-hedging strategies within single clutches of eggs may also occur in the tropics. Ingrisch (1998b) found that whereas most eggs of the Thailand lipotactine, *Lipotactes minutus*, hatch without any dormant period (approximately 50 days after oviposition), some eggs stay dormant for up to 400 days before hatching. Ingrisch suggested that this year-long diapause in some eggs may be an adaptation for some progeny to survive habitat destruction by fire.

Egg-laying and diapause strategies of katydids can be different in warmer climes. The cone-heads *Neoconocephalus* and *Mygalopsis* are two of the taxa that can be found as adults all year-round in central Texas (Isely 1941) and southwestern Australia, respectively (Lymbery 1987; Gwynne et al. 1988). In these seasonal environments with mild winters, the two katydids appear to have two annual bouts of oviposition resulting in two overlapping generations. Two cone-heads, the West Australian *Mygalopsis marki* (Lymbery 1987) and Florida's *Neoconocephalus triops* (Whitesell and Walker 1978), have two seasonally distinct populations emerging each year. The subtropical cone-head *Euconocephalus pallidus* also has two generations, which Ando (1991) was able to relate to two bouts of oviposition, one from March to June and a second from September to December (Fig. 3.4). The first bout is a result of an *adult* diapause; females experiencing the short day lengths from November to February delay oviposition until the days are longer. In contrast, females experiencing the long days of midyear lay eggs without delay. The restricted oviposition periods of this species may be timed to coincide with optimal temperatures for egg development (between 20°C and 25°C: see Fig. 3.4).

In these cone-heads with dual generations, adults from both hatching periods can interact reproductively so there is potential for gene flow between the subpopulations (e.g., Lymbery 1987). Interestingly, each of the two subpopulations of *N. triops* produces a distinctly different male calling song, but the premating signal differences do not act as a behavioral barrier to reproductive interactions (Whitesell and Walker 1978).

Finally, in areas of the world without any cold seasons there are species with no egg diapause at all in their development, for example, the African cone-head *Ruspolia differens* (Hartley and Ando 1988). This developmental

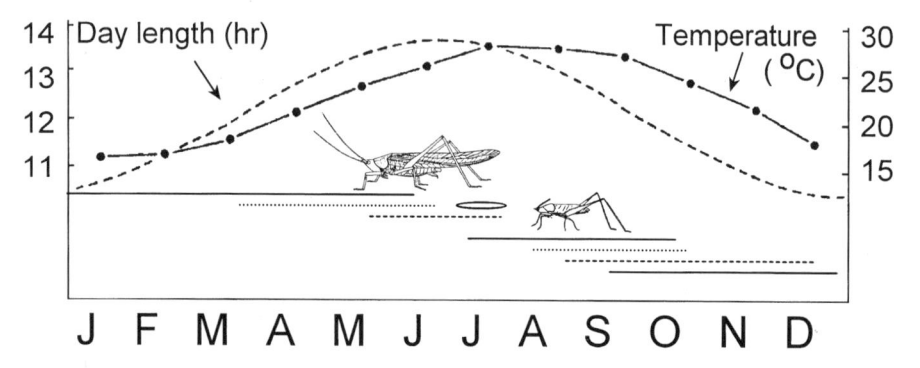

Figure 3.4. The inferred life cycle of the cone-head, *Euconocephalus pallidus*, in Okinawa, Japan. The upper lines show the mean monthly temperatures and day lengths (from sunrise to sunset). Below, the dotted line represents the egg stage; the dashed line, the larval stage; and the continuous line, the adult stage. (Graphs redrawn from Ando (1991). Adult katydid redrawn from Alexander et al. 1972).

pattern in tropical taxa can produce continuously overlapping generations where all stages of the life cycle are present year-round. Examples of genera are the New Guinea *Hexacentrus* (Robinson and Pratt 1975), *Segestidea* (Room et al. 1984), and Costa Rican *Orophus conspersus* (Rentz 1983). But even in tropical areas there are species with only one generation per year, for example, *L. sylvestris*, a lipotactine whose adults appear only at the beginning of the rainy season (Ingrisch 1995), a pattern that indicates an egg diapause associated with drought as demonstrated for the phaneropterine *Stictophaula armata* (Ingrisch 1996).

3.3. Growth

The first task of the callow katydid hatchling is to worm its way out of the substrate in which the egg was laid. Hatching of katydids often occurs at dawn and appears to be cued by the light-dark regime (Reinhold 1998). As in other exopterygote insects, newly hatched nymphs are small but wingless versions of the adults. The degree of nymph-adult resemblance tends to be especially close in species with flightless adults. However, in species with long-winged adults, and particularly those in which nymphal stages mimic distasteful insects such as ants (Poulton 1898; Wickler 1968; Marshall and Haes 1988; Helfert and Sänger 1995), adults and wingless nymphs can be strikingly different in appearance and even behavior (see Chapter 4; Plates 14 to 17).

A katydid prepares for molting by hanging upside down in vegetation. It emerges from the old integument, as do other arthropods, by increasing hemolymph pressure and splitting the old skin and "after much exertion crawls out headfirst," as Feilner (1864) described for the Mormon cricket (Fig. 3.5). Katydid hatchlings molt four to nine times, depending on the species, before reaching adulthood (Ramsay 1964; Ragge 1965, cited in Eluwa 1970). The number of molts shows no obvious taxonomic pattern: there are five instars in the African hetrodine *Acanthoplus discoidalis* (Power 1958) and the mecopodine *Euthypoda acutipennis* (Eluwa 1970); six in the European tettigoniine *Metrioptera roeselii* (Kevan et al. 1962), the Macedonian sagine *Saga natoliae* (Kaltenbach 1990), the subtropical and tropical cone-heads *Ruspolia nitidula* (Hartley 1967) and *Euconocephalus pallidus* (Ando 1991), the European bradyporine *Ephippiger ephippiger (cruciger)* (Hartley and Dean 1974), and the African hetrodine *Acanthoplus speiseri* (Mbata 1992a); and seven in North American shield-back *Anabrus simplex* (Tettigoniinae) (Cowan 1929), African phaneropterine *Zabalius apicalis* (Eluwa 1975b), and the New Guinea mecopodine *Segestidea uniformis*.

Variation in the number of instars occurs in several katydids, including *Pterophylla robertsi* (Barrientos and Jaramillo 1998) and *Conocephalus* species. Instar variation in the European *C. discolor* and *C. dorsalis* may be

Figure 3.5. A Mormon cricket has almost completed its molt to adulthood. The newly shed skin still covers the ovipositor and the end of the abdomen above the insect.

an adaptation to regional climates; the number of instars (six) recorded in areas with long growing seasons are reduced in mountain areas (five) (Ingrisch 1978; Sänger 1980). Variation in the number of instars occurs between the sexes in lipotactines: *L. sylvestris* males require seven molts to reach adulthood while females take eight. *L. minutus* males grow through six instars and females, seven (Ingrisch 1995). In the mecopodine *S. uniformis*, the variation is within males. Only 9 percent of seventh-instar individuals are males, a decrease from the equal number of males and females in each previous instar. This change between instars suggests that most males molt to adulthood in the sixth stage (Room et al. 1984). The seventh-instar adult males undoubtedly pay a cost for the extra larval growth. However, there may be benefits in a larger adult size. Variation in instar number, maturation rate, and adult body size might represent differing mating strategies as suggested for males of a katydid relative, the tree weta (*Hemideina* species: Stenopelmatidae). Because large males fight for females in these ensiferans (Field and Sandlant 1983), faster developing but smaller males may adopt a less aggressive "satellite" tactic of obtaining mates (Spencer 1995) (see Chapter 7).

3.4. Food Habits

The final nymphal molt of the zaprochiline *Anthophiloptera dryas* occurs during spring (October) in eastern Australia in time to take advantage of the newly bloomed flowers on which the adults feed. Individuals of *Anthophiloptera*—literally, "winged flower lover"—are common in gardens around Sydney and will climb to the tops of *Banksia* trees in search of flowers (Rentz and Clyne 1983; D. Rentz, pers. comm. 1993). At the same time on the opposite coast, in bushlands around Perth, the first adults of another zaprochiline, *Kawanaphila nartee*, are emerging. These katydids are also flower lovers, a habit again reflected in the generic name of this flightless western cousin of *A. dryas*: *kawana* is an aboriginal word for "flower". *K. nartee* can be found at night feeding on the red velvet flowers of kangaroo paws (*Anigozanthos manglesii*) (Plate 23). This species is a pollen thief whose flower visits appear to provide no benefit to kangaroo paws, which is pollinated by nectar-seeking birds (Hopper 1993). Both katydids feed exclusively on pollen, a diet that is characteristic of the Zaprochilinae (Rentz and Clyne 1983; Rentz 1993).

Another group that specializes mainly on plant material is the broadwinged Phaneropterinae—Mecopodinae—Pseudophyllinae group, which includes many leaf mimics. These insects "are what they eat" in that they appear to consume much more leaf material than other katydids. Gangwere's (1961) study of the food habits of North American Orthoptera showed that six phaneropterine species ate leaves and some flower parts (see also Young 1985; Gorochov 1988). One phaneropterine, *Microcentrum rhombifolium*, can be raised on an all-leaf diet from first-instar larva to adulthood (Grove 1959). In nature, however, phaneropterines prefer proteinaceous plant parts such as flowers (Gangwere 1961). Many neotropical leaf katydids (pseudophyllines) also appear to be herbivorous (Belwood 1988).

European and North African katydids in the predatory subfamily Saginae also have a very narrow food niche. Kaltenbach (1990) reported that they never eat plant material but instead mainly consume live arthropods, with the occasional morsel of carrion. Sagines are not sit-and-wait predators but move around in the vegetation and jump on prey, grasping it with spined forelegs. Lipotactines are also predators and will wait motionless, following the movements of their prey by turning their head, until finally they jump or run up to 20 cm to grab the victim (Ingrisch 1995). Tympanophorinae (Gorochov 1988), neotropical forest cone-heads

(*Copiphora*) (Belwood 1988), and certain Listroscelidinae (*Hexacentrus*: Togashi 1980; Heller 1986) have similar roving predatory habits. *Phlugis poecila* and *Ancistrocercus inficitus* have adopted a rather risky way of obtaining fresh meat by stealing larvae from wasp nests. O'Donnell (1993) suggested that *A. inficitus* steals brood from the nests of paper wasps when workers are on foraging trips. He found 2 to 11 katydids roosting near 71 percent of the nests censused in Costa Rica. The wasps also may provide protection for the katydids (Downhower and Wilson 1973). North American meadow katydids, *Orchelimum gladiator*, will also steal from social wasps by tackling individual workers (*Polistes* species) and taking chewed-up prey directly from the wasps' mandibles (Feaver 1977).

The narrow range of food preferences seen in zaprochilines, phaneropterines, pseudophyllines, and sagines is unusual among katydids. Most species are best described as omnivores that show preferences for nonleafy items. This was certainly true for the first katydid I ever collected, a great-green grasshopper (*Tettigonia viridissima*) captured in my grandmother's garden in Somerset, England. Following the expert advice of a local gardener I fed my new pet bread and jam, a meal that it voraciously consumed after spending several days in a jar with just grass leaves. Years later I recalled this event while reading Fabre's (1917) account of his captive (French) specimen of the same species: "The Green Grasshopper resembles the English: she dotes on underdone rump-steak seasoned with jam"! Fabre was surprised to find that both his captive *T. viridissima* and the white-faced decticus rejected "the tastiest and tenderest garden stuff," the leaves of grasses and other plants, in favor of more proteinaceous items, including seeds and live prey.

Gangwere (1961, 1967) described most katydids in his study of Orthoptera as having a preference for meat and other protein in a general omnivorous diet. He observed the insects and examined their gut contents. *Atlanticus testaceous*, the single shield-back in his study, had a similar diet to the shield-back subject of Chapter 1, the Mormon cricket. Although primarily carnivorous, *A. testaceous* did eat leaves and fruits. Shield-backs in Greece (*Eupholidoptera megastyla*) also have been observed to eat fruit. Sigfrid Ingrisch related to me that he once saw individuals of this usually hard-to-catch species immobile and apparently drunk after gorging on overripe mulberries! Like shield-backs, many other katydids will consume vegetation but feed opportunistically on proteinaceous items such as seeds, flower parts, carrion, and live prey, including "breakfasting" on the eggs of other insects (Manley 1985) and indulging in gourmet items such as the embryos of terrestrial frogs (Hayes and Rentz 1986).

Conocephalines are also omnivores. While the single cone-head species (*Neoconocephalus ensiger*) observed by Gangwere (1967) favored seeds, the nine meadow katydids (species of *Orchelimum* and *Conocephalus*) ate leaves, flowers, and fruits. Meadow katydids are also opportunistic predators on other insects (Hancock 1904) or their eggs (Manley 1985). Manley implied that *Conocephalus longipennis* might be a blight and a blessing as it destroys the flowers and young grains of rice, but also acts as a "biological control agent" when consuming the eggs of two lepidopteran pests of rice.

Food preferences aside, the everyday diet of meadow katydids appears to include a lot of vegetation. In fact, ecological studies of these insects have shown that their voracious appetites can have quite an impact on the local plant community. Eastern American populations of *Orchelimum fidicinium* consumed 2 percent of the growth of the marsh grass *Spartina* (Smalley 1960), whereas *O. concinnum* and two species of *Conocephalus* convert an impressive 15.9 percent of the biomass of rush species (*Juncus*) into katydid biomass (Parsons and de la Cruz 1980). The impact of the katydids on the plant community is even larger than just the amount of foliage consumed. First, all of the salt marsh katydids clipped off the leaf tips and dropped them uneaten, in order to reach the preferred nutritious growth area several centimeters below the tip (Parsons and de la Cruz 1980; Stiling et al. 1991). Second, feeding by *Conocephalus spartinae* and *O. concinnum* greatly lowered (by 30%–50%) the seed production of rushes and grasses by damaging the seeds developing on the flowers (see also Bertness et al. 1987). These damage estimates do not include indirect reductions in seed set due to pollen loss from katydids' grazing on stamens (Bertness et al. 1987).

The effect of these salt marsh meadow katydids on vegetation also appears to regulate the population size of another salt marsh herbivore, the homopteran *Prokelisia marginata* (Stiling et al. 1991). An increase in the numbers of *Orchelimum* resulting from the experimental addition of fertilizer was correlated with a decrease in *Prokelisia* numbers. This negative effect on homopterans was apparently a result of food competition rather than predation by *Orchelimum* katydids, because experimental clipping of salt marsh grass leaves also reduced homopteran numbers.

The structure of the mouthparts of meadow katydids reflects their diet of both plants and prey; the mandibular structure (Fig. 3.6) includes a hooklike device, apparently an adaptation for holding prey, and well-defined molar "dentes" (small teeth on the inner side of the mandible) for grinding vegetation (Gangwere 1965). The phaneropterines *Amblycorypha* and *Scudderia* have similar dentes for chewing leaves, whereas in the carnivorous *Atlanticus* the dentes are partially fused and "extended into a hook for ripping flesh," a mandibular morphology similar to that of a more famous orthopteroid carnivore, the preying mantis (*Tenodera*) (Gangwere 1965). Finally, the mandibles of the seed-eating *Neoconocephalus* cone-heads have very blunt dentes and a slight concavity on their inner edges, possible adaptations for chewing seeds (Gangwere 1965).

Mastication using mandibles is the first stage in food processing. In the

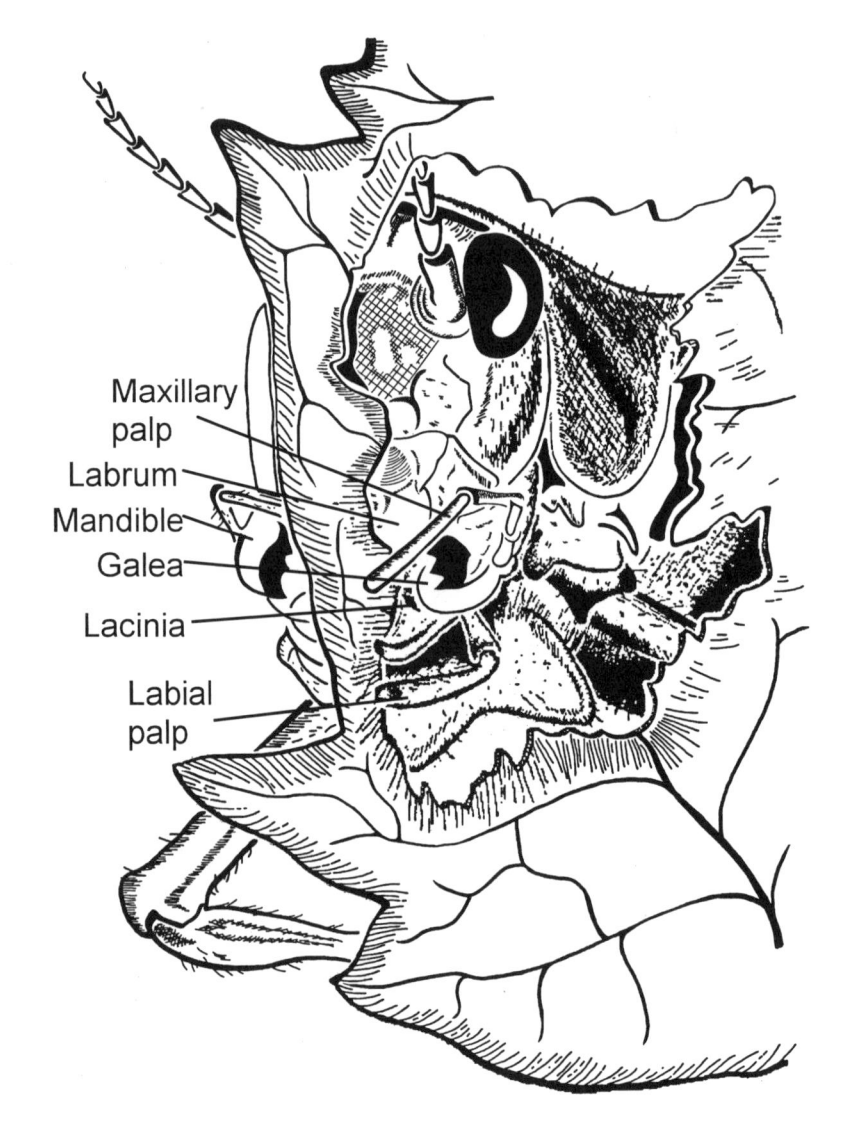

Figure 3.6. Mouthparts of an *Amblycorypha oblongifolia* (Phaneropterinae: North America) feeding on a leaf. From Gangwere (1960). Redrawn by Alison Dias.

next step food is pushed into the preoral cavity, mainly using the maxillae. Maxillary structures, the galeae (the "lateral lips") and bladelike laciniae are used to brush liquid or pulp and to push solid particles, respectively, into the mouth (Fig. 3.6) (Gangwere 1960). After this, food enters the foregut (*a* in "Fig: 6" within Fig. 2.30) and undergoes further mechanical processing in the proventriculus, a muscular structure lined with teeth and spines (Judd 1947; Elzinga 1996) (*b, c, d* in "Fig: 6" within Fig. 2.30). Before food is processed, however, other structures in the mouthparts of katydids are involved in tasting it (Gangwere 1960). Maxillary and labial palps (Figs. 2.11 and 3.6) support the main sensory structures (sensillae) that function in testing the palatability of food as well as sampling other chemicals of the substrate being examined (Zacharuk 1985). Katydids have fewer types of palpal sensillae and a much lower overall sensilla density than do the related ensiferans, gryllacridids, and stenopelmatids, perhaps because katydids rely less on chemosensory reception in reproductive and shelter-seeking contexts (Bland and Rentz 1991). (For example, gryllacridids are known to use scent to locate their silk-lined burrows or nests [Lockwood and Rentz 1996].)

Finally, there is one other important part of the katydid diet to mention because it features prominently in the later chapters of this book. This is a sex-specific item, the mating meal supplied by the male to his mate. As we saw in Chapter 1, the spermatophylax attached to the male's spermatophore is characteristic of most katydids and can be an important part of the female's diet (see Chapter 6), particularly when other sources of food are scarce (Gwynne 1984c; Simmons and Bailey 1990) (see Chapter 9). The spermatophylax is derived from specialized male reproductive glands (*c* in "Fig: 10" within Fig. 2.30; Fig. 6.3), and the prevalence of proteinaceous food items in the diet of katydids is no doubt important to the manufacture of this important nuptial gift.

3.5. Economic Importance and Gregarious Behavior

An increase in population density is expected to exacerbate competition for food. High population densities are uncharacteristic of katydids in general but there are reports of dense aggregations of individuals in Central and South American pseudophyllines. These include small groups of *Nastonotus* species "roosting" together in hollow logs (Montealegre 1997), and *Ancistrocercus* katydids near wasp nests (Downhower and Wilson 1973), as well as large populations of adult *Pterophylla beltrani* and *P. robertsi* aggregating on tree branches (Shaw and Galliart 1987; Barrientos and Montes 1997).

The aggregations of ovipositing *Pterophylla* species can result in economic damage to forest trees (Barrientos and Montes 1997). Other economically important katydids include several genera of Mecopodinae that defoliate young coconut palms and bananas in Papua New Guinea (Room et al. 1984; Young 1985, 1987; Solulu et al. 1998). There are a few pests of grain and crops including the Ethiopian "degasa" or "wollo cricket" (a shield-back, *Decticoides brevipennis*) (Rentz and Gurney 1985; Jago 1997) and certain African hetrodines (Kevan 1982; Mbata 1992a).

At very high population densities, individual katydids within aggregations can show coordinated group behavior. The best-documented cases of this involve flying swarms of cone-heads and the flightless American shield-backs, such as Mormon crickets (see Chapter 1), that migrate in bands. Gregariousness and swarming in *Ruspolia* cone-heads is reminiscent of the infamous crop-destroying locusts (suborder Caelifera: see Uvarov 1977). Huge aggregations of winged adult *R. differens* can reach pest status when they take to the air and fly into crop fields (Bailey and McCrae 1978). The swarms travel mainly at night and can migrate long distances. These insects have even been recorded hundreds of kilometers from land. There is also the occasional report of flying swarms of North American conocephalines, but on a much smaller scale than that for *Ruspolia*. In the autumn of 1983, "millions" of *Neoconocephalus triops* descended onto the streets of Houston, Texas (Reinhold 1983), and large numbers of *Conocephalus fasciatus* can be attracted to street lights (Cantrall 1943). Other katydids that (rarely) form flying swarms are certain shield-backs. In the Canary Islands Owen (1988) reported species of *Decticus* and *Platycleis* in a swarm mixed with a locust species.

Flightless, ground-dwelling species are known for the second type of gregarious migratory behavior, band formation. In Europe such behavior has been noted in short-winged phaneropterines (Bei-Benko 1954, cited in Hartley and Bugren 1986) and also in the bradyporine *Bradyporus*, which was reported in a "colony" of conspecifics mixed with *Decticus albifrons* and, reminiscent of Mormon crickets (see Chapter 1), even being crushed in large numbers by cars on a road (Burr et al. 1923). However, most information on band-forming species comes from North American shield-backs, the Mormon (Fig. 3.7) and coulee crickets (Fig. 3.8) (Snodgrass 1905; Melander and Yothers 1917; Cowan 1929; Wakeland 1959). Both of these flightless species can form bands that migrate, a habit that appears to be derived independently in each species when the phylogenetic relationships between these two genera and other tettigoniines are considered (Rentz and Colless 1990).

As we saw in Chapter 1, the most striking feature of shield-back migrations are the directional headings taken by individual bands. This behavior was even known to native Americans; Kevan (1979) wrote of an old chief of the Dakota Indians who thought the directed movements of these shield-backs "point to the buffalo." We were able to document directional migration in *individual* Mormon crickets by using radiotelemetry of small "radiotags" glued onto several insects (Fig. 3.8). This technique showed that Mormon crickets within a band maintained virtually identical directional headings and covered similar distances (Lorch and Gwynne 2000).

Gregarious Mormon crickets are quite dark in coloration so that individuals and the bands they move in are clearly visible against the back-

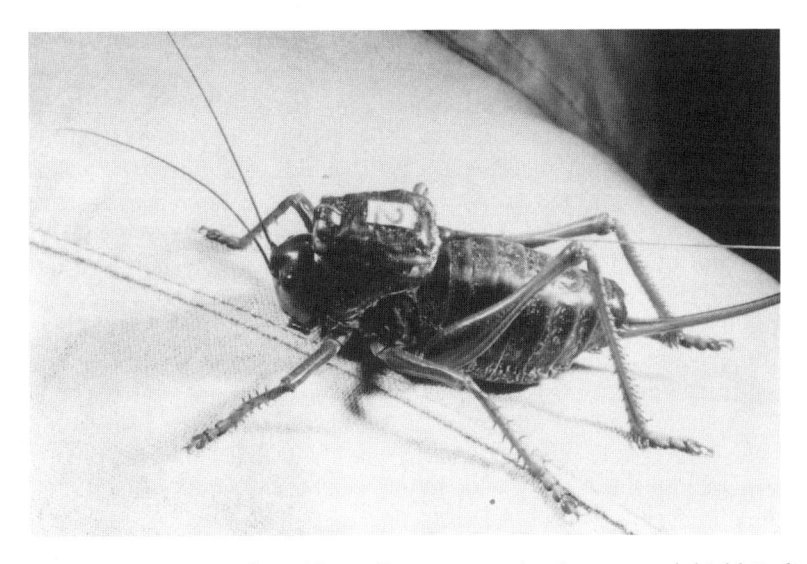

Figure 3.7. A Mormon cricket with a radiotag mounted on her pronotal shield. Radio-tracking of this and other radiotagged individuals gave important information on the distances and directions that individuals moved.

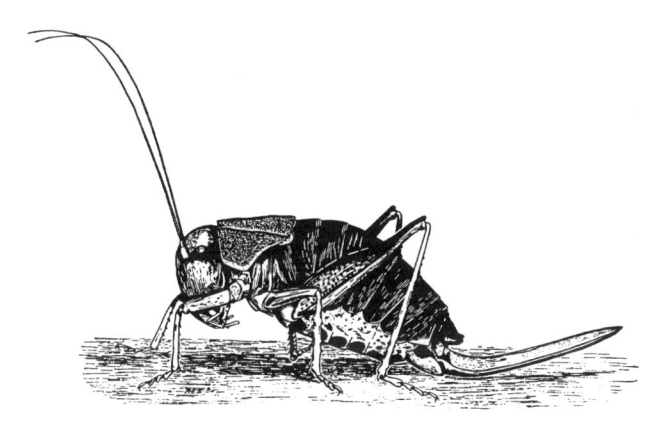

Figure 3.8. A coulee cricket female, *Peranabrus scabricollis*. From Melander and Yothers (1917). By R. L. Snodgrass.

ground substrate (Fig. 1.1). However, at low population densities the insects are smaller and cryptically colored (Plate 9), leading Cowan (1990) to suggest that this species may have solitary and gregarious phases, much like certain locusts. There is evidence that gregariousness causes increased melanization (darkening) or color change in other katydids such as the bradyporine *Ephippiger* (Verdier 1958; Robinson and Hartley 1978, cited in Hartley and Bugren 1986), certain phaneropterines (Hartley 1986), and the pseudophyllines *Pterophylla beltrani* and *P. robertsi*, where the gregarious phase is pink (Barrientos and Den Hollander 1994). In phaneropterines the dark form has also been referred to as a gregarious phase (Bei-Benko 1954, cited in Hartley 1986). A change to a different color phase at high densities is also seen in stick insects (Key 1957), acridids (Sword 1999; Sword et al. 2000), and certain lepidopteran larvae (e.g., see references in Hartley and Bugren 1986). Such a change appears to be a facultative shift to an aposematic display because individuals become distasteful to certain predators. I will have more to say on this topic in Chapter 4 in a discussion of the defense mechanisms of katydids.

A different kind of color polymorphism is seen in the swarming winged katydid *R. differens*, which comes in both green and brown versions (Owen 1965, 1969). Robinson and Hartley (1978) reported that first-instar nymphs are always green; they showed experimentally that isolated individuals remain green through successive molts, whereas about 50 percent of "gregarious" ones, those in contact with conspecifics, turn brown after molting. It is unclear how these results relate to these animals in nature, however, as Owen (1965) and Bailey and McCrae (1978) could find no evidence of a relationship between polymorphism and swarming in *R. differens*. Moreover, color polymorphisms such as green-brown are seen in a number of more solitary katydids and in some, represent adaptive variation in cryptic coloration (see Chapter 4). A striking pink morph can be found in the cone-head *Neoconocephalus ensiger* (Cantrall 1943) and the phaneropterines *Amblycorypha oblongifolia* (Hancock 1916) and *Torbia* species (D. T. Gwynne, unpublished data 1985) (Plate 2). In contrast to the gregarious phase of *Pterophylla* (Barrientos and Den Hollander 1994), pink coloration is rare in cone-heads and phaneropterines and so may represent a maladapted mutant morph in these groups.

Another polymorphism that may be related to variation in population density is wing length (Fig. 3.9). Macropterous katydids are known to fly and settle in new areas (Feaver 1977). The ability of individuals within flightless populations to develop functional wings should be adaptive when high population densities result in intense local competition for limited resources. In locusts, increased social contact between individuals causes an increase in the frequency of long-winged (macropterous) morphs (e.g., McCafferty and Page 1978). Although the appropriate experiments have not been done with katydids, there is some evidence that crowding leads to an increase in the number of macropterous individuals (although species showing actual gregarious behavior have not been examined). In *Platycleis tessellata*, crowding after the fifth instar caused some larvae to become long-winged adults (Sänger 1984), and in *Conocephalus discolor*, crowding in laboratory cultures resulted in a 50 percent increase in adults with wings a third longer than the average wing length (Ando and Hartley 1982). In *Conocephalus* and other katydids, wing length polymorphism

Plate 1. A lone Mormon cricket at the edge of a band crossing a sand dune (see Chapter 1).

Plate 2. Nymph of the pink morph of a *Torbia* species (Phaneropterinae: Australia).

Plate 3. A *Phaneroptera nana* (Phaneropterinae: Europe) female with her spermatophore. Photo by K. Vahed.

Plate 4. *Stetharasa exarmata* (Pseudophyllinae: Colombia). Photo by Fernando Vargas.

Plate 5. A new genus and species of Listroscelidinae from South America. Photo by Fernando Vargas.

Plate 6. *Metholche nigritarsis* (male), an Australian agraeciine (Conocephalinae).

Plate 7. A female *Pholidoptera griseoaptera* (Tettigoniinae: Europe) with her spermatophore. Photo by K. Vahed.

Plate 8. A *Scudderia* female about to oviposit in a leaf (Phaneropterinae: North America). Photo by Maria Zorn.

Plate 9. Possible phase polymorphism in a katydid, the Mormon cricket, *Anabrus simplex*. The "solitary" phase is smaller and more cryptically colored (in this case green) than the bandforming gregarious phase (see Chapter 4).

Plate 10. Green morph of *Promeca perakana*, an Old World leaf mimic (Pseudophyllinae: Malaysia). Photo by K.-G. Heller.

Plate 11. Brown morph of *Promeca sumatrana*, an Old World leaf mimic (Pseudophyllinae: Malaysia). Photo by K.-G. Heller.

Plate 12. Brown morph of a New World leaf mimic, *Typophyllum* (Pseudophyllinae: South America). Photo by D. Klimas.

Plate 13. Green morph of a New World leaf mimic, *Typophyllum* (Pseudophyllinae: South America). Photo by D. Klimas.

Plates 14 to 17. The differing mimicry and color strategies of *Macroxiphus sumatranus siamensis* (Conocephalinae: Agraeciini: Southeast Asia). The black-colored early instars (**Plate 14**) mimic the appearance and behavior of ants, whereas larger individuals from the third stage (**Plate 15**) to the technicolor adult (**Plates 16 and 17**: male and female, respectively) are much more katydid-like in appearance. See Chapter 4. From Helfert and Sänger (1995). Photographs by Brigitte Helfert and Karl Sänger.

14

15

16

17

Plate 18. Aposematic display of the Australian mountain katydid, *Acripeza reticulata* (Phaneropterinae; see Chapter 4). Photo by Howard E. Evans.

Plate 19. This male of an undescribed species of South American conocephaline (Agraeciini) is a "covert caller," referring to the domed pronotal cap that covers the calling wings and enhances the intensity of the song (see Chapter 5). Photo by D. Klimas.

Plate 20. Two great golden digger wasps orienting to a paralyzed phaneropterine katydid that one of them has taken as prey (see Chapter 4). Photo by Maria Zorn.

Plate 21. The black-legged meadow katydid, *Orchelimum nigripes,* is distributed west of the Appalachian Mountains. It hybridizes with the handsome meadow katydid in two locations (see Chapter 7). Photo by David Funk.

Plate 22. The handsome meadow katydid, *Orchelimum pulchellum,* is distributed east of the Appalachian Mountains. It hybridizes with the black-legged katydid in two locations (see Chapter 7). Photo by David Funk.

Plate 23. A female pollen katydid, *Kawanaphila nartee* (Zaprochilinae: Australia), bends to eat her spermatophylax. When food availability (pollen) is scarce, this species shows a reversal in the courtship roles—males become choosy and females compete for access to males (see Chapter 9).

Plate 24. Males of the neotropical *Gnathoclita sodalis* (Pseudophyllinae: South America) have elongated jaws that may be a result of sexual selection (see Chapter 9).
Photo by Fernando Vargas.

Figure 3.9. Wing length polymorphism in *Metrioptera roeselii* (females). The long-winged form of this species was particularly common when it expanded its range in North America after being introduced into Canada from Europe.

involves both forewings and hind wings (Hudson 1972), and in the tettigoniine *Metrioptera roeselii* (Fig. 3.9), only the longest of three wing morphs is capable of flight (Ebner 1951). A particularly high incidence of macroptery occurred in the expanding population of these katydids after they were introduced from Europe into the Province of Quebec, where much habitat was available for colonization (Urquhart and Beaudry 1953; Kevan et al. 1962; Vickery 1965).

Selection for migratory behavior is not, however, without its costs. I observed one of these costs in the late 1970s, after *M. roeselii* had spread several hundred kilometers west from its Quebec introduction point; at the Queens University Biological Station in eastern Ontario a katydid hunter, the great golden digger wasp, *Sphex ichneumoneus*, began to prey on the newly arrived European in great numbers (this wasp has a reputation as an opportunistic predator on adventive Europeans; Sismondo (1978) reported *Meconema thalassinum* as prey some 20 years after this meconematine first turned up on Long Island, New York). Digger wasps (Sphecidae) are just one of a large array of natural enemies that are suffered by katydids. Let us now examine these enemies, how they affect populations of katydids, and the individual defenses that katydids have evolved to thwart their predators and parasites.

4

Survival Strategies: Natural Enemies, Counteradaptations, and Population Regulation

> ... we saw her approaching with a large, light green meadow-grasshopper, which was held in the mouth and supported by the fore legs, which were folded under. On arriving the prey was placed, head first, near the entrance, while the depredater went in, probably to reassure herself that all was right. Soon she appeared at the door of her nest.... Then seizing it ... she dragged it head first into the tunnel.
>
> G. W. Peckham and E. G. Peckham (1898), "On the Instincts and Habits of the Solitary Wasps"

4.1. Natural Enemies

The Peckhams (1898) described the large black, orange, and gold-colored katydid killer, the great golden digger wasp (*Sphex ichneumoneus*), as "one of our most beautiful species" (Fig. 4.1; Plate 20). This was high praise indeed, as these turn-of-the-century naturalists studied some striking animals, including another group of colorful predators, the jumping spiders (Salticidae) (Peckham and Peckham 1895). Digger wasps are sphecids, a group of mainly solitary wasps that hunt and sting arthropod prey, transporting the paralyzed victim to a subterranean cell as provisions for a growing larva. As the name suggests, great golden diggers excavate nests and usually in close proximity to other burrows. The sight of many active wasps together moved another well-known student of wasps, Howard Ensign Evans (1963), to confess that he "know[s] of few things more exciting than sitting beside a flourishing colony ... and watching the females soar in, each with a katydid clutched beneath her: crisp, green songsters, creatures of sunshine and warm moonlit evenings, doomed to be devoured by flabby grubs in dark chambers."

As we saw for *Palmodes* preying on Mormon crickets (see Chapter 1), digger wasps can take a large number of katydids from local populations, particularly when the wasps nest in large aggregations as great golden digger wasps do. A hundred or so wasps in a colony of this species can bring in about 600 prey, whereas several thousand *Conocephalus* can fall victim to a typical colony of an Australian *Sphex, S. cognatus* (Ribi and Ribi 1979).

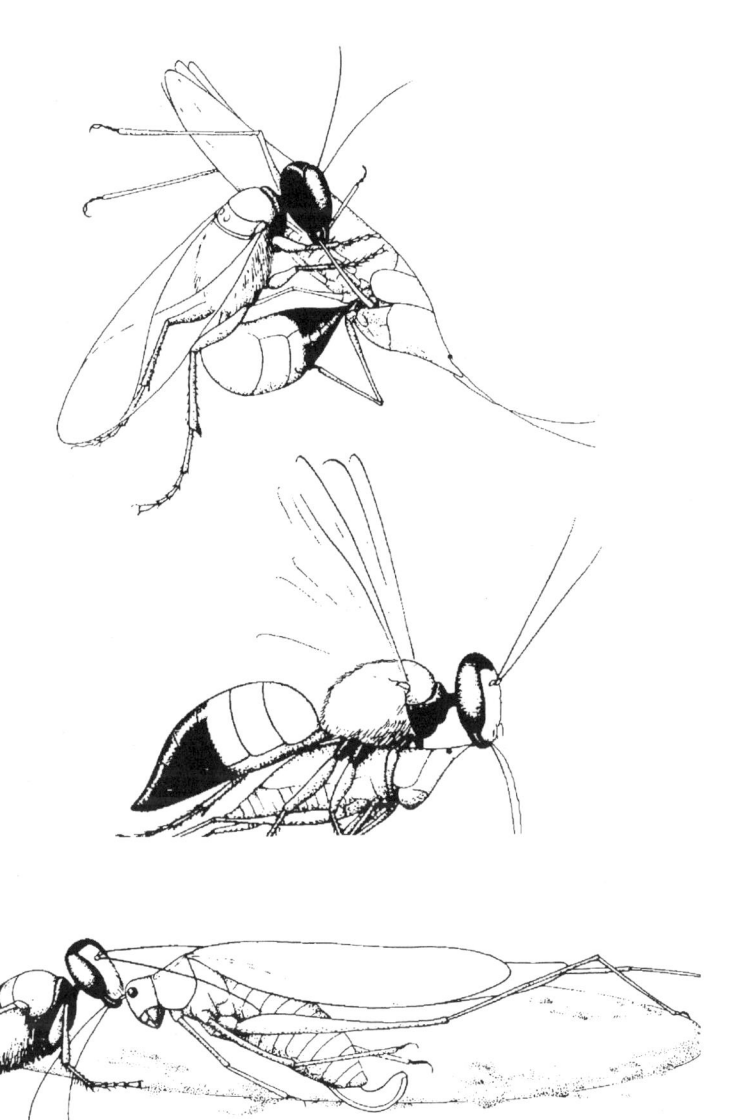

Figure 4.1. The hunting behavior of the great golden digger wasp, *Sphex ichneumoneus*. The female wasp stings (top) a cone-head (*Neoconocephalus*) into paralysis and carries it in flight (middle) back to the nest. Finally (bottom), the prey (here an *Amblycorypha* species) is dragged into the nest as provisions for a larval wasp. From Brockmann (1985). Drawings by Cheryl Hughes and Paloma Ibarra.

Different sphecid taxa use a variety of different arthropods as prey (Evans 1966). *Palmodes*, great golden diggers, and other *Sphex* species, are specialists on katydids (Ristich 1953; Bohart and Menke 1976; Brockmann 1985; Belavadi and Mohanraj 1996). Other Orthoptera hunters are not quite so narrow in their selection of prey and will use a range of prey, including katydids. Of the 33 taxonomic tribes in the family Sphecidae, representatives of the Sphecini, Larrini, and Stizini have been recorded to hunt katydids (Bohart and Menke 1976) (Table 4.1). Katydid hunters in the Larrini are species of *Tachytes* and *Tachysphex*, and in the Sphecini, species of *Sphex* and *Palmodes*. Hunting *Palmodes laeviventris* females can be very abundant in the vicinity of a Mormon cricket band, a virtual supermarket of food for the foragers (Figs. 4.2 and 8.10) (LaRivers 1945). *Isodontia* species are sphecines that do not nest in the soil but instead haul their prey up into holes in wood. Sixteen of 18 species of this genus use Tettigoniidae, and 11 are reported to take katydids exclusively.

Digger wasps are often referred to not as predators but as parasitoids, "parasites" whose feeding activities eventually kill but do not dispatch the "prey" right away so that it is kept fresh and unspoiled. The line between predator and parasite is blurred because the actual capture of the host insect by the adult wasp is certainly akin to predation, whereas the larva's slow consumption of its motionless but living cell mate has all the hallmarks of a parasitic act.

Three other natural enemies of katydids fit the parasitoid category better because they do "capture" prey and their larval dining habits are similar to the digger wasp's "flabby grub" except that the host is devoured from the inside. One group includes flies in the family Tachinidae that parasitize other insects by burrowing as larvae into the host tissues (Adamo et al.

Table 4.1. Records of katydid prey for sphecid wasps

	Con	Cop	H	L	M	Ph	Ps	T
Sphecini								
Sphex	X	X		X	X	X	X	X
Isodontia	X	X		X		X	X	X
Palmodes			X			X		X
Chilosphex								X
Larrini								
Tachytes	X	X				X		X
Tachysphex	X							
Stizini								
Stizus	X							

Data from Bohart and Menke (1976).
Con = Conocephalini; Cop = Copiphorini; H = Hetrodinae; L = Listroscelidinae; M = Meconematinae; Ph = Phaneropterinae; Ps = Pseudophyllinae; T = Tettigoniinae.

Figure 4.2. A paralyzed Mormon cricket female excavated from the nest of a digger wasp, *Palmodes laeviventris*. The wasp's egg (see arrow) can be seen on the abdomen between the second and third legs of its paralyzed victim.

1995). Each species in the tachinid tribe Ormiini specializes on a particular group of katydids, field crickets, or mole-crickets. Host information originally was gathered from flies that emerged from recently collected orthopterans, but the data presented a puzzle—virtually all the hosts were males. For years, the host data were the only information available on ormiines, so just why females were almost never infected remained a mystery. The answer came when William Cade (1975), during his Ph.D. research on cricket mating strategies, found that female ormiines were attracted to the tape-recorded songs of his subjects. These fly parasitoids locate their hosts by homing in on their calls using a unique ear located in the inflated thoracic prosternum (Fig. 4.3) that is tuned to the dominant frequency of the host's calling song (Lakes-Harlan 1992; Robert et al. 1992, 1996).

In response to the song of a *Sciarasaga* katydid, a female ormiine *Homotrixa alleni* faces its victim and "sprays" several larvae up to 6 cm toward its host. Larvae appear to rely on their mother's aim to contact the host, as the youngsters do not move very far and die within an hour or two if she misses (Allen et al. 1999). After arriving safely on the host, one or more larvae will burrow through the intersegmental membranes and into the songster to feed. After one or two weeks of parasite development, the male succumbs and the flies exit from its silenced corpse to pupate (Fig. 4.4).

Pupae and emerged adults of a diversity of ormiines have now been collected from a number of katydid species, and the data show that various

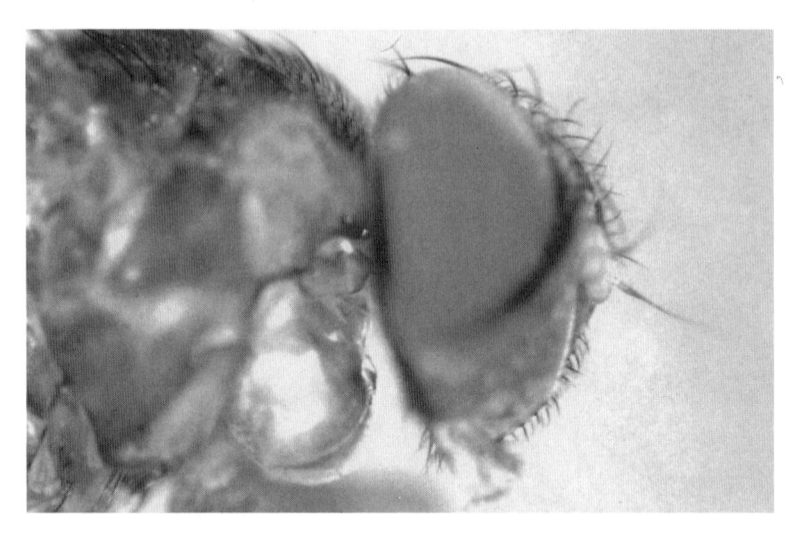

Figure 4.3. An ear that functions only to hear the song of a katydid prey. Side view of the head and thorax of a female parasitoid fly, *Therobia leonidei* (Tachinidae: Ormiini), showing the bulbous ear above the leg. This species hunts *Poecilimon* katydids in Europe (Phaneropterinae). Photo by R. Lakes-Harlan.

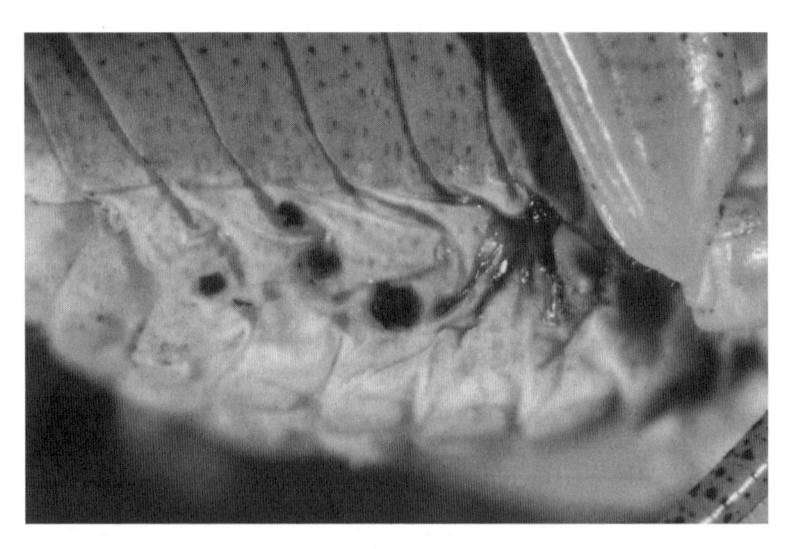

Figure 4.4. The exit holes of parasitoid ormiine flies, *Therobia leonidei* (Tachinidae: Ormiini), in the corpse of a *Poecilimon veluchianus* katydid (Europe). The holes are visible on the abdomen just behind the prey's right jumping leg. Photo by R. Lakes-Harlan.

groups play host to their own fly species. Thus, the fly *Therobia leonidei* parasitizes four species of the European phaneropterine *Poecilimon* (Lakes-Harlan 1992), whereas *Ormia brevicornis* attacks the American cone-head *Neoconocephalus robustus* (Nutting 1953). *O. lineafrons* also uses Conocephalinae, including *N. triops* (Burk 1982) and two species of meadow katydids (*Orchelimum*) (Shapiro 1995). Finally, the only studied ormiine from the southern hemisphere, *H. alleni*, has hosts in three species in two subfamilies, the austrosagines *Sciarasaga quadrata* and *Pachysaga croceopteryx* and the cone-head *Mygalopsis pauperculus* (Allen 1995a, 1995b; Barraclough and Allen 1996). *S. quadrata* in particular is severely hit by fly parasites: about 90 percent of callers end up infected with as many as 16 acoustically orienting parasite larvae per host (Allen 1995a). Moreover, about 11 percent of both males and females also suffer parasitism from a "deaf" dipteran parasitoid, a sarcophagid *Blaesoxipha ragg* that apparently uses nonacoustical cues to locate its host. Geoff Allen (1996) found some unfortunate male victims that hosted both types of fly parasites at the same time. We will return to ormiine parasitoids and their effect on host behavior in Chapter 8.

The second group of parasitoids whose internal feeding and burrowing activities are lethal to orthopterans are the horsehair worms. *Gordius aquaticus* and *G. robustus* infect conocephalines (*Conocephalus* and *Orchelimum*), phaneropterines (*Phaneroptera*), and tettigoniines (Mormon crickets) (Thorne 1940; Baker and Capinera 1997) (Fig. 4.5). These cosmopolitan

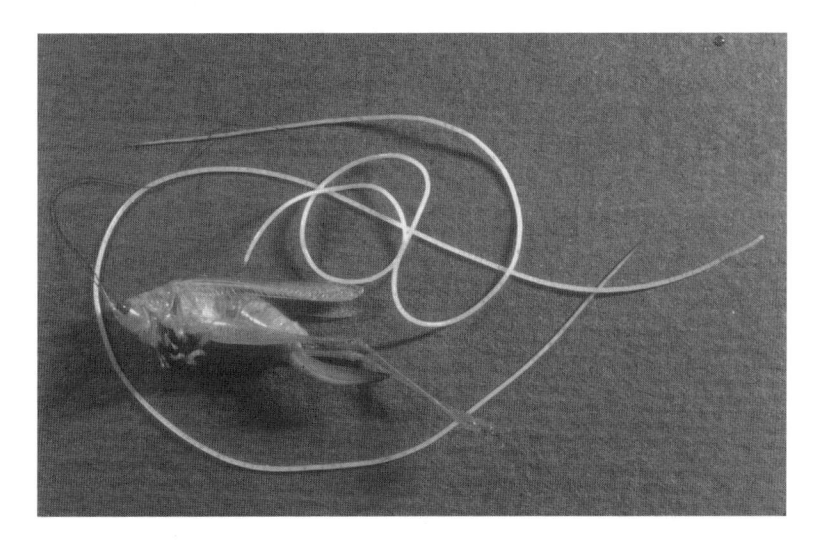

Figure 4.5. This 2-cm-long meadow katydid, *Orchelimum gladiator* (a female), fell victim to two parasitic horsehair worms (phylum: Nematomorpha) that totaled 32cm in length when they emerged from the host.

parasites are so unique that they are placed in their own phylum, the Nematomorpha. Nematomorphs have evolved an uncanny ability to feed while maintaining the host's vital functions, until the host's abdominal cavity is completely filled with almost nothing but the threadlike parasites and its own alimentary tract (Fig. 2.30). I once observed two horsehair worms, totaling 32 cm in length, slowly emerging like spaghetti strands from a 2-cm-long host, an adult *Orchelimum gladiator*. The long, threadlike appearance of the free-living adult worms, and the brown coloration of some species, give them their common name; adults are aquatic and were once thought to have been derived by spontaneous generation from horse hairs shed into livestock watering troughs. In these troughs, as well as ponds and streams, tightly knit groups of mating adults can be observed (I have found them woven around the mesh of minnow traps). These "Gordian knots" (giving *Gordius* its genus name) eventually untangle, and eggs are laid and ingested by an intermediate host such as an aquatic insect. When the intermediate host is eaten by a larval orthopteran, the larval worm infects its final host (Poinar 1991a, 1991b).

Horsehair worm infections in katydids have been reported to kill 40 to 60 percent of young adult *Orchelimum* species (Morris et al. 1975c; Feaver 1977) and up to 59 percent of a band of Mormon crickets (Thorne 1940). A high parasitism rate in *O. gladiator* is not unexpected because this species inhabits riparian areas. On the other hand, the rate for Mormon crickets is surprising given that bands inhabit semidesert environments. However, migrating *Anabrus simplex* do cross rivers and streams (e.g., Tyus and Minckley 1988) and appear to encounter enough watercourses to allow both initial infection and the return of some of the fully developed adult worms to water. Evidence of the latter comes from Thorne's (1940) description of large mats of worms "congregating" along the shore of a reservoir after emerging from their spent hosts within a band of Mormon crickets rafting across the reservoir. Anecdotal evidence suggests that nematomorphs may even facilitate their return to lakes and streams by manipulating the host to show an unnatural attraction to water. Harris (1993) reported a New Zealand ground weta (*Hemiandrus* species, Stenopelmatidae) leaping into a dog's water bowl moments before a lengthy *G. dimorphus* burst from the weta's body!

A third group of worm-like parasitoids are mermithid nematodes. These worms rival the horsehair worms in body length and their ability to fill the host's body cavity. They also kill their hosts and can infect up to 80 percent of the population of *Metrioptera roeselii* (Kevan et al. 1962). Mermithid species infect many orthopterans including katydids: *Mermis nigricens* infects European and North American tettigoniines (*Steiroxys* and *Decticus*) and phaneropterines (*Scudderia*); *Hexameris* species parasitize the tettigoniines, *Tettigonia viridissima* in central Asia and *Platycleis intermedia* in Russia; and two nematodes, *Amphimermis bukari* and *Agamermis catadectaudata*, use *Conocephalus* species in Australia (Grove 1959; Bailey and McCrae 1978; MacVean 1987; MacVean and Capinera 1992; Baker and Capinera 1997).

A number of "parasitic" Hymenoptera are known to attack katydids; the larvae of these tiny wasps, chalcids, scelionids, encyrtids, and eulophids, consume and destroy eggs (Riley 1874; Stiling and Strong 1982; MacVean 1987; Young 1987), killing as many as 20 percent of the eggs of *O. fidicinium* (Stiling and Strong 1982) to half of those laid by *Segestes* (Young 1987).

In addition to parasites, pathogens of katydids include fungal species that can destroy significant numbers of laid eggs (*Segestes decoratus*: Young 1987) or adults (Mormon crickets: Turnbow et al. manuscript), and microsporidian and gregarine protozoans that attack the gut tissues (MacVean 1987; MacVean and Capinera 1992; Lange et al. 1995). Gregarine infections reduce female fecundity and decrease the ability of males to produce spermatophores (Simmons 1993, 1994a). Microsporidians, nematodes, and fungi can kill their hosts, and recent work showed that these natural enemies have considerable promise as biological control agents for Mormon crickets (MacVean and Capinera 1991, 1992).

Katydids are hosts to a number of true parasites, true in the sense that they do not usually kill their hosts. Ectoparasites include tracheal mites (Acarina) that can clog up the auditory tracheae (see Chapter 5) (Bailey and McCrae 1978; MacVean 1987) and mites that attach to the intersegmental membranes and tegmina. There is also a curious group of ceratopogonid flies called *stick ticks* (genus *Forcipomyia*) that attach to the scutella of neotropical pseudophyllines, *Roxelana crassicornis* (Wirth and Castner 1990) and *Polyancistrus* and *Spelaeala* species (Perez-Gelabert and Grogan 1999). Finally, parasitic enemies of the mecopodines *Segestes* and *Segestidea* are the twisted-wing parasites (insects in the order Strepsiptera) that are part ectoparasite and part endoparasite and can infect over 50 percent of the population (Young 1987). Strepsipterans can reduce significantly the feeding ability and ultimately the fitness of their hosts (Solulu et al. 1998). Indeed these parasites show great promise as a biocontrol agent of important katydid pests of the palm oil industry (Kathirithamby et al. 1998).

Female larvae of Myrmecolacidae, a very diverse family of Strepsiptera (Kathirithamby and Hamilton 1995), first attach to the abdominal cuticle and then burrow into and grow within the abdominal cavity. After its adult molt, the parasite female protrudes through the host's abdominal wall (Fig. 4.6). This "partial ectoparasitism" turns out to be critical to the completion of the strepsipteran life cycle; the exposed parts of the female's abdomen

Figure 4.6. Cephalothoraxes (visible as dark spots) of several strepsipteran females, *Stichotrema dallatorreanum*, extruded through the abdominal wall of the host katydid, *Segestidea novaeguineae*. Photo by J. Kathirithamby.

allow copulatory access for the strepsipteran male, who flies in to mate after completing his own endoparasitic life cycle, not in a katydid host but in an ant.

Better-known natural enemies of katydids are the true predators, which, as we might expect, include invertebrates such as spiders, scorpions, mantids, and even other katydids (sagines, see Chapter 3) (e.g., Feaver 1977; Heller and Helversen 1990) as well as many vertebrates, including birds, lizards, bats, and monkeys (Heller and Helversen 1990; Bailey 1991).

Predators of Mormon crickets are particularly well known, as we saw in Chapter 1, because of interest in the biological control of these potential pests. Predators of another katydid pest, the swarming cone-head *Ruspolia differens*, are also well known and include 16 species of birds and seven mammals (Bailey and McCrae 1978).

Homo sapiens is the only species on this list of natural enemies for both of these katydid pests. I mentioned in Chapter 1 that Mormon crickets were eaten by native Americans. As it turns out, native Africans have similar preparation methods for their katydid meals, boiling, roasting, or sun-drying *R. differens* for later consumption (Bailey and McCrae 1978). Another species, *R. nitidula*, is regularly sold in markets in Uganda (Samways 1997). Some Africans allow cone-heads to be eaten only by men, in the belief that they cause sterility or beard growth in women (Bailey and McCrae 1978)! Another belief, this time in China, was that conocephalines and other katydids could be used as aphrodisiacs owing to their perceived "vitality" (fecundity) (Li 1578, cited in Jin 1994).

Katydids with more solitary lifestyles also fall prey to vertebrates. Despite their cryptic coloration leaf-mimicking phaneropterines and pseudophyllines are not always successful in duping their predators, because they end up as a large component of the diets of certain neotropical monkeys. Nickle and Heymann (1996) reported that moustached tamarins (*Saguinus mystax mystax*) specialized on phaneropterines by foraging mainly in the lower to middle parts of the forest canopy. In contrast, saddleback tamarins (*Saguinus fuscicollis nigrifrons*) took mainly pseudophyllines that inhabit understory vegetation. Another major vertebrate predator of neotropical pseudophyllines is bats that listen at night for the sounds of their prey before swooping in and "gleaning" the katydids from vegetation and then returning to a roost where the insects are eaten. Belwood (1988) collected prey remains from these roosts and found that 60 percent of the diet (by weight) of *Micronycteris hirsuta* included katydids. Gleaning bats respond either to male calls or to the incidental sounds produced by insect movements (see Chapter 8 for details). In Europe, Arlettaz et al. (1997) found that lesser mouse-eared bats (*Myotis blythii*) also glean katydids from vegetation, but reported that these bats eat their prey on-the-wing rather than carting the meal back to a roost (R. Arlettaz, pers. comm. 1999). Like the tamarins, mouse-eared bat species appear to use different ecological niches. The lesser mouse-eared bat hunts grass-inhabiting insects at night and so ends up taking far more night-moving katydids (almost 100% of diet by volume) than common daytime grass dwellers such as acridid grasshoppers. By contrast, the closely related greater mouse-eared bat (*M. myotis*) eats mainly ground-dwelling insects. As a result, this hunter takes mainly ground beetles (Carabidae), with katydids comprising only 13 percent of its diet.

4.2. Population Regulation

Although it is obvious that predators and other natural enemies of katydids can have a significant impact on a particular life-history stage, it is not clear whether these biological forces are the main ones that regulate populations in any species. For example, in tropical mecopodines, Young (1987) showed that natural enemies can kill up to 50 percent of the population of *Segestes decoratus*, whereas Room et al. (1984) concluded that physical factors were more important in controlling populations of *Segestidea uniformis*. Room's evidence is that long dry periods tended to correlate with a low abundance of katydids whereas sudden increases in the population size of this important coconut pest occurred after heavy rains. Moreover, other than several predatory lizards, few predators and parasites were seen to exploit this species.

There are few detailed ecological studies comparing the relative importance of different sources of mortality on katydid populations. Long-term research on populations of wart biters (*Decticus verrucivorus*) indicates that many variables are involved. For a relict population in southern England, Cherrill and Brown (1990) estimated that only about 1 percent of hatchlings survived to adulthood, and pointed to a number of likely mortality factors. These include natural enemies, disease, death during molting, and inclement weather, as well as a need for a mosaic of habitat types to support the different ecological requirements of different growth stages (see Cherrill and Brown 1992).

Weather is particularly important, and Haes et al. (1990) provided evidence that population size in English wart biters correlated positively with the amount of sunshine the animals had experienced. Temperature appears to influence the probability of survival of Swedish populations as well; this species was more likely to occupy habitat patches on warm, south-facing slopes (Hjermann and Ims 1996). Long hours of sunshine are probably important to survival, development, and reproduction in the wart biter and other diurnal shield-backs such as Mormon crickets, which can often be seen basking in sun patches (see Chapter 1) where they maintain preferred body temperatures between 35 and 37°C (Turnbow and O'Neill manuscript). In Mormon crickets, elevated body temperature may be important not only to metabolism but also as a form of "fever" that kills fungal pathogens (Turnbow et al. manuscript). Although Mormon crickets thrive best in warm temperatures, they can cope with long periods of inclement weather. Cowan (1990) reported nymphs surviving several cool weeks during which 7 inches (18 cm) of rain fell.

Climate also can influence survival indirectly through its effect on vegetation in the habitat. Populations of the Swedish wart biters (Hjermann and Ims 1996) and another shield-back, *Metrioptera bicolor* (Kindvall and Ahlen 1992; Kindvall 1996), were less likely to go extinct in larger patches of habitat (habitat availability is no doubt influenced by human activity as well). Finally, nutritious food may be a limiting factor for salt marsh meadow katydids, judging from an experimental study in which fertilization caused an increase in foliage nitrogen that in turn led to a significant increase in katydid numbers (Stiling et al. 1991).

4.3. Defenses

Even if the formidable array of predators and parasites is not the main factor controlling katydid numbers, the spectrum of different katydid defenses shows that natural enemies have clearly provided an important selective force. The basic defense for most katydids is to "play possum" by remaining motionless during daylight hours, so that all activity occurs at night. This diel pattern of activity decreases the risk of being spotted by visually hunting mammals and birds, groups that probably have had a long selective influence on katydids (see Chapter 2). Inactivity during daylight is only effective, however, if the potential prey cannot be seen. Ensiferans stay out of sight in different ways: most hide in burrows or crevices (Gwynne 1995), but katydids tend to specialize in blending into the background on which they rest, or crawl into vegetation to hide (Plates 9 to 13).

In many katydids, camouflage is the only line of defense. It is a primary defense, one that is not cued by a predator's approach or attack. Secondary defenses are those that come into play after the predator has made contact with the potential prey (Table 4.2) (Fig. 4.7) (Robinson 1969). The diversity of cryptic morphology and coloration shown in Tettigoniidae is rivaled only by that in stick insects (Phasmatodea) and certain praying mantids. Cryptic coloration is particularly striking in the many tropical leaf mimics, as pointed out by Alfred Russell Wallace (1891), the great tropical biologist (and co-originator, with Darwin, of the natural selection theory): "... most of the tropical Mantidae and Locustidae [= Tettigoniidae] are of the exact tint of the leaves on which they habitually repose, and many of them in addition have the veinings of their wings modified exactly to imitate a leaf."

Striking examples of foliage mimicry in katydids are seen in some Pseudophyllinae (the literal meaning of which is "false leaf") (Belwood 1990; Heller 1995) and Phaneropterinae whose tegmina are dead ringers for leaves, right down to matching the details of fungal spots and insect feeding damage (Plates 12 and 13). Other Pseudophyllinae look like lichen-covered bark (Robinson 1969; Belwood 1990), and the endemic Australian Phasmodinae and Zaprochilinae when at rest are perfect matches to twigs and sticks (Rentz 1991, 1993) (Fig. 4.8). The tegmina of many other katydids bear only a generalized resemblance to grass and other types of

Table 4.2. Katydid defenses against predators

Primary defenses (present in the absence of a predator)
 1. Cryptic coloration, with resemblance to:
    Leaves (e.g., leaf and false-leaf katydids)
    Sticks (e.g., stick and pollen katydids)
    Lichen-covered bark (e.g., some leaf katydids)
    Stones (e.g., *Acripezia reticulata*)
    Generalized color match to background (many species)
 2. Mimicry of distasteful, stinging, or biting arthropods (e.g., *Aganacris*)
 3. Protective spines (e.g., armored ground katydids and some neotropical cone-heads)

Secondary defenses (functioning after attack or approach of a predator)
 1. Fight
    Visual display that startles or confuses the predator (e.g., some sagines)
    Stridulation that startles or confuses the predator (many species)
    Secretions that startle or confuse the predator (e.g., *Bradyporus*)
    Chemicals that are noxious or distasteful (e.g., Crayola katydid)
 2. Flight
    Swiveling on grass stem (e.g., meadow katydids)
    Diving or jumping into water from vegetation (e.g., some *Orchelimum*)

---

leaves. These examples represent several independent origins of leaf and stick mimicry from non-leaf-like ancestors that were active on the ground. This conclusion is derived from a parsimony analysis (see Chapter 2) in which I traced leaf and stick mimicry as characters on an evolutionary tree of katydids (Fig. 2.12, Table 4.2).

Although camouflage is the most common type of adaptive coloration, a few katydids avoid visually hunting predators by mimicking the appearance of distasteful arthropods or those that bite and sting. The larvae of some phaneropterines mimic spiders, tiger beetles, and ants (Fig. 4.7 and Plates 14 to 17) (Poulton 1898; Wickler 1968; Kevan 1982). The adult phaneropterine *Aganacris insectivora* is a South American day-active katydid that is a Batesian mimic (not distasteful) of a spider wasp (Pompilidae), both in its shiny black body and orange wings and in its movements (Belwood 1990; Nickle and Castner 1995).

The early instars of *Macroxiphus sumatranus* (Conocephalinae: Agraeciini) in particular show an uncanny resemblance to ants in color, shape, and behavior (Plates 14 to 17). Diurnally active first and second instars are the mimic stages that forage for their detritus food while running alongside their ant models on or near the ground. The behavior of the young katydids is supported by an almost perfect match to color and shape; they are black and possess morphological features such as long antennae that appear to be short because they are black only at the base. The antennae also move in an antlike manner. When *M. sumatranus* grows out of the size range of ants by the third instar, it suddenly changes to a typical katydid

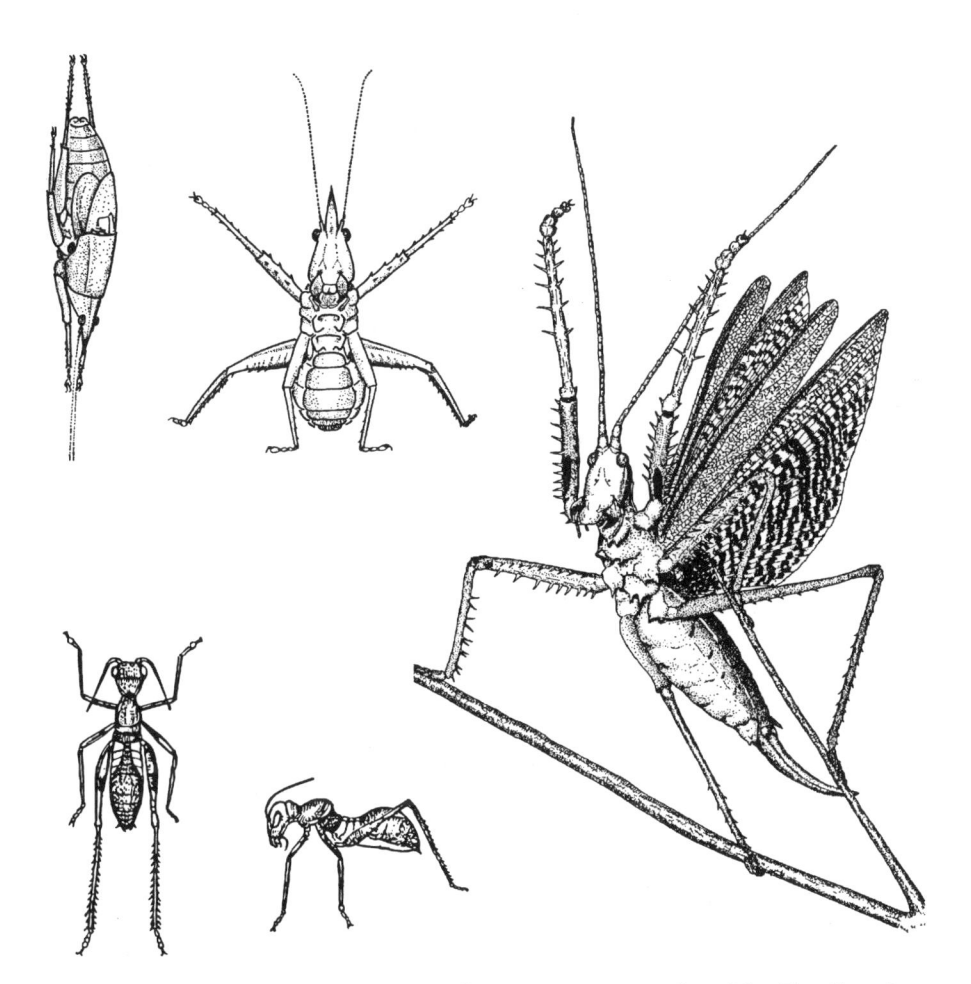

Figure 4.7. Primary and secondary defensive responses in katydids. The diurnal cryptic posture of the conocephaline *Mygalopsis marki* (top left) is a primary defense that turns into a noisy display when the insect is disturbed by a predator. Redrawn from Sandow and Bailey (1978). A more elaborate startle display (right) is seen in *Clonia multispina* (Saginae). From Kaltenbach (1990). The nymph of *Condylodera* (lower left) is an ant mimic. From Poulton (1898).

shape and behavior, after which the insect becomes strictly nocturnal and moves up into the vegetation to feed on leaves. An even more striking metamorphosis takes place with the final molt; the adult takes on aposematic apparel with a scarlet head, long orange forewings, and pink tibiae (Helfert and Sänger 1995).

Nickle and Castner (1995) counted the species of Peruvian katydids that use these different camouflage and mimicry strategies. Of 384 species of Phaneropterinae, Pseudophyllinae, Listroscelidinae, and Conocephalinae,

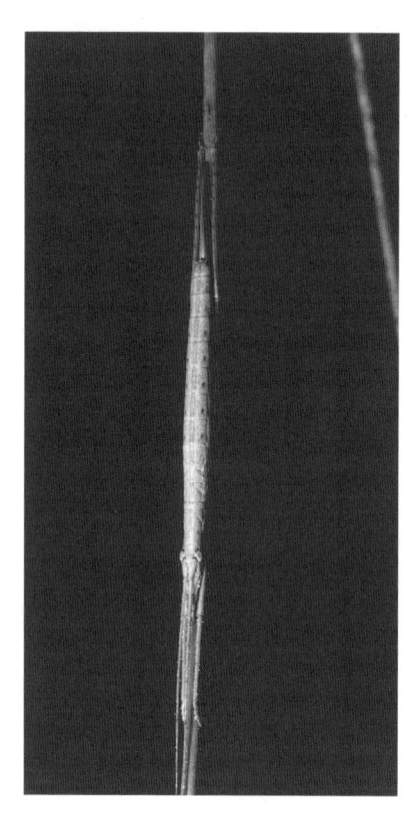

Figure 4.8. The cryptic, daytime resting posture of the sticklike *Kawanaphila nartee*, a pollen katydid (Zaprochilinae). On close inspection, the insect's eyes and antennae are visible at the bottom of the photograph and her ovipositor and hind legs can be seen at the top.

53 (13.8%) used visual imitation. In this fraction, 5 species were bark mimics, 13 looked like twigs, 29 looked like leaves, 4 resembled lichens, and 2 were Batesian mimics of wasps. Most taxa, however, showed a green or brown color match to background vegetation, without any other sort of mimicry (71.8% of species). Only 18 (4.6%) of the species did not use camouflage but instead moved completely out of sight, hidden deep within vegetation during daylight hours (37 species [9.2%] could not be reliably categorized).

Camouflage is only effective when backed up by a good match to the background on which the animal rests (Wallace 1891; Robinson 1969, 1973; Belwood 1990). Among the most perfect leaf mimics, the tropical Pseudophyllinae, "matching" behavior varies. Neotropical species sit on twigs and hold their forewings in a laterally flattened manner and thus mimic a whole leaf (Plates 10 and 11). In contrast, species from the Old World tropics tend to hold their tegmina in a dorsoventrally flattened position while sitting on large leaves, imitating only part of the leaf (Plates 12 and 13) (Heller 1995).

Evidence that cryptically colored katydids actually select the appropriate background comes from a detailed study not of a tropical leaf katydid but of a micropterous Western Australian cone-head, *Mygalopsis marki*, that shows a color match to its background (Lymbery 1984, 1992). Individuals of this species come in brown and green versions, and for day-resting animals (Fig. 4.7) sampled in nature, Lymbery found a significant association of body color with the color of the background vegetation. Background color matching is clearly adaptive because captive magpies (*Gymnorhina tibicen*) selectively ate *M. marki* whose color contrasted with background vegetation. The variable coloration of *M. marki*, in contrast to the color dimorphism seen in facultatively gregarious species (see Chapter 3), is apparently a response to seasonal and spatial variations in background vegetation colors. In fact, *M. marki* has a remarkable adaptive flexibility to its unpredictable and variable environment. Although all individuals are born (hatched) green, they change to brown after molting if the moisture content of the vegetation in the diet is experimentally reduced (Lymbery 1992). Lymbery's work does not rule out the possibility that there may be a genetic influence on color morph in *M. marki*. In fact, a genetic effect on a green-brown polymorphism does occur in another conocephaline, *Conocephalus maculatus*. Oda and Ishii (1998) used genetic crosses and other experiments to show both genetic and environmental influences (temperature and relative humidity) on color.

Bradyporines and hetrodines are exceptions among katydids in that they do not adopt crypsis as a primary defense mechanism. Hetrodines are the armored ground "crickets." As their common name indicates, they employ mechanical defenses against certain predators; species of *Acanthoplus, Enyaliopsis, Eugaster,* and *Gymnoproctus* all possess a pronotal shield with various large spines (Grzeschik 1969; Schmidt 1990; Glenn 1991; Mbata 1992a, 1992c) (Fig. 2.27). Such weaponry probably makes these species a very disagreeable food item for many vertebrates. Spines are also found on several green neotropical cone-heads such as *Panacanthus* species (Fig. 4.9). Here the spines are on the legs as well as the pronotum. The coloration of *Panacanthus* katydids is probably a cryptic defense against diurnal visual vertebrates, whereas the spines probably make it difficult for nocturnal foraging bats to handle these insects (Belwood 1990).

In addition to its cryptic green-brown polymorphism, *M. marki* also shows an intriguing behavior that is an example of a secondary defense mechanism. When confronted with a lizard predator, the insect rears up with mandibles spread (Fig. 4.7) and produces a rasping hiss by rubbing its hind wings on its abdomen. Although the visual component of this startle display increased the probability of surviving lizard attacks, Sandow and Bailey (1978) could find no evidence that the defensive stridulation added to the effectiveness of the display. Perhaps the acoustical component operates against other predators. *Mygalopsis* species turn out to have quite a repertoire of different sounds, including a squeaking noise

Figure 4.9. The defensive display of the cone-head *Panacanthus*. The insect faces the predator, raises its head, spreads its mandibles, and lifts its front legs. Photo by D. Klimas.

produced by expelling air from the mouthparts (Bailey and Sandow 1983), which might explain the literal meaning of *Mygalopsis*—"like a field mouse."

Disturbance sounds produced by an abdominal-tegminal mechanism have evolved independently in a very different katydid, the African pseudophylline, *Pantecphylus cerambycinus* (Heller 1996). An abdominally produced sound also occurs in the Haglidae (*Cyphoderris*), the sister group to Tettigoniidae (Fig. 2.5). But here it is produced when the first abdominal segment contacts the hind part of the thorax (Ander 1938). Male and female disturbance stridulation also has been reported for a number of neotropical false leaf katydids (Belwood 1988). Another type of disturbance stridulation is produced by the mecopodine *Sexava* katydid when it rubs its labrum on the mandibles (Lloyd and Gurney 1975). However, the most widespread defense noise in katydids is when the forewing calling mechanism of males (or females, in bradyporines with female song: Dumortier 1963 [see Chapter 5]) is co-opted for a disturbance function. Gregarious male Mormon crickets in bands almost always stridulate when disturbed, whereas solitary crickets only do it on occasion (Table 4.3). Although this sort of disturbance stridulation crops up in some species of Conocephalinae, Tettigoniinae, Mecopodinae, Pseudophyllinae, Phaneropterinae, and Bradyporinae (Heller 1996), it is quite rare. After handling some 90 tettigoniid species (mainly North American), Alexander (1960) reported tegminal disturbance sounds only in two pseudophyllines, a conocephaline and a tettigoniine.

Table 4.3. Differences in the defensive behavior of gregarious and solitary Mormon crickets (males only): a significantly greater number of gregarious than solitary males both stridulated and spat "tobacco juice" when disturbed (X^2, $P < 0.05$)[a]

Behavior	Gregarious (band) on ground	Solitary	
		On ground	Calling
Proportion stridulating when handled	20/20	3/16	1/18
Proportion spitting "tobacco juice"[b]	20/20	8/16	No data

[a] I made these observations during August 1999 at one site where the crickets were gregarious (Dinosaur National Monument, Colorado) and one where the species was solitary (Poudre Canyon, Colorado).
[b] Gregarious crickets also produced much larger amounts of regurgitate than did solitary individuals and often fouled the vials into which they were placed.

These examples of acoustic disturbance occur when the katydid is touched or handled in some way, so there are two possible functions of this mechanism: to startle the predator into dropping the prey or to fool the attacker by mimicking the disturbance stridulation of dangerous or bad-tasting insects such as reduviid bugs and cerambycid beetles (Heller 1996). A third function of disturbance stridulation, to confuse the predator, may occur in gregarious Mormon crickets; males stridulate while jumping away from a potential predator approaching from several meters away. I have seen this reaction when trying to sample insects from a migrating band; a wave of insects jumping left and right is always accompanied by the frantic "zips" of leaping males, resulting in "Those nearest the source jump[ing] away from it, thus exciting others near by, and the motion spreads out in waves resembling those produced by casting a stone into a quiet pool of water" (Cowan 1929).

Startle displays can also be visually conspicuous. Some sagines and listroscelidines (Cohn 1965; Kaltenbach 1990) have a particularly dramatic response when they open the mandibles and raise the forewings and hind wings as well as the spiny forelegs (Figs. 4.7 and 4.9). Flashy secondary-defense displays also occur in leaf-mimicking pseudophyllines: when discovered by a predator, *Acanthodis curvidens* breaks out of its cryptic pose with a sudden visual flash of colored hind wings (Robinson 1969), whereas the Peruvian *Pterochroza ocellata* spreads a brilliant pair of hind wings equipped with large eyespots (see also Belwood 1990; Castner 1995). The Australian mountain katydid, *Acripeza reticulata* (Phaneropterinae), adds a third defensive component. Its primary defense is a cryptic stone-like coloration, but when disturbed, adults raise the forewings to reveal brightly colored blue and red abdominal bands, and they puff out a yellow band behind the head that exudes drops of fluid (Plate 18) (Rentz 1991, 1996).

The technicolor display of *A. reticulata* is probably less a startle display than it is a warning that the insect will make a disagreeable meal; the droplets of body fluid that ooze out of the intersegmental membranes are likely to be distasteful (Rentz 1996). A colorful Peruvian cone-head that everts an odoriferous gland on its abdominal tip is almost certainly displaying warning coloration. The "bright waxy" colors of the "Crayola katydid," an undescribed species in the genus *Vestria*, are never concealed by the tegmina and so are always on display, much like those of the adult stage of the ant-mimicking agraeciine we discussed earlier (Plates 14 to 17). This *Vestria* species produces a strong odor by emitting pyrazine alkaloids, but only when disturbed. Nickle et al. (1996) found this scent to be effective in causing two naive woolly monkeys (*Lagothrix lagotricha cana*) to reject the *Vestria* species as prey. This occurred after previous encounters in which the katydids were dropped after being sniffed or tasted by the monkeys. Following the initial learning experience, one monkey sniffed its hands, and the other (the unfortunate individual that tasted its multicolored prey) "became agitated" and stuffed its mouth with leaf litter. Encounters with Crayola katydids in nature presumably teach woolly monkeys to dine on less colorful prey.

Less obvious cases of warning coloration in distasteful katydids appear to include some of the gregarious species I reviewed in Chapter 3. Sillén-Tullberg and Leimar (1988) pointed out that distastefulness is an important condition for the evolution of gregariousness. Distasteful, colorful prey benefit from having others close by, because predators quickly learn to avoid these prey when a number of them are sampled quickly. Thus, the typical "dilution" benefits of being a member of a prey group are enhanced by distastefulness. Similarly, Key (1957) hypothesized that a change in color correlated with distastefulness in stick insects; certain species become bad tasting when densities increase and stick-mimicking is no longer effective. A recent example involves *Schistocerca* locusts from Africa and the United States (including the infamous desert locust *S. gregaria*). In these species crowding induces a conspicuous black and yellow coloration as nymphs feed on distasteful native plants and thus become unpalatable to lizards (Sword 1999, Sword et al. 2000).

There is some evidence for a similar phenomenon reflected in two morphs of Mormon crickets (Plate 9). Cryptic, solitary individuals do not eat sage leaves (*Artemisia* species), a foliage high in toxic compounds, even though the plant is common in their habitat (Hansen and Ueckert 1970; Ueckert and Hansen 1970). However, the foreguts of dark, noncryptic individuals in bands contain many sage leaves (over 90% of the adult diet) (MacVean 1987, 1990; Redak et al. 1992) that probably make the katydids distasteful when they regurgitate by "spitting tobacco juice." This behavior is common in Orthoptera and can be effective against predators (Lymbery and Bailey 1980). Thus, it is adaptive for individual Mormon crickets to group together in the dark-colored bands that are so obvious to visually hunting predators (see Chapter 1). Regurgitation appears to be the only reason why gregarious Mormon crickets make disagreeable prey. Other parts of their bodies appear to be palatable; I have often seen small birds picking out "choice parts" from road-killed band members.

The distasteful-gregariousness hypothesis predicts that Mormon crickets in bands should be more likely to regurgitate than solitary individuals should. Some preliminary experiments indicate that this is the case (Table 4.3). Bands of Mormon crickets do attract predators such as birds, as Mormon pioneer history shows (see Chapter 1). The gulls viewed by Mormon pioneers as heroic crop savers that gorged on and then regurgitated Mormon crickets may well have been naive predators undergoing an important learning experience.

Another way in which katydids can use exudates to fight predators involves the bizarre habit of "autohemorrhage" or reflex bleeding. This has been noted in some bradyporine and hetrodine katydids. Burr et al. (1923) observed that *Bradyporus dasypus* responded to handling by squirting blood from the tergites and from under the thorax. A few years later, Boldyrev (1928) reported similar behavior in *B. multituberculatus* following "a slight cracking noise" from the opening of pores on the pronotal ridges, which caused jets of hemolymph to shoot 15 cm, "striking the hand for 1–2 seconds." Hetrodines also squirt blood. Glenn (1991) stated that *Enyaliopsis nyala* can squirt a jet of "light green liquid" up to 25 cm. For a *Eugaster* species, Grzeschik (1969) showed that blood squirting is caused by a contraction of abdominal muscles and that discharges are localized to the "nozzles" (pores in the cuticle) nearest the disturbance stimulus. The hemolymph ejected from *Eugaster* is not toxic, and Grezeschik concluded that the advantage of reflex bleeding probably lies in confusing the predator with a sudden faceful of sticky fluid (Fig. 4.10).

Finally, in addition to "fight" responses, designed to slow or deter attacks, a few katydids use "flight" as a secondary defense against predators. In response to ultrasound (such as those produced by rustling vegetation), *Neoconocephalus ensiger* first stops calling (Faure and Hoy 2000), then swivels its body around the plant stem away from the direction of the disturbance so that the whole insect (except for its tarsal tips) is concealed from the sight of the approaching predator (Cantrall 1943). Anyone who has attempted to collect meadow katydids (*Conocephalus* and *Orchelimum*) by hand has encountered the frustrations of this swiveling behavior as the insects dodge most attempts to grasp them. As a "last-resort" response to an approaching hunter, *N. ensiger* and other katydids jump from their perches in vegetation. I have commonly observed calling male Mormon crickets do this as I approached them. Several species of the riparian genus *Orchelimum* (Blatchley 1920; Smalley 1960), and the bromeliad-inhabiting *Agraecia pulchella* in the neotropics (G. K. Morris, pers. comm. 1999), even

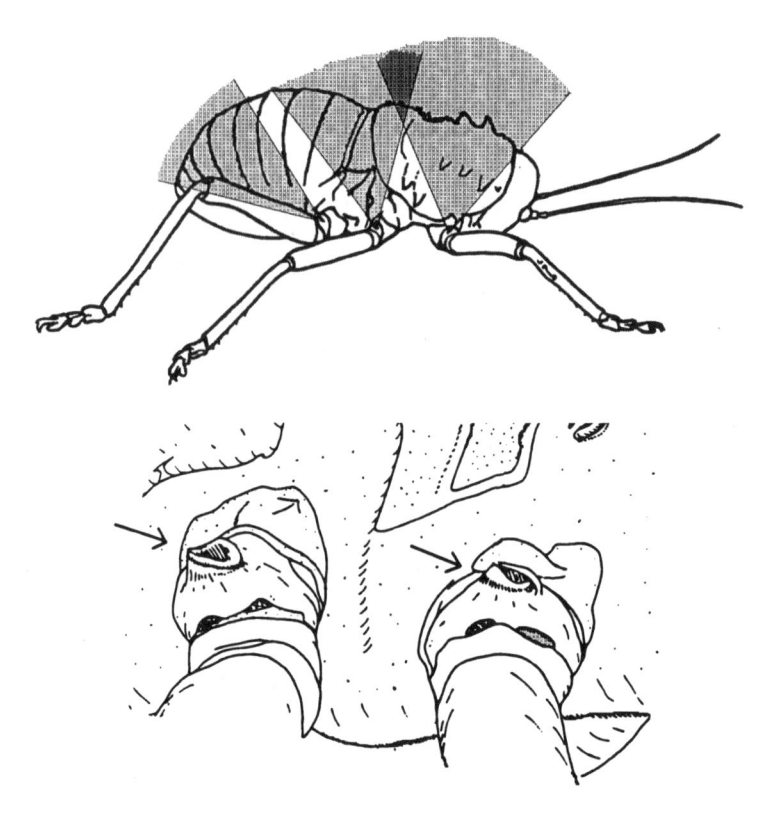

Figure 4.10. Reflex bleeding as a defensive response in *Eugaster* species (Hetrodinae). The shaded areas in the upper drawing show the range of blood squirts from "nozzles" at the bases of the legs (lower drawing). Redrawn from Grzeschik (1969).

dive from their perches into water. This can be a risky "out of the frying pan and into the fire" maneuver, as several *O. nigripes* observed to leap from their perches ended up in the jaws of turtles (Feaver 1977)!

With the basic survival and reproductive biology of katydids in mind, we can now turn to the main focus of this book—sexual behavior and mating. We start in Chapter 5 with the singing behavior for which the group is so well known. Katydids, along with birds, anurans, cicadas, grasshoppers, and true crickets, are nature's songsters. What are the mechanisms than enable katydids to produce and hear sound signals? And how do these communication mechanisms function in reproduction?

5

Entomological Choristers: Song and Mate Attraction

> *I love to hear thine earnest voice,*
>   *Wherever thou art hid,*
> *Thou testy little dogmatist,*
>   *Thou pretty Katydid!*
> *Thou mindst me of gentlefolks,—*
>   *Old gentlefolks are they,—*
> *Thou say'st an undisputed thing*
>   *In such a solemn way.*
> *Thou art a female, Katydid!*
>   *I know it by the trill*
> *That quivers through thy piercing notes,*
>   *So petulant and shrill;*
>
> Oliver Wendell Holmes, "To an Insect"

5.1. The Choristers of Summer

Clarence Valentine Riley, Missouri State entomologist, introduced his 1874 insect report featuring the katydids with the first four lines of Holmes's poem. Riley acknowledged these insects as the dominant songsters of late summer and autumn in North America: "Then do grass and wood resound with song—not so much of feathered tribes, as of insect tribes, and especially of the Katydids, green vaulters from leaf to leaf and from branch to branch—essentially American. The song of these entomological choristers may not compare in melody with that of the ornithological warblers; but though it grate at times, it has a merriness all its own, and, as it comes from bough or spear, can never be unpleasant, for there is no sadness in the earth's minstrelsy."

Whether sad or not, "minstrelsy" is a crucial component of successful reproduction in almost all katydids. It is a signal that guides the female to the male. Katydid sound signaling has been the subject of a great deal of study, both at the level of the underlying machinery of sound production and reception and at the level of evolutionary causation. The evolutionary research suggests that song is more than just an acoustical beacon; it probably contains important information on the quality of the caller that is assessed by both mate-seeking females and eavesdropping rival males. This aspect of song research is covered in Chapter 7. A main purpose of the present chapter is to describe what we know about the basic function and evolution of katydid calling. But first, how are calls produced and how are they detected by potential mates and rivals?

5.2. Producing and Hearing Songs

5.2.i. An *instrumentum stridoris*

A widespread entomological myth is that crickets sing by rubbing their legs together; "legs" is the usual answer I get from schoolchildren when I ask, "How do crickets and katydids sing?" However, no orthopterans—indeed no arthropods—stridulate using leg structures alone (Ewing 1989). There are orthopterans that rub the legs on their abdomens or even on their wings, but as mentioned in Chapter 2, legs are not involved at all in the stridulatory movements of either crickets or katydids; both groups produce their distinctive calling sounds using their forewings (tegmina) (Fig. 5.1).

The myth about leg stridulation is rather puzzling because stridulation using forewings in katydids and crickets has been known for hundreds, and probably thousands, of years. The earliest account dates back to ancient Egypt, where a hieroglyphic for an "initiated" or "mystic" man uses a grasshopper as a symbol (Greek = τέττιγα), "for he does not utter sounds through his mouth, but chirping by means of his spine, sings a sweet melody" (deciphered by Egyptian scribe Horapollo in the fifth century A.D. and later translated into Greek by Philip) (Horapollo and Cory 1840, cited by Myers 1929). The reference to sounds from the "spine" may well point to the dorsal forewing mechanism of the long-horned grasshopper rather than leg-abdomen stridulation of the short-horned variety. Chinese naturalists had recognized the forewing mechanism by 250 B.C., as they had determined that various parts of the exoskeleton in insects, including the wings, could be organs of sound production (Chou 1960).

In western scientific circles, an appreciation of wing singing came much later. In *Theatre of Insects*[1] Thomas Moffett et al. (1634) referred to the cricket-like sound that resulted when the detached forewings of a field cricket were rubbed together (cited in De Geer 1752, and Kirby and Spence 1818, p. 397). Also, in his delightful "monography" on crickets in the

---

[1] Kirby and Spence (1826) described this Elizabethan text as the "fruits of the successive labours of several men of talent." The works of Dr. Edward Wotton and Conrade Gesner were expanded by physician and botanist Thomas Penny. Penny died in 1589 before completing the manuscript. Moffett, another physician, purchased the work but also died before publishing. Finally, Sir Theodore Mayerne, a court physician to Charles I, published it. Moffett's daughter is better known today than her father, as she is thought to have inspired the rhyme "Little Miss Mouffett sat on a tuffet eating her curds and whey. Along came a spider . . ." (Evans 1985).

Figure 5.1. A comparison of tegminal calling in two ensiferans from southwestern Australia. **a**. The forewings (tegmina) of a calling katydid, *Conocephalus upoluensis*, are barely raised. **b**. Those of a calling cricket, *Eurepa marginipennis* (Gryllidae: Eneopterinae), are almost vertical when the male is calling. At the base of the katydid's left forewing is the square-shaped mirror on which the file (on the underside) is clearly visible as a white mark.

Natural History of Selborne, Gilbert White (1789) deduced that field crickets sang using a "brisk friction of one wing against another."

So how do the forewings of male katydids actually generate sounds? Baron De Geer, "observant historian of the manners and economy of insects" (Kirby and Spence 1826), provided the main details in his massive seven-volume *Mémoires pour servir à l'Histoire des Insectes* (1752). De Geer noted that great-green grasshoppers, *Tettigonia viridissima* (Fig. 2.29 and 2.30), "... in that part of the right wing case which is folded horizontally over the trunk, have a round plate, made of fine transparent membrane [top right *a* in "Fig. 1" within Fig. 2.30], resembling a little mirror or piece of talc, and as tense as a drum. It is surrounded by a strong and prominent nervure [vein], but is concealed under the fold of the left wing case, where there are strong nervures corresponding to what may be called the hoop of the drum." He concluded that "it is exceedingly probable that the quick motion with which the insect rubs these nervures against each other, produces a vibration in the membrane, whence the sound is augmented" (cited in Kirby and Spence 1818 and Russell et al. 1831) (Fig. 5.2). Seventy-five years after De Geer's work was published, the Rev. Lansdown Guilding (1827) had a closer look at the underside of the left tegmen's nervure, an "instrumentum stridoris," which is "... supplied with hard and regular teeth over which, on the right [forewing] a bony process is placed so as to act on the serrated projection of the [forewing] which lies above it...." In modern terminology the toothed vein is the file, and the "projection" is the "scraper" or "plectrum." Goreau (1837), who appears to have coined the term *stridulate* (*The Oxford English Dictionary*), also gave a detailed account of the katydid's tegminal organs.

Finally, in his *Manual of Entomology*, Hermann Burmeister (1836) proposed an elegant but erroneous theory for the katydid's song mechanism (mysteriously he did not even mention the wing-stridulation theory even though he cited the works of both De Geer and Kirby and Spence). Burmeister's was a harmonica-like mechanism in which air jets from the katydid's two large (meta-) thoracic spiracles blast upward to strike De Geer's "little mirror" on the forewing. As we shall see, the large spiracles do have an acoustic role, but in receiving rather than in producing sounds.

5.2.ii. Behavior and mechanics of song production

When raising his forewings to begin a calling bout (Fig. 5.1), the male katydid first warms his thoracic muscles, using rhythmic contractions very similar to those used during flight (Heath and Josephson 1970; Heller 1986). Singing involves a sequence of tegminal openings and closings, with the "power stroke" coming during closing when the right wing's scraper is pushed over the left wing's file (Fig. 5.3) above it (Fig. 5.1, top, and Fig.

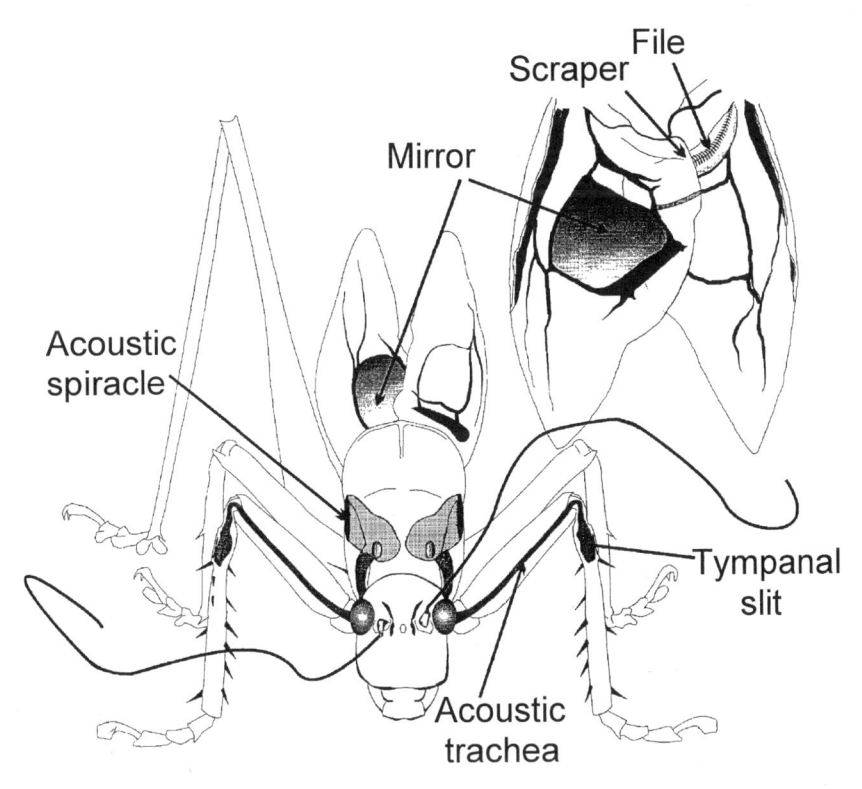

Figure 5.2. Mechanisms of calling and hearing in katydids. The main part of the figure shows a calling male and both the location and position of his tegminal calling structures and foreleg-thoracic hearing system. Details of the calling structures are shown on the tegminae in the top right figure (anterior parts at top). To produce the calling song, the scraper on the left wing moves over a file on the right, and sound energy is transduced into the frame surrounding the resonating mirror. Original drawing by G. K. Morris.

5.2); often, when the forewings are drawn open, the file and scraper do not make full contact so that little sound is produced (Morris and Pipher 1972; Walker 1975). Sequences of wing closures can produce complex song patterns in which the scraper on the right wing touches different parts of the file on the left wing at different speeds in a series of movements that T. J. Walker and D. Dew (1972) aptly described for one phaneropterine as "fiddling finesse."

The left-over-right tegmen rule in katydids means that the wings are asymmetrical because the file on the right tegmen is vestigial. But there are exceptions to this rule. In the shield-back *Neduba* males have a functional file on both forewings, and some males in the population can be found singing in the typical left-over-right position, whereas others show the

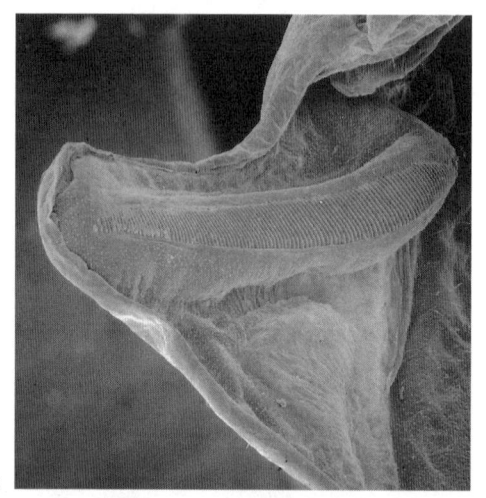

Figure 5.3. Scanning electron microscope photographs of the stridulatory files (on the underside of the forewings) of two ultrasonic katydids. Compare the many-toothed file (typical of most katydids) of a *Haenschiella* species (**a**) to the simple file of *Kawanaphila nartee*, a species with only a dozen or so teeth in the stridulatory file (**b**). **a**: Photo by G. K. Morris. **b**: Photo by author.

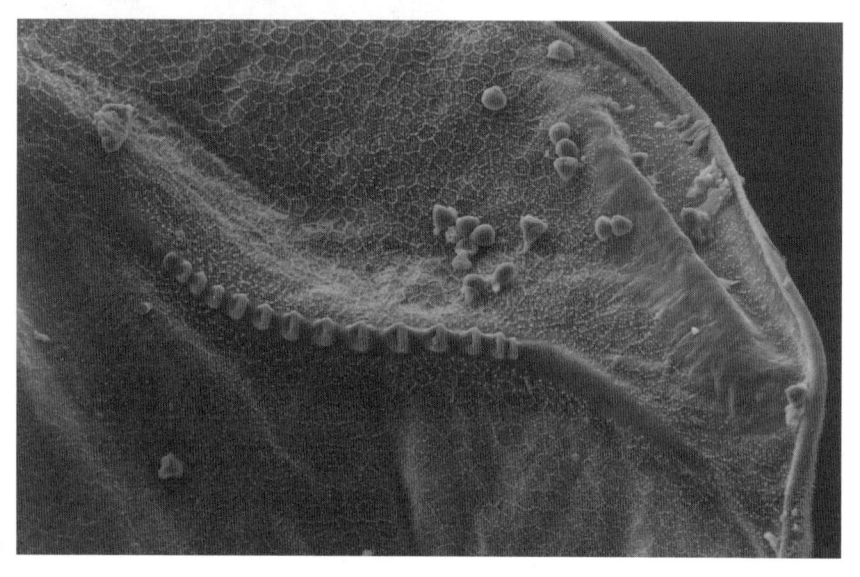

right-over-left wing configuration (Morris et al. 1975a) that is more typical of calling gryllids (Rentz 1991). However, it is unknown whether a song produced by right-over-left calling is as effective as one produced by the usual position. For example, in the pollen katydid *Kawanaphila nartee*, 3 of 67 sampled males had a reversed wing overlap but produced calls with lower frequencies than did all other males (Fig. 5.4) (Gwynne and Bailey 1988). Ambidextrous singing involving the use of both types of overlap, even by the same male, appears to be normal in the haglid genus *Cyphoderris* (Spooner 1973). Haglids appear to be ancestral to Tettigoniids (Sharov

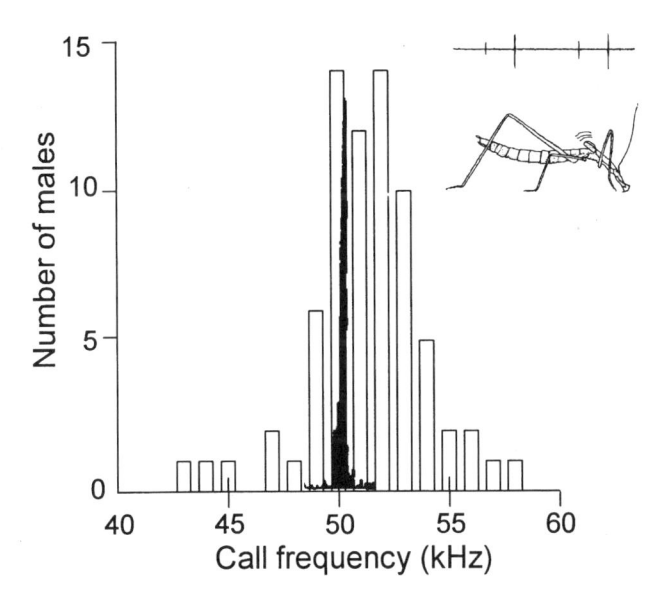

Figure 5.4. The calling song (top right) frequencies of a pollen katydid, *Kawanaphila nartee*. The histogram (white bars) shows the distribution of the peak song frequencies produced by different males sampled in nature. The three lowest frequencies (43–45) represent individuals that had a reversal of the left-over-right overlap of wings that is typical both for this species and for katydids in general. The black part of the figure is the calling song spectrum of one male (with relative sound energy being represented in the vertical axis), showing the very narrow distribution of frequencies used by this ultrasonic katydid (compare to the broad spectrum songs of the katydids in Figs. 5.9 and 5.13). Redrawn from Gwynne and Bailey (1988).

1968; Gwynne 1995), so it is interesting to note that a 55-million-year-old fossil katydid *Pseudotettigonia amoena* recently was reported to have symmetrical wings, an indication that *Pseudotettigonia* songsters used both wing overlaps, thus representing a possible ancestral behavior for katydids (Rust et al. 1999).

We can be fairly certain that when these ancient katydids closed their wings, as in today's singers, vibrations were produced by scraper-on-file friction and transferred through the scraper into the main resonating region of the tegmen—"... the stretched membrane of the mirror... to which vibration is communicated by the shaking of the surrounding frame" (Fabre 1896, 1917) (Fig 5.2, top). The surface of the mirror itself radiates vibrations to the air much like the cone of a speaker, and this small generator can produce some very loud songs: the Rev. Guilding (1827) claimed that his tropical pseudophylline males could be heard up to a mile away! Morris and Pipher (1967) and Bailey (1970) experimentally con-

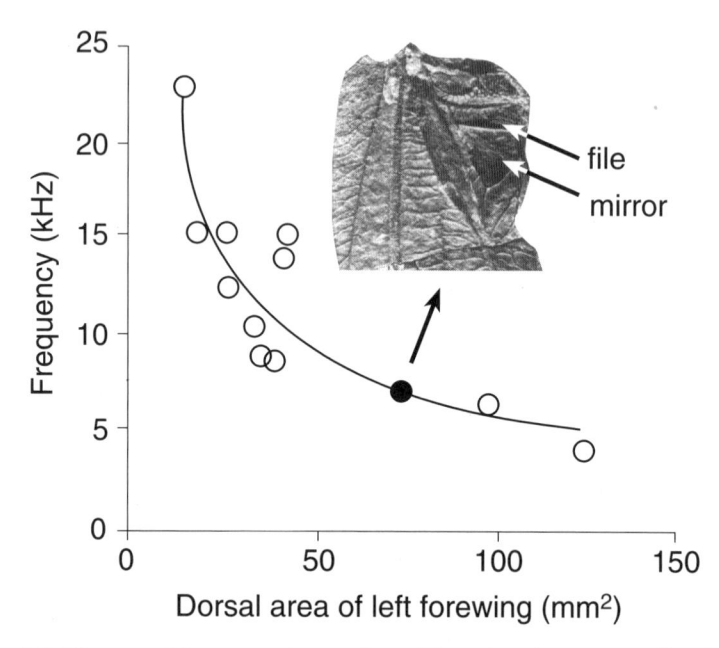

Figure 5.5. The area of the resonating surface of the wings (as measured by the area of the left forewing) of 11 species of katydids (open points) correlates with the main frequency of the song. This allowed Rust et al. (1999) to predict the song frequency of an extinct katydid using the area of its wing (shown in the top right of the figure) (filled point). Plot redrawn from Rust et al. (1999).

firmed a radiation function for the mirror[2] (Fig. 5.2, top). However, other parts of the forewings also must radiate sounds because not all species have a distinct mirror region (e.g., the zaprochiline *K. nartee*: Gwynne and Bailey 1988).

It is the frame around the mirror of the katydid's left tegmen that determines the pitch (frequency) of the call (Morris and Pipher 1967; Bailey 1970; Sales and Pye 1974). The size of this region of the wing correlates with call frequency, allowing Rust et al. (1999) to describe the likely frequency at which the extinct *Pseudotettigonia* males pitched their songs (Fig. 5.5).

Most katydid calls are nonmelodious combinations of frequencies in the range of 10 to 100 kHz (e.g., Ahlén 1981) producing the "buzzes, zips or clicks" (Alexander et al. 1972) that do "grate at times" (Riley 1874), although we should not forget that "harsh sounds [do not] always displease" (White 1789). These noisy calls of most katydids contrast with the narrow-band, pure tones typical of the melodious calls of birds, anurans,

---

[2] Chinese owners of katydid pets supposedly treated the mirror with a mixture of brass and rosin "to refine and heighten the volume of the insect's voice" (Jin 1994).

and gryllid crickets, songsters that use a frequency range audible to humans (Sales and Pye 1974).

A number of katydids do produce more melodic, pure-tone calls either in the audible (Morris and Beier 1982; Morris et al. 1989; Heller 1995) or in the ultrasonic range of 50,000 Hz and above (Suga 1966; Sales and Pye 1974; Gwynne and Bailey 1988; Mason et al. 1991; Morris and Mason 1995). In fact, a few katydid groups are exceptional in specializing in the production of mainly pure-tone songs. Montealegre and Morris (1999) reported that 75 percent of the described songs of the (mainly tropical) Pseudophyllinae are of this sort; the narrow frequency bands and the rapid attenuation of high frequencies may function as a way of avoiding the plethora of eavesdropping bats in the neotropics.

The songs of katydids certainly span a huge range of frequency bands, from a mere 600Hz produced by the Malaysian pseudophylline *Tympanophyllum* to a high-pitched 135kHz of *Arachnoscleis* (a listroscelidine from western Colombia) that rivals many bats in its use of very high frequencies (Morris 1999). These exceptional katydids, and most crickets, can produce pure-tone calls because the rate at which the scraper strikes individual file teeth matches the resonant frequency of the tegminal structures (in the case of gryllids, this is the "harp" area of the left forewing). As a result, the vibrations of these structures do not decay between tooth strikes. The decaying of mirror-frame vibrations in most katydids supplies the additional "side-band" frequencies that combine to produce a broad-band, buzzy call.

In addition to the tegminal surfaces, which are retained in males in abbreviated form even when females are wingless (e.g., some lipotactines: Ingrisch 1995), there can be modifications of other body parts that resonate sounds. The boxlike pronotal shield that protects the front part of the forewings in many Orthoptera (Fig. 2.11) is co-opted into a call-amplification device by some katydids (Morris et al. 1975a). One of the most elaborate examples of this effect involves "covert stridulation" in an undescribed species of South American conocephaline. The "covert" component refers to the fact that the male's forewings are completely concealed beneath an enlarged pronotum, which acts like a speaker box in radiating the call (Morris and Mason 1995) (Fig. 5.6, Plate 19).

5.2.iii. Other katydid signals

In the last four lines of the verse that opens this chapter, O. W. Holmes was dead wrong about the singer's sex; the "trill" of the true katydid is from the male. There are, however, some species in which females produce low-intensity, and usually very brief, sounds in response to the male's call.

In general, the tegminal mechanisms used by females are different from

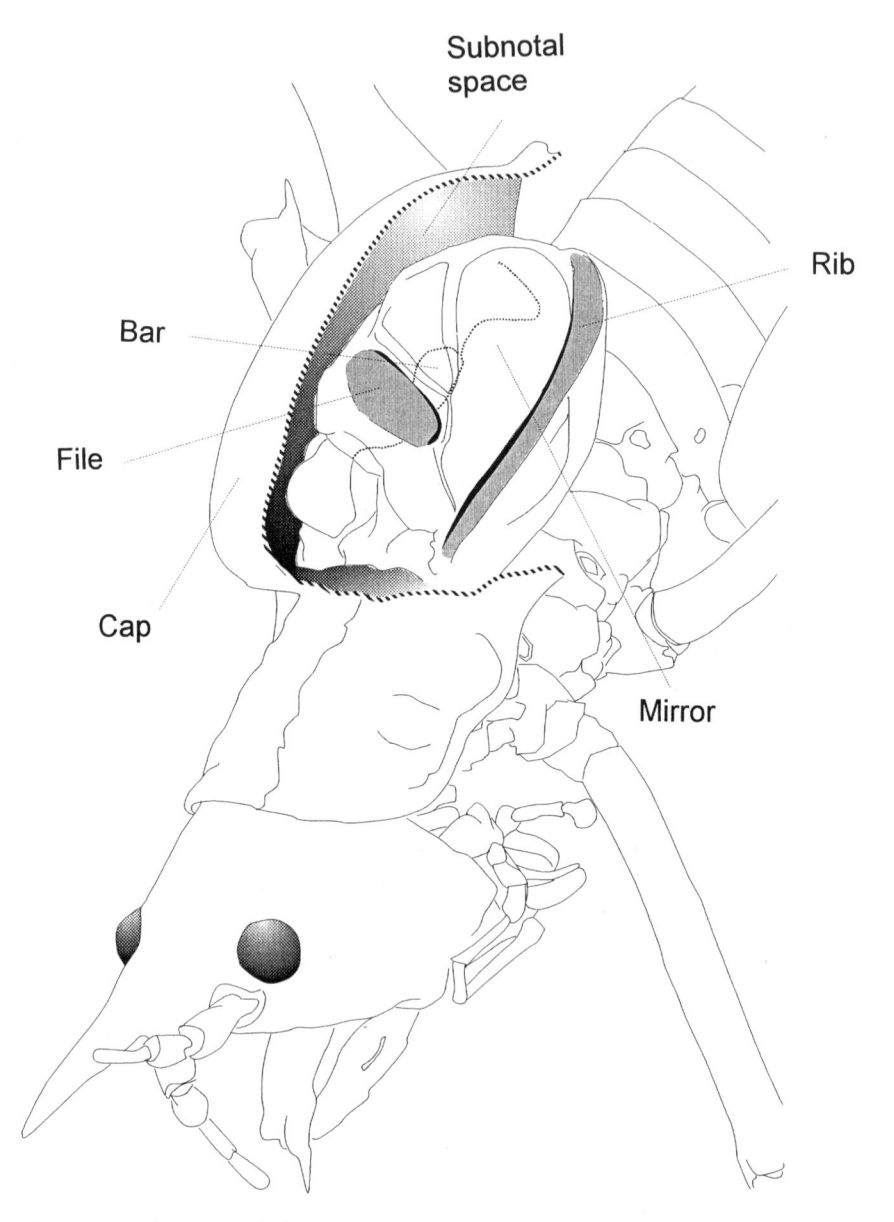

Figure 5.6. Covert stridulation in an undescribed agraeciine katydid from South America (see Plate 19). The head, thorax, and part of the abdomen of a male are shown, with the calling wings hidden beneath the domed pronotal "cap." From Morris and Mason (1995). Drawing by G. K. Morris.

those of the males. The answering "clicks" of some female phaneropterines (e.g., Fig. 7.15) are produced by spines moving over anal veins of the dorsal surface of the forewing. In other female Phaneropterinae, the anal veins are on the wing's ventral surface. In contrast, female *Meconema thalassinum* rub opposing spines on both surfaces of the tegmina, although the function of sound production by these females is unclear (Nickle and Carlysle 1975; Robinson 1990). In most ephippigerine Bradyporinae, and apparently in the pseudophylline *Panoploscleis* (Beier 1960) (Fig. 9.21; see Chapter 9), females have a forewing organ that is derived independently from the male's because the female file veins are on the right wing rather than the left and are toothed on the dorsal rather than the ventral surface. As files are on the right wing, females still show the typical katydid "left-over-right" forewing overlap (Goreau 1837; Hartley et al. 1974; Nickle and Carlysle 1975).

Although katydids are best known for the distinctive male calling songs produced by forewing stridulation, as we saw in Chapter 4, they can also produce biologically relevant sounds to defend themselves against predators. Phyllophorines, for example, stridulate using the coxae (Fig. 5.7) and do not show tegminal stridulation (Carl 1906; Dumortier 1963; Lloyd 1976). Because this subfamily retains tibial ears, could it be possible that coxal sounds have replaced tegminal stridulation as a calling signal in this group? More likely, however (as there have been no reports of phyllophorine calls), the ears of these katydids detect predators.

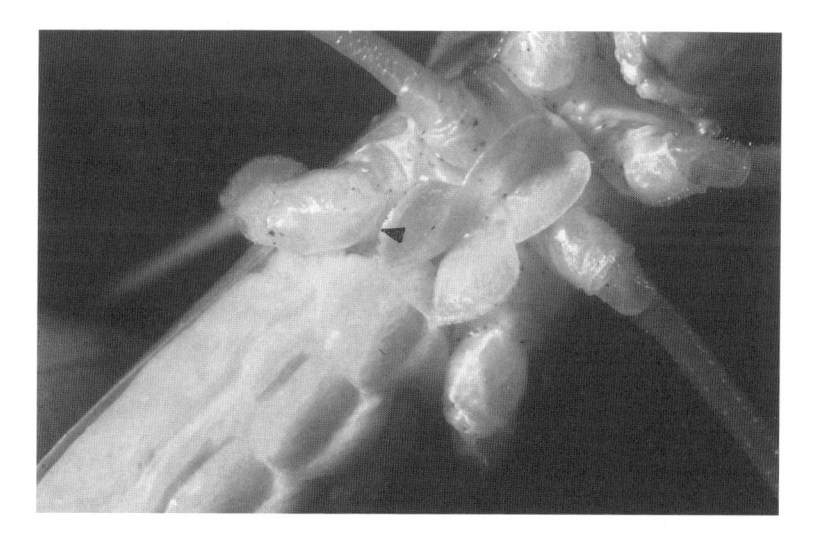

Figure 5.7. Ventral view of a phyllophorine katydid (*Phyllophora* species), showing the stridulatory mechanism on the thorax. A scraper on the sternite (see arrowhead) is rubbed onto a file on the hind coxa. Sounds produced probably function in a disturbance context. Photo by D. Klimas.

Whereas sound communication brings the sexes together over long distances, signaling through the vibration of substrates such as plant stems can be an important mode of communication when individuals are closer (Bell 1980; Morris 1980; Markl 1983; Dambach 1989). Insects can vibrate the substrate by drumming, which produces longitudinal compression waves in the substrate, and by producing transverse substrate waves (Morris 1980). *Meconema thalassinum* males drum one hind leg against a leaf, producing both substrate vibrations and an airborne component that is audible several meters away (Ragge 1965). Drumming may be the only way of attracting mates in this species, one of the few katydids that has lost tegminal singing (Gwynne 1995).

One source of transverse vibrational waves is the by-product of forewing stridulation (Kalmring et al. 1990). These vibrational cues may be useful when a katydid seeking a calling conspecific cannot localize airborne sounds as might occur in dense vegetation: in *Tettigonia cantans*, vibratory cues do make it easier for the female to locate the singing male (Latimer and Schatral 1983). At distances of less than a meter, *Ephippiger ephippiger* females can locate a singing male using just the vibratory components of the call (Stiedl and Kalmring 1989).

A second type of transverse wave is produced in the absence of stridulation when a katydid signals by shaking its perch (Busnel et al. 1956; Morris 1980; Morris et al. 1994). Shaking behavior, called *tremulation*, can produce very elaborate signals (Fig. 5.8) and has been observed in a number of species, particularly in neotropical Pseudophyllinae and Conocephalinae (Belwood 1990; Morris et al. 1994). Tremulation is used to call mates in a few species, especially some neotropical species in which males can be seen tremulating in the absence of females (e.g., *Choeroparnops gigliotosi*: Fig. 5.8). Many of these species tremulate a lot but have very low "duty cycles" of singing; they spend less than 1 percent of the time stridulating. Mates are often attracted by alternating tremulation with drumming, where the abdomen is tapped onto the substrate. Other katydids, such as the pseudophylline *Myopophyllum speciosum*, tremulate both to attract and while courting a female (Fig. 5.8) (Morris et al. 1994).

By their nature, vibrational signals are confined to the plant in which the signaling katydid is perched. As such, they function over closer ranges than communication by airborne sound. A potential function of the more reclusive substrate signals is to reduce eavesdropping by two types of enemies: acoustically orienting bats (Belwood 1990) (see Chapter 8) and competing males (Alexander 1975; see Morris 1980; Morris et al. 1994). Alternatively, tremulation signals may be more highly localizable signals directed at females (especially in dense vegetation) or represent a response to other acoustically orienting katydid predators such as lizards (Bateman 1995). The various hypotheses for tremulation have yet to be tested, although some comparative data are consistent with an eavesdropping-bat

Choeroparnops gigliotosi

Myopophyllum speciosum

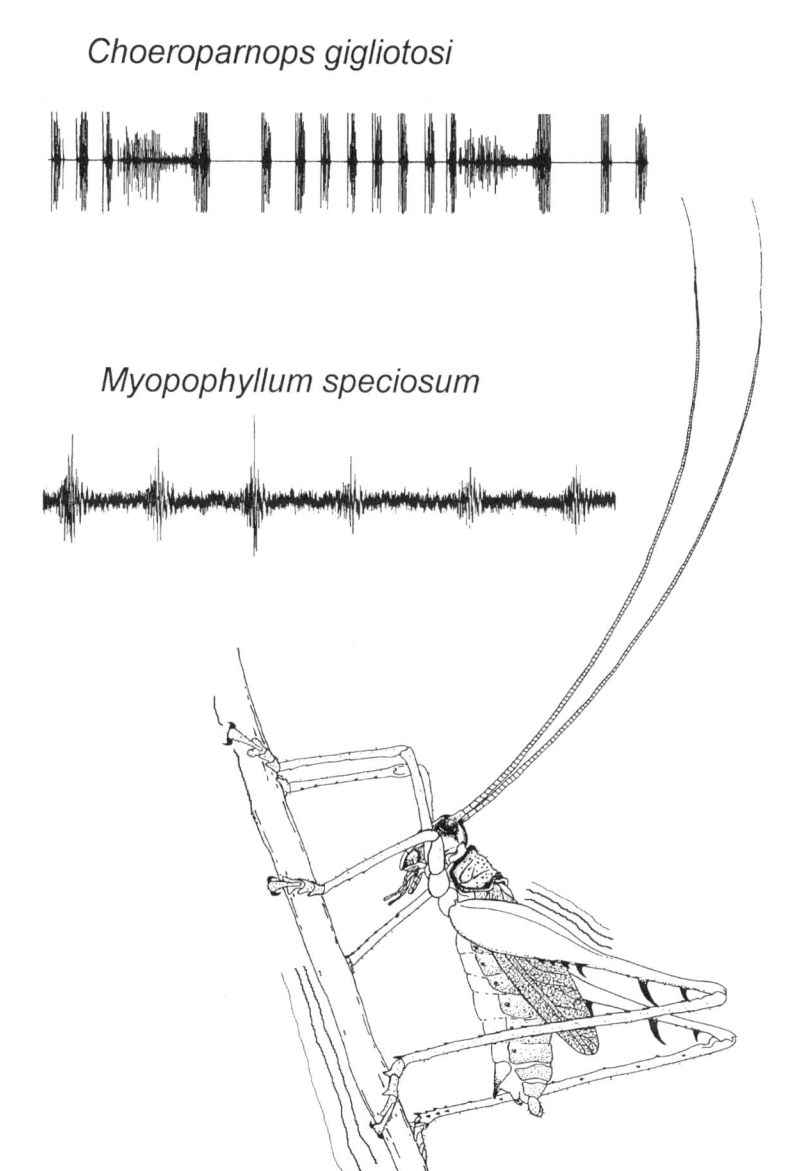

Figure 5.8. Male tremulation signals of two neotropical pseudophyllines. *Choeroparnops gigliotosi* repeats a complex tremulation "calling" signal (in the absence of females) made by shaking his perch then striking his perch with his abdomen. Two signals totalling about 50 seconds are shown in the upper trace. Males of *Myopophyllum speciosum* (bottom part of figure including drawing) tremulate when courting the female. A 5-second trace of the signal is shown. Redrawn from Morris et al. (1989). Artwork of katydid by Heather Proctor.

hypothesis: Heller (1995) compared acoustical calling patterns in species of New and Old World tropical Pseudophyllinae. Malaysian katydids have never been observed to tremulate, and the average calling times of males in this part of the world, where there appear to be few bats that orient to the songs of prey, are significantly higher than those in Ecuador or Costa Rica.

5.2.iv. Hearing and finding other katydids

At sunset on the coastal heathland on the edge of the Pinnacles Desert in Western Australia, a male cone-head, *Mygalopsis marki*, begins to call from the upper branches of a bush. A few other conspecific males are calling in the distance, but our male cannot hear them at first because their sounds are masked by the loud call of another katydid, *Hemisaga denticulata*, singing in the next bush. It is only when the *Hemisaga* song stops a few seconds later (inhibited by the *Mygalopsis* song: Fig. 5.9) that the cone-head male can hear his rivals, including a large (and loud) male about 2 m away. In response to his vocal neighbor, our male moves silently down to the ground and walks away from the other caller. After a 10-m journey, the neighbor's call is a distant sound of about 65 dB. The male ascends another bush to finally begin an uninterrupted bout of signaling.

This description of the nocturnal acoustical world of the *Mygalopsis* male is distilled from some 15 years of published research on the hearing and behavior of this species (Thiele and Bailey 1980; Bailey and Thiele 1983; Dadour and Bailey 1985, 1990; Römer and Bailey 1986; Römer et al. 1989; Römer 1993). These studies reveal details of the acoustic environment of these insects and find that *Mygalopsis* males maintain a specific calling distance so that neighboring conspecific calls are heard at a sound level of approximately 65 dB. This rule of thumb means that the distance between calling males averages 11.5 m in the open heathland of the Pinnacles site but only about 6 m at a site several hundred kilometers to the south where there is greater sound attenuation owing to the more dense woodland vegetation (Römer and Bailey 1986).

The *Mygalopsis* research is unique in revealing what the katydid actually hears in nature, not by extrapolating from the usual laboratory neurophysiological results but from "tapping" into and recording from the neural auditory circuitry of live physiological preparations in the field. In the words of Heiner Römer and his colleagues, the ears and neural circuitry of the *Mygalopsis* species are used as a "biological microphone" (see Roeder 1966). Recordings are taken directly from single cells, auditory interneurons located in the prothoracic ganglion, a neural center positioned between the neural connections from the brain to the ears (Fig. 5.10). These "omega" interneurons are particularly useful to scientists because they are sensitive to a range of sound frequencies found in the calling song,

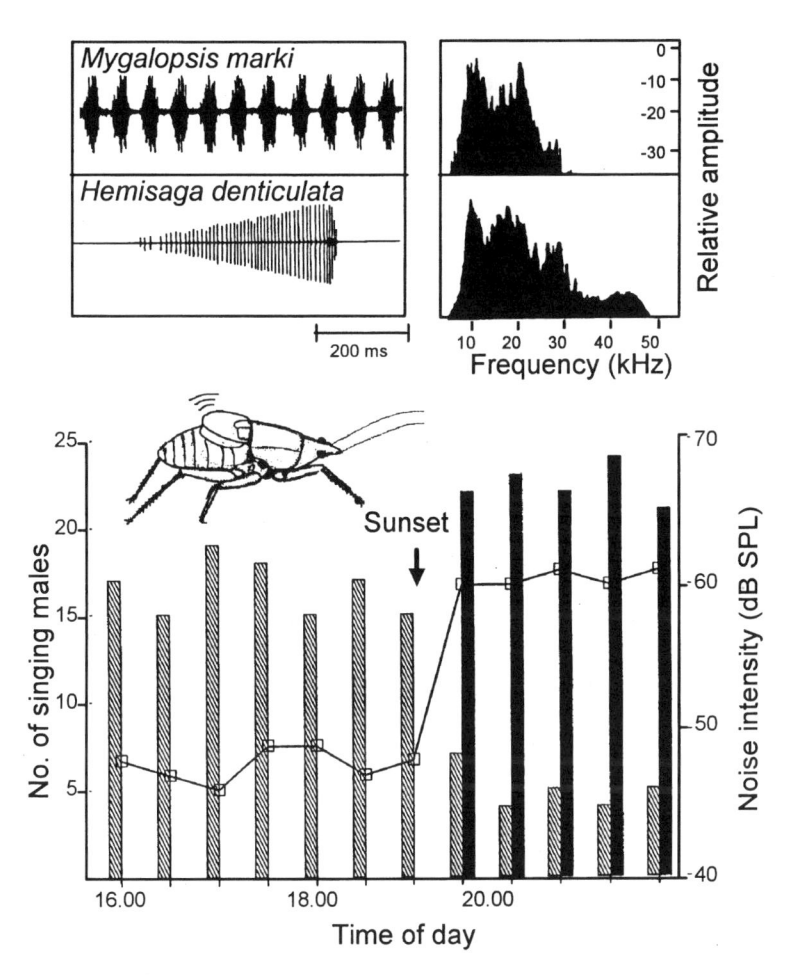

Figure 5.9. Inhibition of the austrosagine *Hemisaga denticulata* song by the cone-head *Mygalopsis marki*. Approximately half a second of the songs of both species, showing the patterns, is illustrated in the top left panel. Although the song patterns are different, they share similar frequency spectra (top right panel; in both, the vertical axis represents relative intensity or *amplitude* of the song components). The bottom graph shows the number of calling males of both species, illustrating the decrease in singing activity of *H. denticulata* (hatched bars) after the onset of singing by *M. marki* (filled bars). SPL = sound pressure level. Redrawn from Römer (1993).

thus matching the sensitivity of the ear itself (see below) (Rheinlaender and Römer 1990).

The auditory neural network of a katydid extends in two anatomical directions from the omega interneuron in the prothoracic ganglion (Lakes and Schikorski 1990). Toward the anterior of the body, acoustical information is sent for further processing by the brain and from the posterior it is

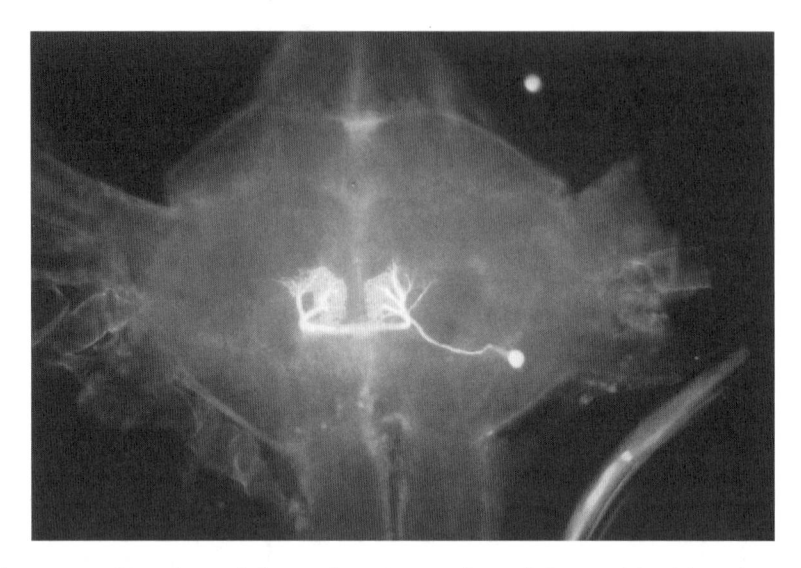

Figure 5.10. Top view of the prothoracic ganglion of *Gampsocleis glabra*, showing the position and shape of the omega interneuron. This single nerve cell can be "tapped" and used as a biological microphone in field preparations. Photo by Heiner Römer.

received from a pair of tympanic organs, the ears, each positioned on the tibia of each foreleg just below the "knee" (Fig. 5.11). Every katydid has four eardrums; each ear consists of two tympanic membranes connected to sensory structures that comprise the "tympanal organ" (Lakes and Schikorski 1990), a structure found in smaller form in the middle and hind tibiae (Kalmring et al. 1994, 1996; Sickmann et al. 1997) (Fig. 5.11).

One part of the tympanal organ is a subgenual organ (Fig. 5.11) that is also found in earless ensiferans such as camel crickets (Jeram et al. 1995) and appears to function as a detector of the substrate vibrations that might be made by predators or tremulating suitors. The rest of the katydid tibial organ includes a linear arrangement of receptor cells, the crista acustica, a structure analogous to the cochlea of the vertebrate ear in that each receptor cell responds to a specific airborne sound frequency (Oldfield 1982; Rheinlaender and Römer 1990) (Fig. 5.12). The sum of these sensitivities (as recorded, for example, from the whole tympanic nerve in the femur of the foreleg) typically reveals that the frequencies to which the ear is most sensitive match the frequency spectrum of the calling song (Fig. 5.13, but see also Fig. 8.5). Interestingly for some species with ultrasonic calls, the most sensitive frequency of the ear tends to be slightly lower than the main frequency of the call. Suga (1966) suggested that this slight difference compensates for the loss of the higher frequencies of the calls in nature; the

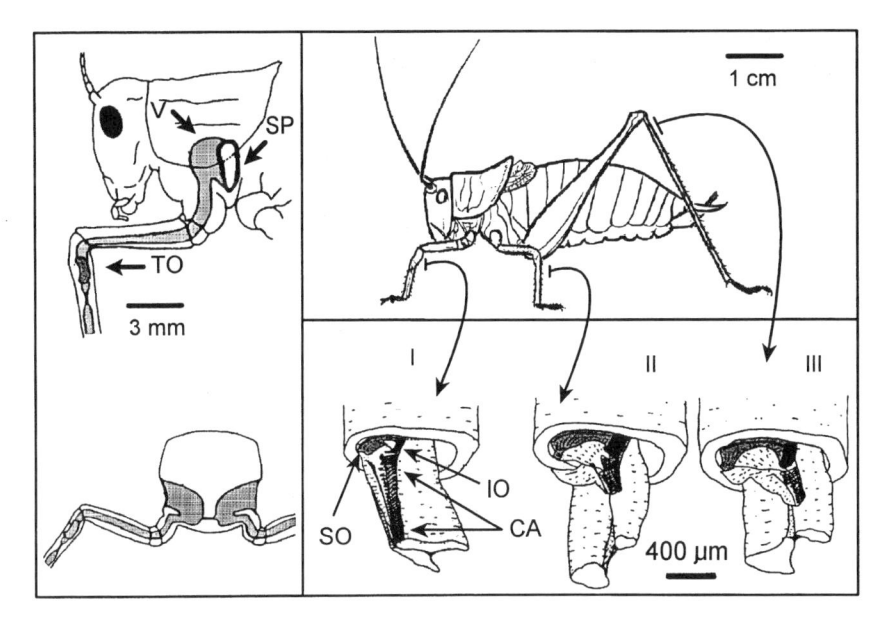

Figure 5.11. The hearing and vibration-sensing system of a katydid, *Polysarcus denticauda* (Phaneropterinae), showing a side view (left panel, top) and front view (left panel, bottom) of the tracheal auditory connections in the front left leg from the tibial hearing organ (TO) to the auditory spiracle (SP) via the auditory vesicle (V) or bulla in the thorax. The lower right panel shows the tibial organ in all three legs (note that there are no tympana in legs II and III), illustrating the crista acustica (CA), vibration-sensing subgenual organ (SO), and the intermediate organ (IO). Diagrams redrawn from Sickmann et al. (1997). Katydid redrawn from Harz (1969).

spectrum of sounds arriving at the ear from the distant caller would be lower than at its origin, owing to atmospheric attenuation.

For many katydids ultrasonic hearing is also important for avoiding predators. In response to ultrasound that may represent an echolocating bat or a predator breaking through vegetation, singing *Neoconocephalus ensiger* males stop calling (Faure and Hoy 2000) and if in flight, fold their wings and nose-dive toward the ground (Libersat and Hoy 1991).

The diversity of functions of the katydid ear may account for the great between-species variation in its external structures (Fig. 5.14). The tympanum of ensiferans is a "pressure-sensitive" structure (Bailey 1993; Greenfield 1997), one that, like the vertebrate ear, can move easily because it is backed on the inside by an air-filled cavity connected to the outside of the body via tracheal tubes (Fig. 5.11). In some katydids, the tympana consist of just two membranous areas on each tibia. This is probably the primitive state of the ear because a similar setup is found in closely related hearing tettigonioids in the Haglidae and Stenopelmatidae (Table 5.1).

In other katydids each tympanum is recessed to some degree in often

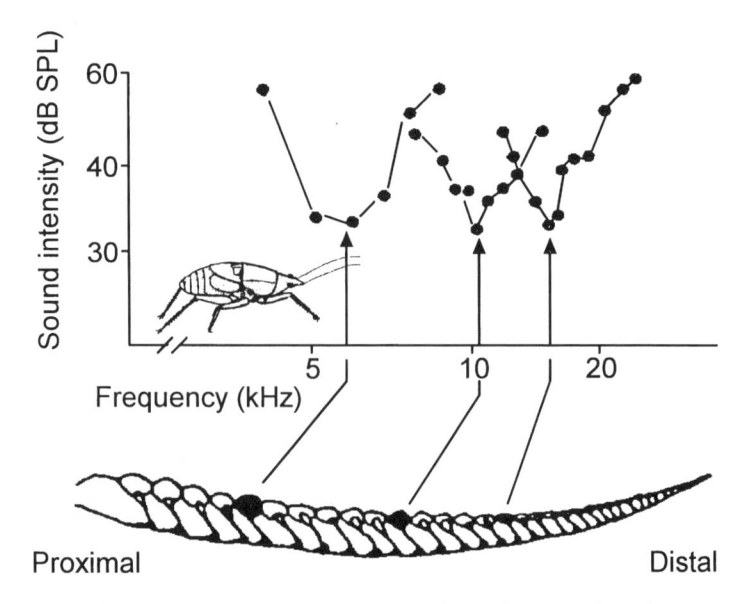

Figure 5.12. The crista acustica (CA in Fig 5.11) of *Mygalopsis marki* (below), showing the specific response of individual cells to sounds of different frequencies. The lowest part of each curve is the frequency at which the cell is most sensitive. Sound intensity is on the vertical axis (Oldfield 1982). SPL = sound pressure level. Redrawn from Rheinlaender and Römer (1990).

enlarged chambers behind cuticular folding (Table 5.1). Cuticular structures can be quite elaborate, such as in the listroscelidine *Phisis pallida*. Even greater elaboration is found in some *Mimetica* species where the tibial cuticle forms mammal-like "pinnae" on each side of the tibia (Fig. 5.15) that probably function to guide sound waves onto the ear surfaces (Bailey 1990, 1993). Finally, the tympana can be concealed behind cuticular folds so that the external part of each ear is barely visible as two small vertical slits on each foretibia.

Variation between species is also seen at the opposite extremity of the katydid's hearing system. An enlarged mesothoracic spiracle is connected to each tibial organ via tracheae from the foreleg into the thorax (Goreau 1837) (Fig. 5.11). This auditory spiracle can enter into an enlarged, and often horn-shaped, chamber. These elaborations of the thoracic tracheal system are not found in other acoustic Ensifera (Haglidae and Stenopelmatidae) (Bailey 1993). The size and elaboration of this tracheal cavity vary greatly between species (Bailey 1990; Heinrich et al. 1993) and even within species. Bailey (1998a) found that within-species variation in auditory spiracle size correlates with auditory sensitivity (Fig. 5.16). In the zaprochiline *K. nartee*, spiracle size varies between the sexes; there is an elaborate spiracle in the female but not in the male (Fig. 9.17) (Bailey and Römer 1991).

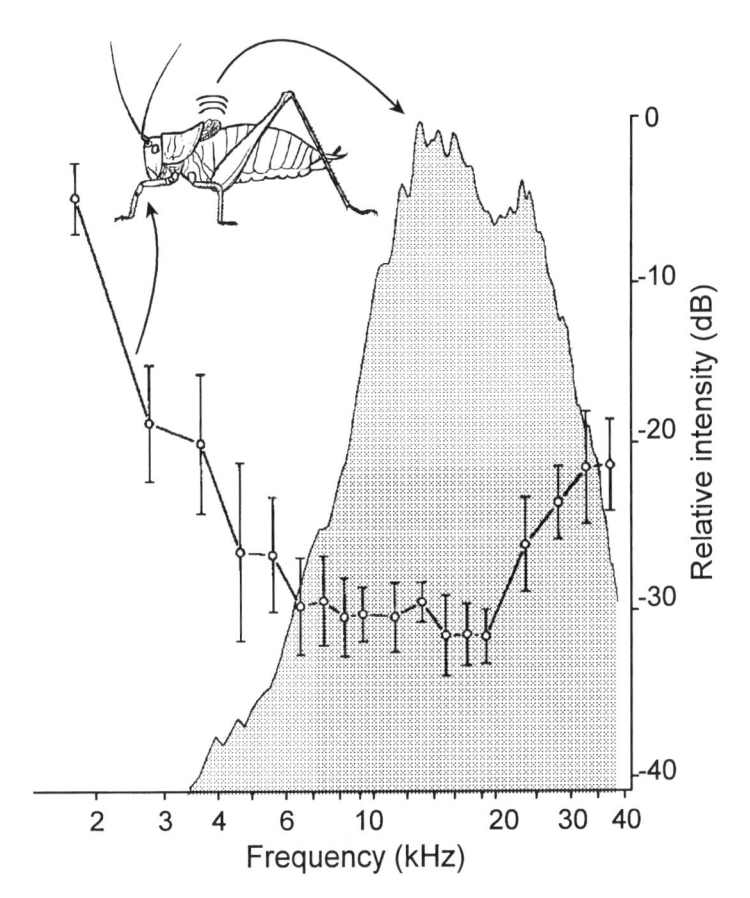

Figure 5.13. Song and response of the ear in *Polysarcus denticauda* (Phaneropterinae) (see Fig. 5.11). The frequency response curve of the ear (showing means +/− standard error bars) is most sensitive (the lowest part of the curve) to the loudest frequencies in the spectrum of the call (shaded area). Redrawn from Kalmring et al. (1996).

This unusual sexual dimorphism may be a result of sexual selection (see Chapter 9).

In katydids the auditory trachea and associated thoracic spiracles have been modified from their original respiratory function as first deduced by French entomologist M. Goreau (1837). Goreau's conclusion followed some rather callous experiments in which he "plunged" several katydids "head-first" into water, holding them under until they drowned. He noted small bubbles of air exiting from all the spiracles except the mesothoracic and concluded that the latter did not have a respiratory function.

Although the tibial-thoracic hearing system is found in both crickets and

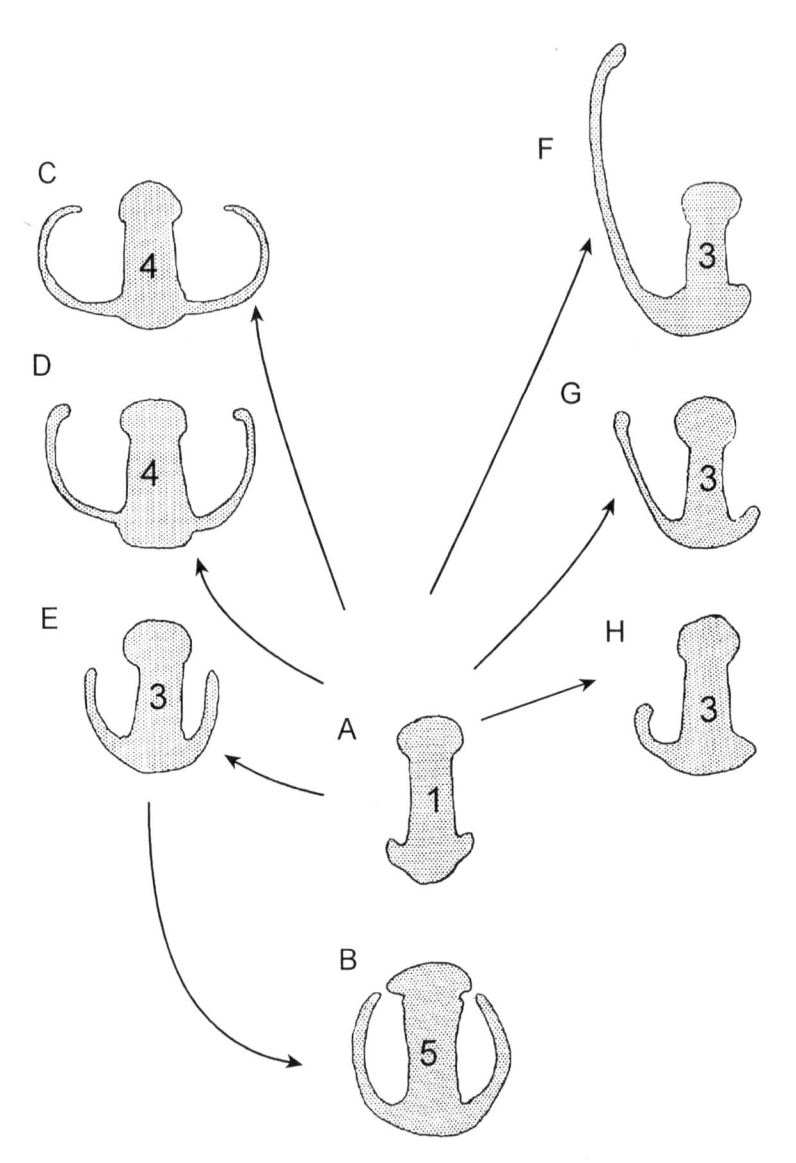

Figure 5.14. Cross sections of the foretibial ears of various katydid species, showing different degrees of elaboration of cuticular folding over the two tympana from the ancestral "open tympana" (Type 1: see Fig. 2.10). In each of these figures the tympana are located, one on each side of the domed central structure. The ear types (numbers) are listed in Table 5.1. A. *Elephantodeta*. B. *Decticus verrucivorus*. C. *Phisis pallida*. D. *Lacipoda immunda*. E. *Horatosphaga meruensis*. F. *Oxycous lesnei*. G. *Zeuneria melanopeza*. H. *Gravenreuthia saturata*. Rredrawn from Bailey (1990).

Table 5.1. Variation in cuticular structures associated with the ears (tibial tympana) of katydids

Type of ear	Representative taxa
1. Simple, "open" tympana with no cuticular elaborations	Bradyporinae, Conocephalinae, Hetrodinae, Listroscelidinae, Meconematinae, and many Phaneropterinae
2. Both tympana open but with raised cuticular fold over the anterior tympanum	Conocephalinae (Agraeciini) and many Phaneropterinae
3. Cuticular folds around tympana but not large enough to form cups	Conocephalinae (Agraeciini), Hetrodinae, Phyllophorinae, and Phaneropterinae
4. Cuticular folds well developed to form cuplike cavities	Conocephalinae (Agraeciini and Copiphorini), Listroscelidinae, Pseudophyllinae, and many Phaneropterinae
5. Cuticular folds forming narrow slits (cavities below slits not bulbous)	Bradyporinae, Conocephalinae (Copiphorini), Listroscelidinae, Hetrodinae, Phaneropterinae, Pseudophyllinae, and Tettigoniinae

From Bailey (1990).

katydids, members of the two families, and different katydid subgroups, use the equipment in quite different ways when determining the direction of a male's song (Bennet-Clark 1984; Bailey 1993; Michelsen and Löhe 1995; Greenfield 1997). Many vertebrates, including humans listening for the direction of sounds in the frequency range of speech (100–1,000 Hz), use differences either in the phase of the relatively long sound waves arriving at the two ears (phase-difference mechanism) or in the arrival time of the sound (time-of-arrival mechanism) (Stevens and Warshofsky 1965). However, insects such as orthopterans cannot detect differences in the time of arrival of sounds because the distance between their ears is too small relative to the length of the sound waves in the signal. Some crickets (Gryllidae) use a type of phase-difference mechanism in which sound waves arriving at the inner face of the tibial tympanum are out of phase with those on the outer face, because sounds moving internally are delayed by both the length of the auditory tracheae and the septum that divides the left and right parts of this system (Löhe and Kleindienst 1994; Michelsen and Löhe 1995).

Many katydids appear to take advantage of a third mechanism, one that can operate only when the sounds are of a sufficiently short wavelength compared to the animal's body size. The body diffracts the sound waves so that the ear in the "sound shadow" perceives a sound of lower intensity than the ear facing the sound source. A sound-shadow mechanism is used by some vertebrates, including humans, to analyze sounds above

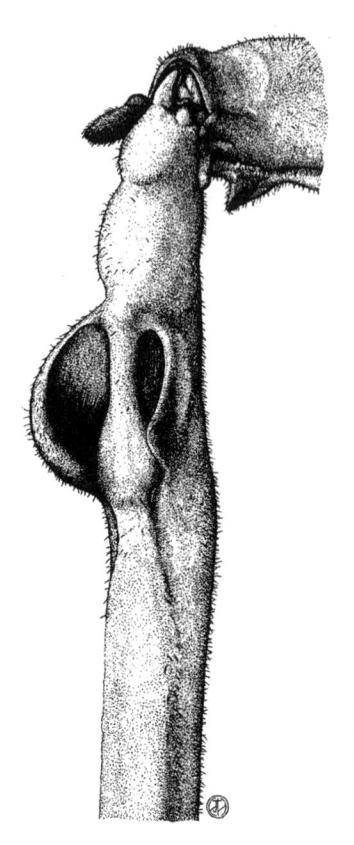

Figure 5.15. Front view of the top of the right foreleg tibia and tip of the right femur of a *Mimetica* species, showing the cuticular "pinnae" that surround the tibial tympana. Drawn by Ellen Hickman, commissioned by Bailey (1993).

4,000 Hz (the frequency of many bird and frog calls). However, this mechanism is unavailable to most insects because only wavelengths of high frequencies are sufficiently small to be diffracted around their small bodies. But a number of katydids are exceptions to this rule because, as we have seen, they tend to specialize in songs containing high frequencies. Several neotropical pseudophyllines can obtain directional information from conspecific calls because their very high frequencies (above 80,000 Hz) are sufficient to cast a sound shadow over the large tibial tympanal chamber, the main acoustical input into the auditory systems of these insects (the pseudophylline auditory spiracle and trachea in the thorax are much reduced relative to the structures seen in other katydids: compare Fig. 5.17H to 5.17A–G) (Mason et al. 1991).

Katydids with enlarged thoracic tracheae use a sound-shadow mechanism involving acoustical input into this "back door" to the inner side of the tibial tympanum (Fig. 5.17A–G) (Lewis 1974; Hill and Oldfield 1981). The intensity of this sound input is enhanced by the hearing trumpets, the horn-shaped trachea just inside each auditory spiracle entrance (Figs. 5.11,

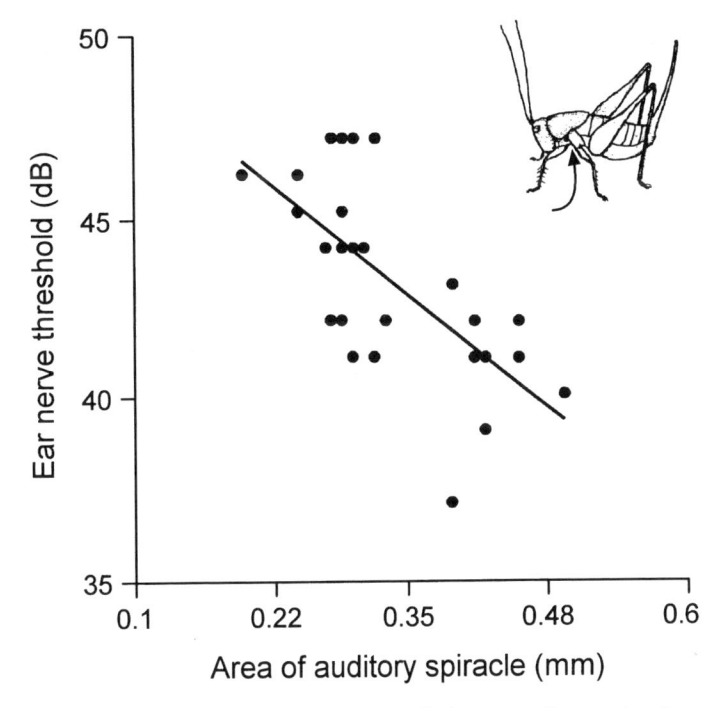

Figure 5.16. In the listroscelidine *Requena verticalis*, larger auditory spiracles are more sensitive to the sound frequencies typical of male song. This is revealed by the threshold at which the auditory (ear) nerve responds to male song. Redrawn from Bailey (1998a). Katydid redrawn from a figure by D. Scott, in Clutton-Brock (1991).

5.17A–G) (Lewis 1974; Nocke 1974, 1975; Hill and Oldfield 1981; Michelsen et al. 1994). Directionality comes from the high-frequency call casting a shadow over the spiracle facing away from the sound source (Fig. 5.18).

A more accurate direction-finding mechanism may operate in certain katydids whose main song frequencies are below 20,000 Hz. In these species, sound input is via the tibia but through much narrower, slitlike cuticular openings (Fig. 5.14B, type 5 in Table 5.1) to the tibial tympana than what is found in the ultrasonic neotropical pseudophyllines studied by Mason et al. (1991). Air in the chamber between the slit and tympanum resonates only at the main frequency of the call (Autrum 1940, cited in Bailey 1993) so that maximum stimulation occurs at the point at which each of the four slits (surrounding the two tympana on each tibia) directly faces the oncoming sound waves (Fig. 5.18) (Bailey and Stephen 1978; Stephen and Bailey 1982). However, this mechanism works only if the animal orienting to the conspecific call has the ability to shut off much of the sound entering into the tracheae via the hearing trumpets. There is some evidence

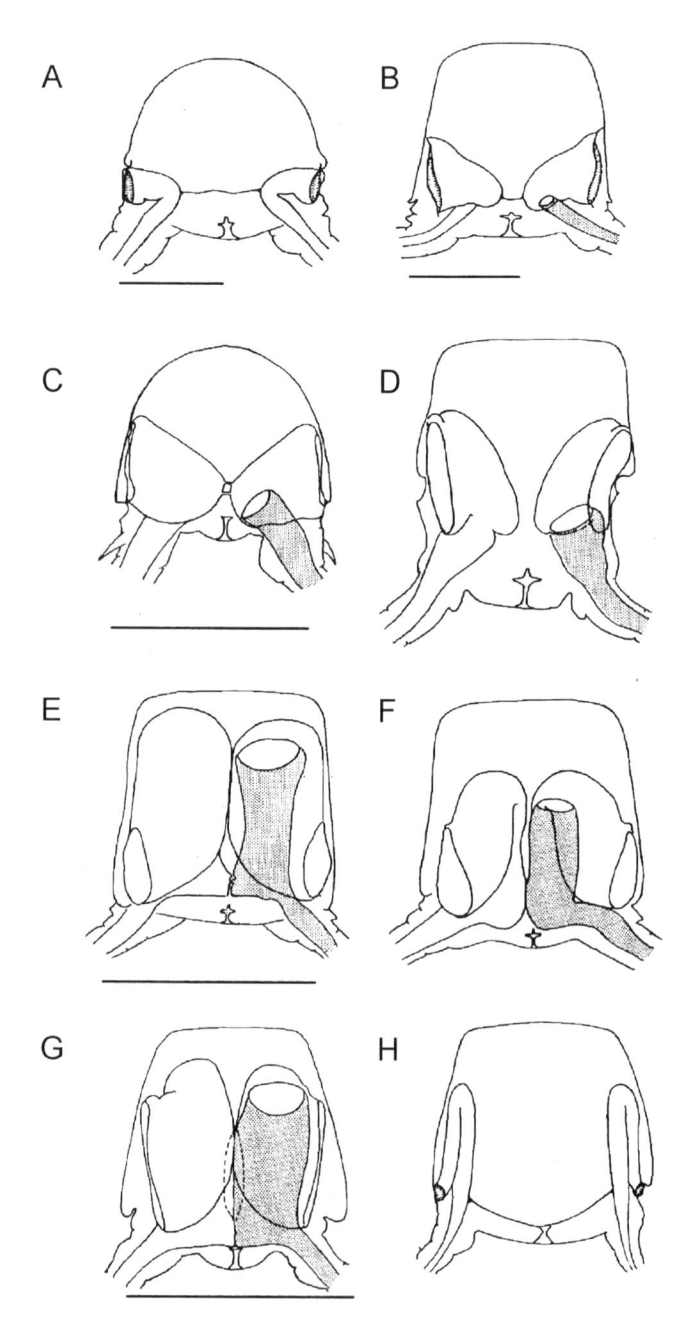

Figure 5.17. Front views of the thorax of various katydid species, showing the diversity of shapes and sizes of the acoustic tracheal vesicles (bullae). A. *Bliastes* species (Pseudophyllinae). B. Mormon cricket, *Anabrus simplex* (Tettigoniinae). C. *Hexacentrus mundus* (Listroscelidinae). D. *Metaballus litus* (Tettigoniinae). E. *Elephantodeta* male. F. *Elephantodeta* female, showing sexual dimorphism (Phaneropterinae). G. *Oxycous lesnei* (Phaneropterinae). H. *Oxyaspis* species (Pseudophyllinae). Scale bars = 0.05 cm. From Bailey (1990).

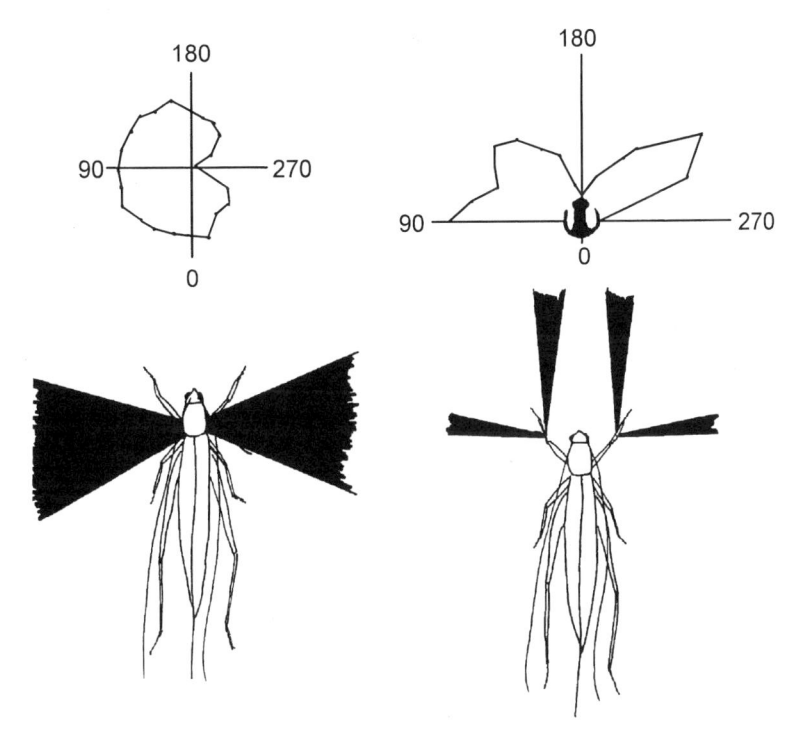

Figure 5.18. Sound-shadow (diffraction) (left) and slit (right) mechanisms of direction finding in katydids. The black areas represent the total angles from which there is sound input into the auditory systems. For the shadow mechanism (left: based on data from Hill and Oldfield 1981), the right side (270°) is away from the sound source (the 90° position) and therefore receives a lower-intensity input, as shown on the "polar plot" of sensitivity to sounds of the left ear. A greater direction-finding accuracy is found when the main source of sound entry is via the four tympanal slits (right: based on data from Bailey and Stephen 1978). The polar plot of one ear (shown in cross section in the center of the plot: see Fig. 5.14, type 5) shows the greater sensitivity when the source of the sound is directly opposite each slit opening into the tibial tympana. Redrawn from Bailey (1993).

that the tracheal input can be closed off while the katydid is orienting to calls in two species of austrosagines, *Pachysaga australis* (Bailey 1990) and *Sciarasaga quadrata* (Römer and Bailey 1998). The slit mechanism of directionality (in contrast to the sound-shadow mechanism) predicts that the female should show highly accurate orientation movements to the male call. Such a high degree of "auditory acuity" was confirmed in *P. australis* (Bailey and Stephen 1984).

Finally, the possibility remains that like humans, some katydids may use a combination of tibial and thoracic sound inputs to determine the direction of a male caller, particularly if a wide range of frequency bands is being detected (e.g., *M. marki*, Fig. 5.9) (Bailey 1993).

5.3. Evolution and Adaptive Significance of Katydid Song

5.3.i. Origin and evolution of singing

As we saw in Chapter 2, tibial ears are found not only in Tettigoniidae but also in the other ensiferan families: Haglidae, Stenopelmatidae, Gryllidae, Mogoplistidae, and Gryllotalpidae. Tegminal stridulation is found in all of these groups except Stenopelmatidae. Hearing and singing organs are sufficiently similar in these different ensiferans that they usually are assumed to have evolved just once in a common ancestor of the suborder (Alexander 1962; Michelsen 1992; Otte 1992). This single-origin hypothesis can be tested by tracing the presence and absence of ears and tegminal organs onto phylogenies of Ensifera. Although several schemes of ensiferan relationships support the single-origin hypothesis (Zeuner 1939; Ragge 1955; Sharov 1968), the trees of Ander (1939) and Gwynne (1995) (Fig. 2.3; see Chapter 2) indicate a dual origin as the most parsimonious explanation, one origin within tettigonioid clade and the other in the grylloids (Fig. 5.19).

The idea that acoustical organs evolved twice seems implausible because these calling and hearing structures are so similar in the two ensiferan groups. However, there are some differences between the organs in the two groups (see Gwynne 1995), as Ander (1939) pointed out: "the [ears of grylloids and tettigonioids] are so different that one cannot regard them as homologous. Doubtless they can be held to be derivatives of the subgenual organs [structures found even in earless orthopterans, e.g., Jeram et al. 1995] . . . but independently developed in the two groups." One example is that grylloids do not have a crista acustica (Fig. 5.12) in the tibial organ (Eibl 1978). Shared neural preadaptations have led to the independent origin of ears on the same body part in some parasitoid flies that we met in Chapter 4. The thoracic ear that acoustically orienting tachinids use to find their singing orthopteran prey (see Chapter 4, Fig. 4.3) is also found in the same location in a subgroup within another family of flies, the sarcophagids (Robert et al. 1996; Lakes-Harlan et al. 1999). This subgroup has developed prey-locating habits similar to those of the tachinids; they listen for and track down their victims, but rather than targeting orthopterans, the sarcophagids hunt calling male cicadas (Soper et al. 1976).

Once singing and hearing evolved in katydids, a few groups subsequently lost these abilities. Losses can be detected when the presence and absence of tegminal and tibial organs are traced onto a phylogeny of Tettigoniidae (Fig. 5.19) (see Chapter 2, Fig. 2.3) (Gwynne 1995). Remarkably, there have been very few losses in Tettigoniidae compared to the grylloids;

5.3 Evolution and Adaptive Significance 115

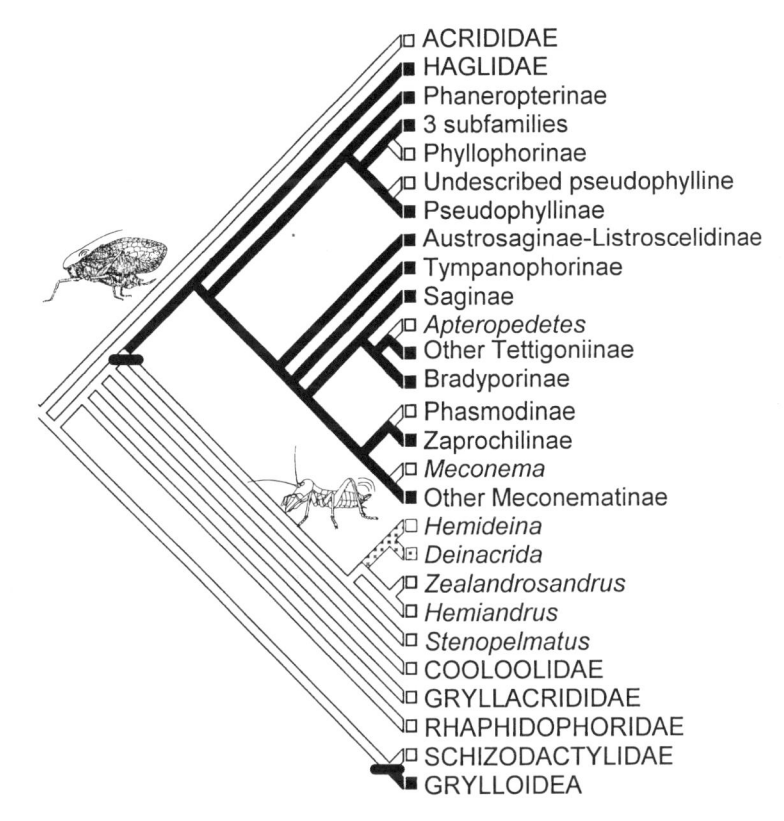

```
╱▫ ACRIDIDAE
■ HAGLIDAE
■ Phaneropterinae
■ 3 subfamilies
▫ Phyllophorinae
▫ Undescribed pseudophylline
■ Pseudophyllinae
■ Austrosaginae-Listroscelidinae
■ Tympanophorinae
■ Saginae
▫ Apteropedetes
■ Other Tettigoniinae
■ Bradyporinae
▫ Phasmodinae
■ Zaprochilinae
▫ Meconema
■ Other Meconematinae
▫ Hemideina
▫ Deinacrida
▫ Zealandrosandrus
▫ Hemiandrus
▫ Stenopelmatus
▫ COOLOOLIDAE
▫ GRYLLACRIDIDAE
▫ RHAPHIDOPHORIDAE
▫ SCHIZODACTYLIDAE
■ GRYLLOIDEA
```

Figure 5.19. A suggested hypothesis for origins and losses of singing in katydids and their kin. A phylogeny of Ensifera from Fig. 2.3 showing an expanded phylogeny of katydids (from Fig. 2.12), and several stenopelmatids (*Hemideina* to *Stenopelmatus*, where taxonomy is used as a proxy for phylogeny). Traced onto the tree is the most parsimonious explanation for the origins and losses of forewing stridulation and tibial ears (black bars). The figure shows (1) the origin of male (tegminal) singing in katydids and in Grylloidea (lower black bar) and leg-abdomen singing (stippled) in weta (Stenopelmatidae); (2) five independent losses of forewing singing in katydids (*Apteropedetes* and Phasmodinae have also lost ears); and (3) two independent origins of tympanal organs. Katydid redrawn from Vickery and Kevan (1983) (See Fig. 2.1.).

in the approximately 6,000 living katydid species, tegminal stridulation has been lost independently in *Meconema thalassinum* (Ander 1939), *Apteropedetes* species (Tettigoniinae) (Rentz 1979, pers. comm.), an undescribed pseudophylline from Ecuador (apparently the genus *Tabaria*, originally described as a mecopodine: F. Montealegre, pers. comm. 2000), stick katydids (Phasmodinae: Lakes-Harlan et al. 1991), and giant leaf katydids (Phyllophorinae: Carl 1906; Dumortier 1963; Lloyd 1976). Tibial ears have been lost in only two of these groups, the Phasmodinae (Lakes-Harlan et al. 1991) and *Apteropedetes* (Tettigoniinae) (Rentz 1979). The other three taxa may retain ears to hear other sorts of sounds: *M. thalassinum* males drum on the substrate (Ragge 1965) and phyllophorines have a coxal stridulatory mechanism (Lloyd 1976). In contrast to katydids, of the 3,000 or so species of crickets (Grylloidea), many, representing almost every subfamily, have lost acoustical function (Otte 1992). The reasons for the large difference in number of losses of sound-communication in crickets and katydids are unknown.

5.3.ii. Songs and sex

About 4 to 5 days after the final molt of the male katydid he begins to sing (e.g., Feaver 1977). Males in most species cease signaling only when a female is attracted, when he has just copulated (e.g., Gwynne 1990b), or when it is too cold, or at a time of the day when the species is typically inactive (Nielsen and Dreisig 1970; Nielsen 1971, 1972). The compulsion to sing is well reflected in anecdotes of males continuing to call while eating prey and even when disturbed by an entomologist intent on capturing the specimen (Alexander 1960).

Observations of male katydids and crickets calling incessantly with no obvious response from conspecifics led to some interesting speculation on why they sing. H. A. Allard in *Our Insect Instrumentalists and Their Musical Technique* (1928) concluded that "the theory ascribing sound to sex in insects has probably been much overdone," preferring instead the argument that insects "love sound" and "find it a means of self-expression" that represents "the beginnings of an esthetic sense" (Allard 1929). Earlier, Fabre (1917) had "not gone so far as to refuse [song] as part of the pairing" but nevertheless concluded that the principal purpose of singing was "to express its joy in living, to sing the delights of existence with a belly well filled and a back warmed by the sun. . . ." Of course whether these insects feel pleasure is irrelevant to whether or not singing serves a reproductive function!

So what is the adaptive significance of singing? Most behavioral biologists today agree that the continuous songs of orthopterans, as well as those of frogs and cicadas, function in reproduction. The Rev. Guilding (1827) came to this conclusion about katydid song in his note on a neotropical pseudophylline's "instrumentum stridoris": "[song is a way] by which the mute female is invited by the male to celebrate their nuptials." About the same time Lord John Russell and his colleagues (1831) came to a similar conclusion in *Insect Miscellanies*, stating that the sounds of males "are intended as signals for their companions." The book cites observations of Italian naturalist Gabriel Brunellius on the great-green grasshopper (Figs. 2.29 and 2.30), the very species De Geer (1752) had first used to study the katydid's singing mechanism. Brunellius (1791) demonstrated that females released on one side of his garden moved to singers placed in small reed cages on the other side. When his first captive male began to sing, the released female "began leaping forth, jumping from one plant to another, and thus flitting about quickly reached the male and came to a halt on his cage." Brunellius repeated his experiment by moving the cage to the opposite side of the garden to watch the frustrated female retrace her movements back through the vegetation.

Over 100 years later, Leo Baier (1930) continued the investigation of pair formation in great-green grasshoppers, by controlling for nonacoustical cues that may have been involved in the attraction of females to singers. Baier attracted females to a telephone speaker piping in the song of a live male singing in a distant room. Moreover, Baier demonstrated that katydids responded to airborne, not substrate vibrations. Using another tettigoniine, *Pholidoptera griseoaptera*, he was able to rule out communication through the substrate by showing that males still responded acoustically to each other when calling from disconnected substrates—cages suspended beneath balloons floating in his laboratory.

Subsequent studies of katydids in a variety of subfamilies demonstrated that song attracts females (Table 5.2). The attraction of females to the call is no doubt an ancestral characteristic of katydids because the male song in the sister group Haglidae also attracts mates (Snedden and Irazusta 1994). Haglid song also can be involved in male-male rivalry (Mason 1996),

Table 5.2. The songs of male katydids attract females in a wide variety of species representing a number of different subfamilies

Subfamily	Species	Reference
Tettigoniinae	*Tettigonia viridissima*	Brunellius 1791*
		Baier 1930
	T. cantans	Latimer and Schatral 1983
	Metrioptera sphagnorum	Morris et al. 1975b
Conocephalinae	*Orchelimum gladiator*	Morris et al. 1975c
	Conocephalus nigropleurum	Morris and Fullard 1983
	C. brevipennis	Bailey and Morris 1986
	C. upoluensis	Bailey 1985
	Neoconocephalus spiza	Greenfield and Roizen 1993
	Ruspolia (several species)	Bailey and Robinson 1971
Bradyporinae	*Ephippiger ephippiger*	Busnel et al. 1956
Phaneropterinae	*Poecilimon*	Heller and Helversen 1993
Hetrodinae	*Acanthoplus speiseri*	Mbata 1992b*
Austrosaginae	*Pachysaga australis*	Bailey and Stephen 1984
Listroscelidinae	*Requena verticalis*	Bailey and Yeoh 1988
Zaprochilinae	*Kawanaphila nartee*	Gwynne and Bailey 1988

*Studies in which females were observed to move to a singing male so there were no controls for nonacoustical cues.

Table 5.3. The male songs of a diversity of katydids attract or repel rival males or, in some instances, cause the listening rival to change his song

Effect of song	Subfamily	Species	Reference
Attraction	Conocephalinae	*Orchelimum gladiator* and *O. vulgare*	Morris 1972
Attraction	Conocephalinae	*Neoconocephalus affinis*	Brush et al. 1985
Repulsion and song change	Conocephalinae	*Mygalopsis marki*	Dadour and Bailey 1985, 1990
Attraction	Hetrodinae	*Acanthoplus speiseri*	Mbata 1992b[a]
Attraction and song change	Tettigoniinae	*Metrioptera sphagnorum*	Morris 1970, pers. comm 1999
Song change	Tettigoniinae	*Pholidoptera cinerea* and *P. griseoaptera*	Baier 1930; Jones 1966
Attraction, repulsion, and song change	Tettigoniinae	*Tettigonia cantans*	Latimer and Schatral 1983; Latimer and Sippel 1987
No attraction, but song change[b]	Zaprochilinae	*Kawanaphila nartee*	Simmons and Bailey 1992
Song change	Listroscelidinae	*Requena verticalis*	Schatral and Bailey 1991

[a]Studies in which males were observed to move to a singing male but there were no controls for nonacoustical cues.
[b]*K. nartee* males have a very low hearing sensitivity.

as is katydid song. As we saw earlier, sound signals can attract or repel rivals (Table 5.3), as observed in the response of a male *M. marki* to the call of a loud neighbor. I examine the significance of male-male competitive interactions further in Chapter 7.

For katydid males, messages to rivals and to females are carried within the same signal, the "calling" song. In contrast, many true crickets (Gryllidae) separate calling, aggressive/rivalry, and courtship messages into quite different acoustical signals (Alexander 1960; Otte 1992). Canada's bog katydid, *Metrioptera sphagnorum*, is a tettigoniid exception; a signaling male produces a calling song with two parts, one that contains mainly audible frequencies ("audio") and the other that is mainly ultrasonic. The male changes his tune, prolonging the audio of the calling song, when a rival male comes within earshot (Morris 1970). There is also some evidence that this continuous audio part of the song is used when courting females (Morris et al. 1975b).

As previously mentioned, Rev. Guilding's (1827) finding of loud calling males and mute females is not universal in katydids. In some katydids, females have tegminal sound organs. These organs are used by some

Phaneropterinae and Bradyporinae in an acoustical dialogue between the sexes prior to pair formation (Hartley et al. 1974; Nickle and Carlysle 1975; Robinson 1980; Robinson et al. 1986). In some of these species, both sexes move during pair formation. In others the male moves toward the female's sound while she remains stationary (Spooner 1964, 1968b, 1995; Heller and Helversen 1986; Robinson 1990). In phaneropterines the female answers the male with a simple click, whereas in some bradyporines the answer is a well-developed song (Hartley et al. 1974). The willingness of females to respond acoustically to the male call is a convenient experimental measure of the call preference of females, and can be exploited by eavesdropping rival males (Fig. 7.15), as we will see in Chapter 7. For species in both subfamilies the female's answer usually comes at a species-specific time interval after the male's signal.

The attraction of male katydids to the clicks of conspecific females can aid in the collection of males. A number of naturalists have attracted singing males by producing artificial sounds resembling the sounds of an answering female. I once duped an *Amblycorpha* songster this way by clicking my fingernails each time he emitted his short call. Spooner (1968a) outlined methods for attracting female phaneropterines to artificial sounds. He reported that the best results (for North American species) come from striking the tip of a small knife blade against glass between 0.1 and 0.5 second after the end of the male's song.

5.4. Nuptials and an "Extraordinary Wallet"

Once the song of the male katydid has attracted a mate, the two insects typically end up in a position facing each other and "fencing" with their antennae (Fabre 1917). The courting pair can break up at any time from antennation, to just before sperm transfer, should one individual choose to reject the other, as we saw in Mormon crickets (see Chapter 1). If mating proceeds, the copulating pair almost always ends up in a position in which the male is curled behind the female with each facing in opposite directions (Boldyrev 1915; Alexander and Otte 1967; Rentz 1972a). This mating position is an ancient one, having evolved early in a distant camel cricket–like ancestor of the clade containing the Tettigoniidae (Fig. 2.3) (Gwynne 1995). The position (Fig. 5.20) is achieved either by the sexes simply moving past each other in the vegetation or by the female mounting the male, with the male eventually dropping back and beneath the female and bending to grasp her ovipositor (Boldyrev 1915; Alexander and Otte 1967). In Fabre's (1917) description of his white-faced decticus, the male at this point is

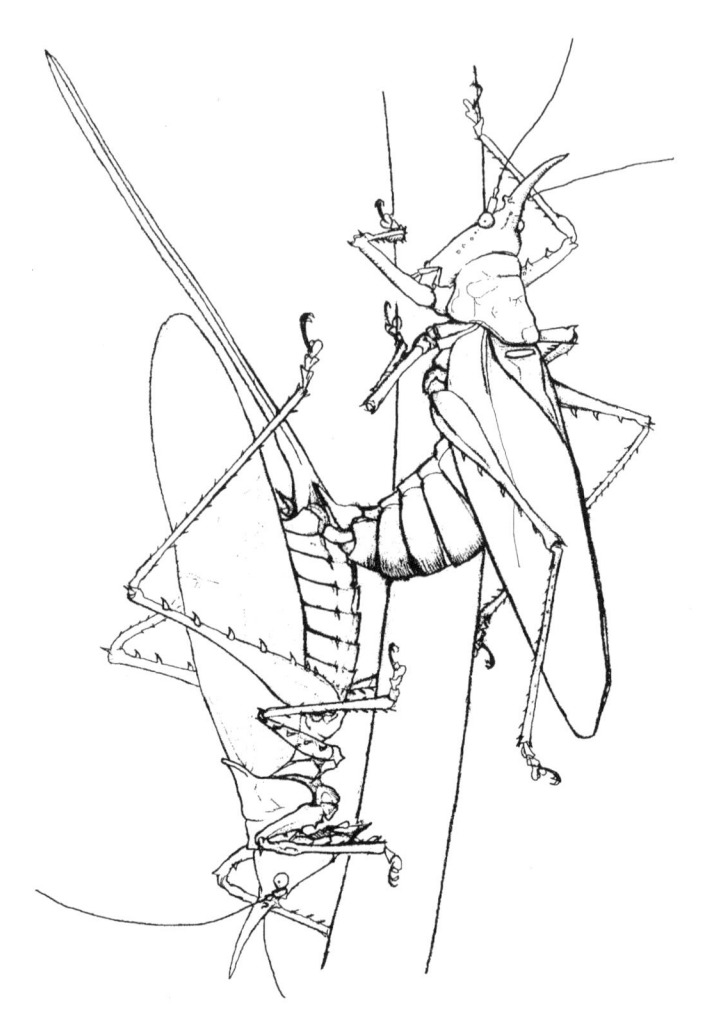

Figure 5.20. A mating pair of *Copiphora rhinoceros* (Copiphorini), showing the typical mating position of most katydids. From Morris (1980).

> ... *tumbled over on his back. Hoisted to the full height of her shanks, the other, holding her sabre almost perpendicular, covers her prostrate mate from a distance. The two ventral extremities curve into a hook, seek each other, meet; and soon from the male's convulsive loins there is seen to issue, in painful labour, something monstrous and unheard-of, as though the creature were expelling its entrails in a lump. ... [After the pair separates] The strange concern remains hanging from the lower end of the sabre of the future mother, who solemnly retires with the extraordinary wallet, the spermatophore.... At intervals she draws herself up on her shanks,*

curls into a ring and seizes her opalescent load in her mandibles. . . . [Females] tear at it, work at it solemnly with their mandibles for hours on end and finally gulp it down.

The evolution, function, and consequences of this "extraordinary wallet" comprise most of the rest of this book.

6

A Nuptial Banquet and Seminal Sac: Evolution of the Spermatophylax Meal

> ... a small bag—evidently the ovary—... attached to the body of the female close to the tail; this is extracted from the other without the tail ...
>
> Captain John Feilner (1864), Report to the Smithsonian Institution

6.1. A Mating Mystery

In 1859, the year that evolutionary theory burst onto the scene with the publication of Darwin's *The Origin of Species by Means of Natural Selection*, Captain John Feilner was engaged "under the auspices of the Smithsonian Institution" to conduct biological observations in northern California. In the summer of that year he explored the Pitt River valley where he observed an insect to "deposit the eggs ... in the ground ... by means of its tail, which is shaped like a bayonet" and in which "immediately before uniting sexually, the insect without the tail (which [he] presume[s] to be the male), utters a shrill whistling sound, as if to call his mate" (Gwynne 1997b). Feilner's subjects were katydids and almost certainly Mormon crickets, given the geographic location and because they were found in "such numbers as actually to cover the ground" (see Chapter 1). He goes on to describe copulation in these animals and describes a "bag" transferred from male to female (see this chapter's opening quotation). The final report (Feilner 1864) was published by the Smithsonian Institution following the captain's untimely end; he was killed by natives while on a subsequent expedition (Feilner 1864; Caudell 1908a).

A half century after Feilner, when field studies of katydid mating were considerably less hazardous, three well-known naturalist-entomologists, J. H. Fabre (1896, 1917), C. P. Gillette (1904), and R. E. Snodgrass (1905), presented further details on the strange mating "bag" in their studies of Mormon crickets and other shield-back katydids. Both Feilner and Gillette noted that the structure eventually disappeared, but it was Snodgrass and, as we saw at the end of Chapter 5, Fabre who determined the bag's fate by noting that it was eaten by the female (Fig. 6.1). Curiously, Snodgrass did not view this meal as beneficial, referring to it as an "albuminous mass" that caused the female "much annoyance" so that she "attempts to

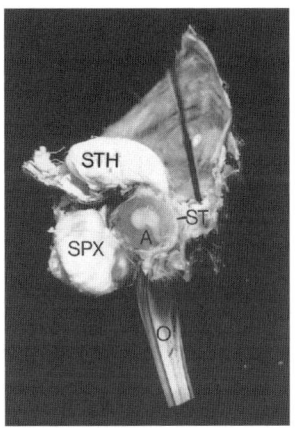

Figure 6.1. **a.** A newly mated Mormon cricket female (*Anabrus simplex*) tucks her head between her forelegs and eats her spermatophylax meal (SPX) from the base of her ovipositor (O). **b.** The parts of this spermatophore are shown in a dissection of the side of a female's abdomen. The spermatophylax is attached to the rest of the spermatophore, which includes a sperm ampulla (A) that is positioned inside the female's genital chamber. The side of the genital chamber has been removed to show the ampulla and one of two comma-shaped sperm cavities it contains. Sperm exits along an ejaculatory canal (the bottom part of the "comma") into the female's spermathecal tube (ST) and into her large spermatheca (STH). From Gwynne (1997a).

rid herself of it by bending her head beneath the abdomen and chewing it off." Fabre seemed ambivalent about the female's response. Although he referred to the "horrible banquet" of the female white-faced decticus (*Decticus albifrons*), he also noted that she "enjoyed" it and called the male donation of another katydid, an *Ephippiger* species, a "nuptial dish." Both Fabre and Gillette recognized that the mating bag served a second role, not as an ovary as Feilner (1864) had surmised, but as a "fertilizing capsule" or spermatophore, a structure that Fabre recognized from his knowledge of reproduction in invertebrates as diverse as centipedes and cephalopods (see Mann 1984) (Fig. 6.2).

The details of the dual role of the mating bag were worked out independently by German biologist Ulrich Gerhardt (1913; 1914; 1921) and a Russian, B. T. Boldyrev (1915). Their observations were presented in lengthy monographs describing not only mating behavior but also the anatomy of the spermatophores of katydids, true crickets (Grylloidea), and camel crickets (Rhaphidophoridae). Boldyrev, in particular, was an avid observer of orthopteran behavior who kept cages of nocturnal species next to his bed for convenient midnight observations. These observations showed that the elaborate spermatophore eaten by females was a standard mating feature in virtually all katydids and in some other ensiferans. As

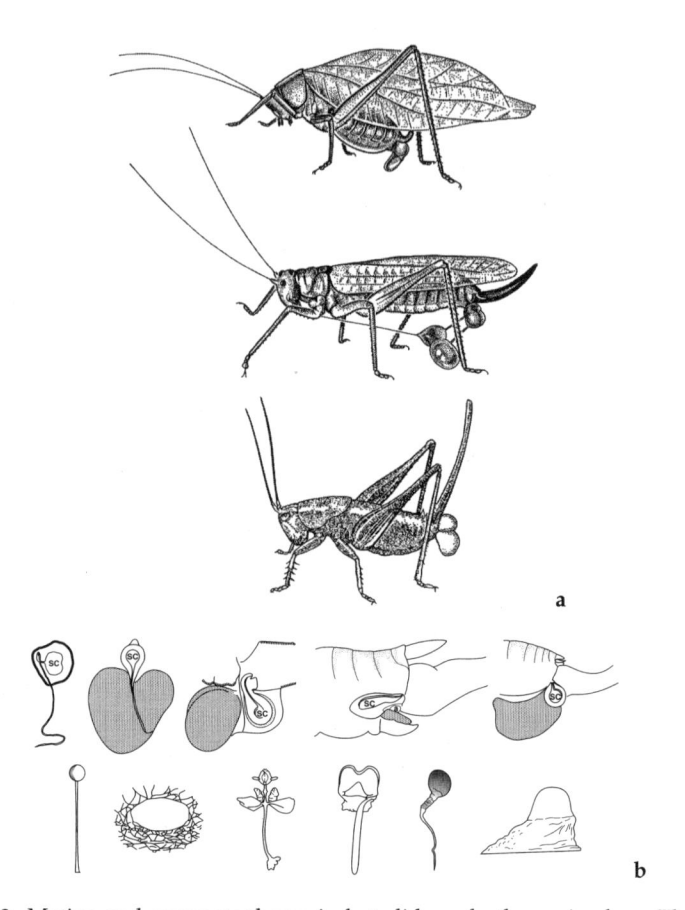

Figure 6.2. Mating and spermatophores in katydids and other animals. **a**. The top three figures are female katydids with spermatophores: *Stilpnochlora marginella* (Phaneropterinae) (top), *Bliastes insularis* (Pseudophyllinae) (middle), and *Requena verticalis* (Listroscelidinae) (below). **b**. Top: Diagrams of spermatophores showing (from left) three spermatophores of other ensiferans: a mole cricket, *Gryllotalpa gryllotalpa* (Gryllotalpidae) spermatophore showing the long threadlike sperm canal, and the sperm cavity (SC); another cricket, *Gryllomorpha dalmatina* (Gryllidae), in which the spermatophore is surrounded by a spermatophylax (stippled); and a rhaphidophorid showing the position of the spermatophore and spermatophylax (stippled) in the mated female (side view). The two top right drawings are similar side views of a mated female *Ruspolia nitidula* and a phaneropterine katydid, an *Isophya* species. a = accessory reservoir of the spermatophore. Bottom: Examples of spermatophores in other animals. From left, the springtail *Sminthurus viridis*, where the spermatophore is placed on the end of a stalk, which is picked up by the female; the centipede *Scolopendra cingulata* (Chilopoda: Scolopendromorpha), where the spermatophore is surrounded by a weblike structure (Arthropoda: Chilopoda); another stalked spermatophore, in the mite *Halacarellus basteri* (Arthropoda: Arachnida, Acarina); the spermatophore of a calanoid copepod, *Labidocera aestiva* (Crustacea); sperm package of a polychaete worm, *Kinbergonuphis simoni* (Annelida: Polychaeta); and a urodele amphibian spermatophore, *Pseudotriton*. **a**. top and middle: From Leroy (1969) by permission of publisher. **a**. bottom: Drawing by D. Scott, from Clutton-Brock (1991). **b**. top: Redrawn from Boldyrev (1915, 1927). **b**. bottom: From drawings in Proctor et al. (1995).

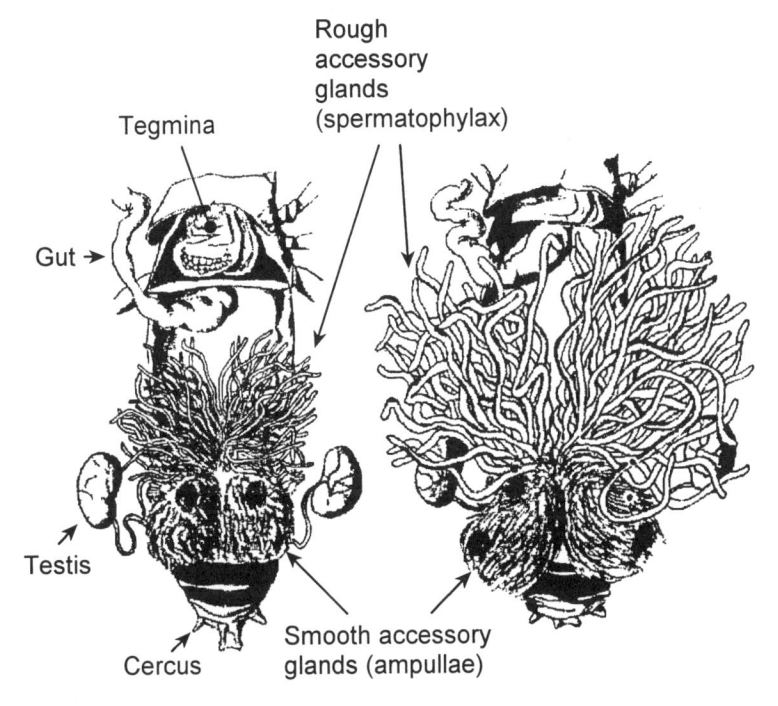

Figure 6.3. A dorsal view of a mated (left) and unmated (right) male's abdomen (*Ephippiger*) showing the testes, the rough accessory glands that produce the spermatophylax, and the smooth glands that produce the sperm ampulla. From Busnel and Dumortier (1955) by permission of publisher.

we saw in Chapter 1, the meal's main course is a sperm-free mass that Boldyrev called the *spermatophylax*. This mass is produced when muscle contractions in the abdomen squeeze out the contents of the "rough accessory glands" (c in "Fig: 10" of Fig. 2.30; Fig. 6.3), structures which, along with the smooth glands (q in "Fig: 10" of Fig. 2.30; Fig. 6.3) that produce the sperm ampulla, virtually fill the male's abdominal cavity (Brunellius 1791; Boldyrev 1915; Busnel and Dumortier 1955).[1]

Boldyrev noted that the female eats the spermatophylax before she removes and eats the ampulla, which in ensiferans is partially inserted into the female. The ampulla is unusual in most ensiferans in that it protrudes outside of the genital tract; in most other pterygote insects that use spermatophores, the sperm-containing vessel is placed entirely inside the female's genital tract ("genital chamber") (Khalifa 1949; Davey 1960; Mann

---

[1] Gabriel Brunellius, the student of katydid phonotaxis whom we met in Chapter 5, encountered the exudates of the rough accessory glands while dissecting male great-green grasshoppers in his search for the penis, a structure he was unable to locate (katydids lack a true intromittent organ).

1984) (Fig. 6.2). Sperm transfer from the ensiferan ampulla is a rather drawn-out process. Katydids can ejaculate millions of sperm, a high number for insects (Reinhold and Helversen 1997). These sperm clump together in "spermatodesms" (Boldyrev 1915), which are ejected slowly through the ejaculatory canal into the female's sperm-storage organ, the spermatheca (Snodgrass 1905; Cappe de Baillon 1922) (*o* in "Fig: 15" of Fig. 2.30; Figs. 6.1 and 6.2). In some species the small spermatodesms in the male reproductive tract transform into larger feather-like aggregations when they reach the female tract (Viscuso et al. 1998), where in some katydids (e.g., tettigoniines and phasmodines) they are stored in individual pouches—spermatodoses—within the spermatheca (*e* in "Fig: 15" of Fig. 2.30). A new spermatodose forms after each mating (Boldyrev 1915; Gwynne 1993). Boldyrev also noted "spurts" of a sperm-free liquid from the spermatophore following the transfer of sperm.

The temporal association of meal consumption and insemination led both Boldyrev and Gerhardt to suggest that the spermatophylax meal functioned as a device to prevent the female from eating the ampulla before sperm transfer had been completed. Boldyrev viewed the spermatophylax as just one of a number of protective methods used by ensiferans. Other protective devices he suggested included prolonged copulation and male vigilance over the newly mated female (guarding) to prevent her from eating the spermatophore. As in other animal species, the transfer of a full ejaculate is likely to be important to the paternity of the mating male (Smith 1984; Birkhead and Møller 1998). The full ejaculate in katydids includes not only sperm but also substances (perhaps in the clear liquid that Boldyrev observed to spurt out of the ampulla following sperm transfer) that appear to suppress female receptivity to the advances of other males.

In modern evolutionary terminology, Boldyrev and Gerhardt's ejaculate-protection hypothesis argues that the spermatophylax is a sexually selected trait that has evolved via an evolutionary "arms race" between the sexes. Early in the race, ancestral nonvirgin females seeking additional matings would have gained the edge in an initial conflict as they interfered with sperm transfer, benefiting either from nutrition in the proteinaceous ampulla eaten or perhaps from manipulation of the ejaculates by destroying only some of the ejaculates while permitting the sperm of preferred males to enter the sperm-storage organ. After the spermatophylax had evolved as a countermeasure to such female interference, its size might be further increased if needed to protect a larger ampulla (ejaculate) resulting from increased sperm competition (Wedell 1993b).

About 70 years after Boldyrev and Gerhardt, the ejaculate-protection hypothesis (see also Fulton 1915) was finally tested in katydids, by me (in collaboration with Barbara Bowen and Chris Codd) (Gwynne et al. 1984) and by Nina Wedell and Anthony Arak (1989) (see also Sakaluk 1984). The experimental manipulation necessary for the test is a fairly obvious one—reduce the size of the spermatophylax meal and determine whether there is a reduction in the male's ability to inseminate females or to fertilize eggs. My first efforts along these lines used my first katydid subject, the Mormon cricket (see Chapter 1), but I quickly found that this species was a difficult laboratory subject (one problem was that in order to develop, the eggs of shield-backs often require several alternating warm and cold periods; see Chapter 3). The species that I finally settled on is a common singer I encountered in gardens around the University of Western Australia shortly after I arrived there as a research fellow in 1981. David Rentz identified the singer as a listroscelidine, *Requena verticalis*, and my first few weeks of lab work showed it to be ideal for the experiments I had in mind. It is small and easily raised in the lab, and its spermatophore is perfect for experimentation; unlike most other katydids (including Mormon crickets), the *Requena* spermatophylax detaches completely from the sperm ampulla when it is grasped by the mated female (Fig. 6.4). In most other katydids the entire spermatophylax is difficult to remove intact, either by an experimenter or by the hungry female katydid that must work her way slowly through the meal by placing her head between her hind legs to take successive mouthfuls (as Fabre described for his white-faced decticus: see Chapter 5).

The results with both *Requena* (Gwynne et al. 1984) and the wart biter *Decticus verrucivorus* (Wedell and Arak 1989) showed some support for Boldyrev's hypothesis; females experimentally deprived of a spermatophylax quickly removed and ate the sperm ampulla. However, some interesting differences between the two species became apparent when the

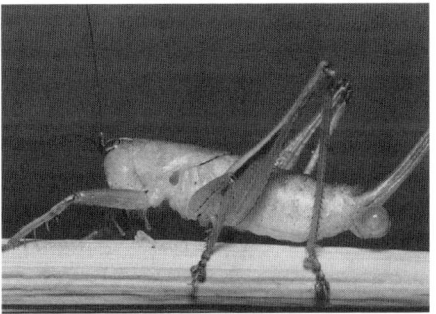

Figure 6.4. a. A female katydid, *Requena verticalis*, with a spermatophore just after mating. Her mate is visible in the background. sc = sperm cavity of ampulla; sx = spermatophylax. **b.** The female is completing her main meal—the spermatophylax. Its remains are visible beneath her mandibles, and the sperm ampulla is attached at the base of her ovipositor. In contrast to Mormon crickets and other katydids, the ampulla of this species is not placed in the female's genital chamber, and the spermatophylax meal detaches completely from the sperm ampulla.

prediction about ejaculate protection was examined more precisely. According to the hypothesis, the male should provide only enough material to distract the female until ejaculate transfer is complete; thus any spermatophylax material not used can be saved toward a full insemination with another female. For the wart biter, the average meal kept the female busy for about 3 hours before she ate the ampulla. The 3-hour meal length correlated precisely with the ampulla-attachment time necessary to induce a full 5-day period during which females both were unreceptive to further matings (Fig. 6.18) and increased their egg-laying rate (Wedell and Arak 1989). Many males produced meals smaller than the average and would have induced shorter refractory periods, thus fertilizing fewer eggs if the female's subsequent mate provided a larger meal to the female (Wedell 1991).

In contrast, the average meal supplied by the *Requena* male kept the female busy for over 5 hours, a much longer time than necessary for the ampulla to fully inseminate the female and to induce a full sexual refractory period (Gwynne et al. 1984; Gwynne 1986). The reason for this long meal may lie in Fabre's (1917) suggestion that the meal is "the source of life for the ovules." Perhaps the large *Requena* spermatophylax is a nutritious meal (Thornhill 1976b; Eluwa 1978) that evolved to nurture the male's own progeny. Under this paternal-investment hypothesis for nuptial gifts (Trivers 1972), the fitness interests of males and females are complementary as both parents clearly benefit from a nuptial meal that increases the fitness of their offspring.

This chapter examines the ejaculate-protection and paternal-investment hypotheses for the adaptive significance of the katydid spermatophylax. Before we examine the hypotheses (Table 6.1) and evidence for and against them, two additional points need to be made. First, the two hypotheses are not necessarily mutually exclusive; the meal may enhance both the mating male's paternity *and* the quality of his offspring. Second, understanding why the meal evolved really involves questions about the meal's original and past functions as well as the selection pressures that continue to *maintain* the male donation in various species. It is entirely possible that the spermatophylax has changed in function in present-day populations of some species. Questions about adaptive maintenance are answered mainly using manipulative experiments like the ones just mentioned. However, historical questions about origins and past evolution of traits are addressed, as we have seen in previous chapters, using the method first championed by Darwin (Ghiselin 1969), by comparing different species.

The significance of the answers to the main questions posed in this chapter goes beyond simply an understanding of the adaptive significance of the katydid spermatophore meal. First, the questions are general ones that are applicable to all animals, but especially arthropods in which males

Table 6.1. The adaptive significance of courtship feeding: potential benefits received by mate-feeding males

Hypothesis	Sexual conflict?
Precopulatory offerings such as prey:	
Ia. A nutritional offering that attracts a female and entices her to mate and stay in copula (e.g., scorpionflies)	Low: both sexes benefit
Ib. A "symbolic" low-nutrient offering that attracts a female and entices her to mate and stay in copula (possible in some balloon flies)	High if female expects nutrition
II. A naturally selected offering that prevents the female from eating the male	Low: both sexes benefit
Offerings produced during copulation:	
IIIa. A plug that prevents sperm leakage	Low: both sexes benefit
IIIb. A plug that prevents rival males from copulating	High: female prevented from adaptive remating
IVa. Ejaculate protection: a nutritious offering that prevents premature consumption of the sperm-containing ampulla	Low: both sexes benefit
IVb. Ejaculate protection: a low-nutrition offering that prevents premature consumption of the sperm-containing ampulla	High if the offering's sole advantage to the female is nutrition
V. Paternal investment: a nutritious offering that benefits the mating male's offspring	Low: both sexes benefit
VIa. Cryptic female choice: a male-produced ejaculate substance comprising a nonnutritional, hormonal substance that the female assesses to judge mate quality	High only in matings with a low-quality male
VIb. Cryptic female choice: a male-produced ejaculate substance comprising a nonnutritional, hormonal substance that manipulates a female by turning on egg production	High: female prevented from adaptive remating

feed their mates. I begin by briefly reviewing the menu of different courtship meals offered by katydids and other arthropods (for a detailed review, see Vahed 1998a). I conclude by arguing that a sexual selection hypothesis, such as that proposed by Gerhardt and Boldyrev for katydids and other ensiferans, provides a general explanation for mate feeding in most species, although a couple of katydid species seem to be notable exceptions.

Second, the current function of the nuptial gift has important implications for Darwinian sexual selection. One concerns female mating preferences. Females are not expected to show preferences for certain meals over others if males have been selected to provide a low-nutrient sticky mass purely as an ejaculate-protection device. On the other hand, if the mating meal has food value, either in an ejaculate-protection or in a paternal-investment role, females should be adapted to prefer the males that can serve the best meals (see Chapter 7). Moreover, if the gift is highly nutritious, there is potential for females to compete with each other for access to males and their meals, and thus an opportunity for biologists to understand how the male investment might control sexual competition by males and females (see Chapter 9).

6.2. The Courtship Gift Menu

Arthropods provide a variety of courtship meals (reviewed by Parker and Simmons 1989 and Vahed 1998a), and within this phylum it is the insect order Orthoptera that tops the list for gift diversity (Gwynne 1983b). As in many butterflies, some short-horned grasshoppers (Acrididae) have a male contribution that is absorbed in the female's genital tract. Such a cryptic nutritive contribution from the ejaculate can have a positive effect on female fecundity by increasing her life span or egg-laying rate (Butlin et al. 1987; Rutowski et al. 1987; Svard and Wiklund 1988; Wiklund et al. 1993; Kaitala and Wiklund 1994; Pardo et al. 1994; Belovsky et al. 1996; but see Eberhard 1996). In crickets, katydids, and other ensiferans, the male contribution is always taken orally by the female and can have very positive effects on her fitness (Table 6.2). The type of meal ranges from the partial cannibalism suffered by a male hump-winged cricket (*Cyphoderris* species) when the female eats his specialized fleshy hind wings (Fig. 2.5), to an array of substances manufactured by males such as the spermatophylax of katydids and the many types of external glandular secretions exhibited by various true crickets (Gryllidae) (Alexander and Otte 1967; Otte 1992). An example of the latter are the life span–increasing secretions lapped by the female tree cricket (*Oecanthus nigricornis*) from a tiny "soup bowl" positioned on the male's thorax (Brown 1997a) (Fig. 6.5, Table 6.2).

Nuptial meals that are manufactured by the male have cropped up independently in a number of other insect orders. Species of Zoraptera (Choe 1995) and Coleoptera (Eisner et al. 1996b) feed on head-gland secretions, whereas male scorpionflies (Mecoptera: Panorpidae) (e.g., Thornhill 1976b; Thornhill and Alcock 1983; Bockwinkel and Sauer 1994) and certain dipterans (e.g., Pritchard 1967; Berg and Valley 1985) literally spit out piles of food for their mates from specialized salivary glands. Males of some Neuroptera (e.g., Withycombe 1922) and dobsonflies (Megaloptera) present females with "a large gelatinous part covering the sperm-containing package" that is remarkably similar to the katydid spermatophylax (Hayashi 1992, 1993, 1998) (Fig. 6.6, Table 6.2). In theory (Buskirk et al. 1984) the male's entire soma could be donated as a nuptial meal. Although sexual cannibalism is well known in mantids and spiders, the

Table 6.2. The effects of glandular meals on females

Species	Gift type	Effect on female	Reference
Requena verticalis (katydid)	Spermatophylax (13% protein)	Affects egg size, hatching, growth rate of sons	Gwynne 1988a
Kawanaphila nartee (katydid)	Spermatophylax	Increases fecundity and egg size	Simmons 1990
Poecilimon veluchianus (katydid)	Spermatophylax (11%–15% protein)	Helps offspring survive starvation	Reinhold 1999
Conocephalus nigropleurum (katydid)	Spermatophylax	Increases life span of mated female*	Lorch 1999
Decticus verrucivorus (katydid)	Spermatophylax (4% protein)	None detected	Wedell and Arak 1989
Gryllodes supplicans (grylline cricket)	Spermatophylax	Increase number of eggs laid soon after mating and supplies water	Kasuya and Sato 1998; Ivy et al. 1999
Oecanthus nigricornis (oecanthine tree cricket)	Thoracic gland secretion	Increases life span of mated female	Brown 1997a
Protohermes grandis dobsonfly (Megaloptera)	Spermatophylax	Increases life span of mated female*	Hayashi 1998

*Effects determined by varying female mating (gift-receiving) frequency rather than by experimentally manipulating amount of gift while controlling the amount of ejaculate a female receives.

male complicity in cannibalism that is necessary if the meal is adaptive has been confirmed only in some spiders (Forster 1992; Andrade 1996).

Finally, some of the best-studied nuptial meals in arthropods are not body parts or secretions but items collected by males, such as carrion defended by panorpid scorpionflies and some sciomyzid flies (Thornhill 1981; Berg and Valley 1985) or prey of bittacid scorpionflies (e.g., Thornhill 1976b) and empidid flies (Fig. 9.1) (Svensson et al. 1990; Cumming 1994).

In contrast to the Orthoptera as a whole, there is little variety in the courtship meals served among katydids. There are some reports of females mouthing possible secretions on the dorsum of the male (Boldyrev 1915, Gerhardt 1914, as cited in Vahed 1998a; and Fabre 1917), including reports of apparent feeding from dorsal glands in *Isophya acuminata* and *Pterophylla robertsi*, similar to the metanotal gland feeding of tree crickets (Gryllidae: Oecanthinae: Fig. 6.5) (Englehardt 1915; Barrientos 1998). However, it is

Figure 6.5. A female black-horned tree cricket, *Oecanthus nigricornis*, feeding on the dorsal thoracic gland of male following copulation. Photo by Maria Zorn.

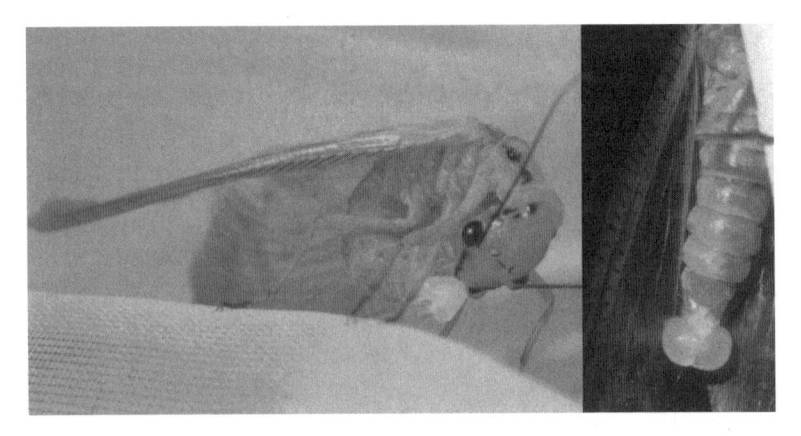

Figure 6.6. Katydids and related orthopterans are not the only insects with a spermatophylax nuptial gift. Here a female megalopteran, *Protohermes*, feeds on her spermatophylax after mating. Photo by F. Hayashi.

not really clear whether the secretions of *I. acuminata* are a source of nutrition or some sort of attractant or "alluring" substance (Englehardt 1915).

Another possible mating meal involves body parts. Rentz (1963) described *Decticita* females biting off and consuming the male's hind femur while copulating. In cases such as this, it is important to demonstrate that this partial cannibalism is a frequent part of mating, as Rentz showed, rather than simply an opportunistic act.

The standard fare for nuptial meals in katydids is the spermatophylax, a part of mating found in most species (Figs. 6.1–6.4) (Gwynne 1983b, 1997a; Vahed 1998a) and consisting of water, protein, some carbohydrates, but few lipids (Heller et al. 1998). This male gift varies greatly in size (Table 6.3, Fig. 6.7), with the current record holders being a few bradyporines. In genera such as *Ephippiger*, the massive meal (ampulla and spermatophylax) can comprise about 40 percent of the male's body weight (Busnel et al. 1956; Vahed 1997). After first reading Busnel's paper on *Ephippiger* some 20 years ago, I wanted to see an enormous katydid mating meal for myself. My first observational foray (to the bottom of my garden) proved rather a disappointment, however, when my subject, a cone-head species (*Neoconocephalus*) in southern Canada, turned out to have one of the smallest nuptial contributions in the family (Gwynne 1977). In this cone-head, as well as the related *Ruspolia* and *Belocephalus* species, the spermatophylax is a mere postcopulatory snack equivalent to about 2 to 4 percent of the male's weight that the female eats in a matter of minutes (Boldyrev 1915; Whitesell 1969; Gwynne 1983b) (Fig. 6.7, Table 6.3).

6.3. Why Katydids Feed Their Mates

6.3.i. Hypotheses

Male animals in general have the potential to copulate frequently because ejaculation is usually an inexpensive activity (although never without some cost: Dewsbury 1982). Thus, the potential for males to mate with many females is not usually limited by copulation costs but by sexual competition with other males (Trivers 1972; Emlen and Oring 1977). Consequently, katydids are unusual because the mating potential of males can be limited by copulation alone, not by sperm costs but by the costs of spermatophore production (see Reinhold and Helversen 1997). "Limp and exhausted, as though shattered by his exploit, he remains where he is, all shrivelled and shrunk" is how Fabre (1917) described the postcoital appearance of one of his captive male white-faced decticus. The male's shrunken abdomen is an indication of the price he pays for a single mating. The price can be measured in the time it takes for a male to recharge his abdominal glands so that he can mate with another female. A male *R. verticalis* donates about a fifth of his body mass and cannot remate for about 5 days (Gwynne 1990b). Even longer is the refractory period of the male hetrodine *Eugaster spinulosa*, who apparently endures a 10-day wait before he can mate again (Bleton 1942).

Why have males evolved such a costly copulation? We have already

Figure 6.7. The large variation in the size of the spermatophylax meal. **a.** A cone-head, *Belocephalus* (Copiphorini) (arrow points to the spermatophylax). **b.** A meadow katydid, *Conocephalus nigropleurum* (Conocephalini). **c.** The Mormon cricket, *Anabrus simplex*.

Table 6.3. Variation in male investment in the nuptial gift and in the female refractory periods of katydids

Species	Subfamily	Relative meal (spermatophore)* size (% of male body mass)	Spermatophylax protein (% wet mass)	Female refractory period (days)
Ephippiger ephippiger	Bradyporinae	22.1 28.1 23.3	9.1	
E. terrestris	Bradyporinae	30.3		
Ephippigerida taeniata	Bradyporinae	27		
E. saussuriana	Bradyporinae	28		
Uromenus stali	Bradyporinae	35		4.8
U. rugosicollis	Bradyporinae	12		
Austrosalomona	Conocephalinae	14.7		
Conocephalus discolor	Conocephalini	9.7		
C. nigropleurum	Conocephalini	10.8		
C. semivittatus	Conocephalini	8.2	25.1	2
Orchelimum (2 spp.)	Conocephalini	(10)		Permanent
Belocephalus sp.	Copiphorini	2.5		
Copiphora brevirostris	Copiphorini	12.8		
Ruspolia nitidula	Copiphorini	0.7		
Coptaspis sp. 2	Agraeciini	6.4	27.2	3.6
Coptaspis sp. 5	Agraeciini		22.4	
New genus # 12 sp 6	Listroscelidinae	27.9	4.6	11.5
Requena sp. 4	Listroscelidinae	17.7	5.8	4.5
Requena sp. 5	Listroscelidinae		20	
R. verticalis	Listroscelidinae	20	13.2	5
Mecopoda elongata	Mecopodinae	0.7		
Cyrtaspis scutata	Meconematinae	9.1		
Meconema meridionale	Meconematinae	1.8		
M. thalassinum	Meconematinae	5.9		
Acripeza reticulata	Phaneropterinae	1.9	4	2.5
Ancistrura nigrovittata	Phaneropterinae	7.4		
Barbitistes ocskayi	Phaneropterinae	31.6	8	
B. serricauda	Phaneropterinae	28.6		
Caedicia simplex	Phaneropterinae	9.3	6	1.8
Leptophyes albovittata	Phaneropterinae	7.6		
L. boscii	Phaneropterinae	7.0		
L. laticauda	Phaneropterinae	25.9		
L. punctatissima	Phaneropterinae	4.0		
Phaneroptera falcata	Phaneropterinae	5.0		
P. nana	Phaneropterinae	14.2		
Poecilimon affinis	Phaneropterinae	14.6 & 15.5	17	
P. jonicus	Phaneropterinae	8.6		
P. macedonius	Phaneropterinae		11.2	
P. schmidtii	Phaneropterinae	13.8	15.5	
P. veluchianus	Phaneropterinae	22.7 & 25.6	10	

Table 6.3. (continued)

Species	Subfamily	Relative meal (spermatophore)* size (% of male body mass)	Spermatophylax protein (% wet mass)	Female refractory period (days)
Polichne parvicauda	Phaneropterinae	13.3	10.7	1.6
Polysarcus scutatus	Phaneropterinae	16		
Metaplastes ornatus	Phaneropterinae	16 & 19.8		
Tinzeda eburneata	Phaneropterinae	6.2	19	2.6
Tinzeda sp. 9	Phaneropterinae		17	
Phasmodes ranatriformis	Phasmodinae	4		1.8
Balboa tibialis	Pseudophyllinae	21.9		
Docidocercus gigliotosi	Pseudophyllinae	15.4		
Pristonotus tuberosus	Pseudophyllinae	9.5		
Undescribed genus "A"	Pseudophyllinae	25.3		
Anabrus simplex	Tettigoniinae	19		
Anonconotus alpinus	Tettigoniinae	2		
Antaxius pedestris	Tettigoniinae	16.1		
Austrodectes monticolus	Tettigoniinae	19.4	15.6	5
Chlorodectes montanus	Tettigoniinae	16.3	6	7
Decticus albifrons	Tettigoniinae	9.1	9.7	4.6
D. verrucivorus	Tettigoniinae	11.1	4.2	
Eupholidoptera sp. 1	Tettigoniinae	13		
Eupholidoptera sp. 2	Tettigoniinae	16.4		
Gampsocleis glabra	Tettigoniinae	11		
Metaballus litus	Tettigoniinae	24		
Metrioptera bicolor	Tettigoniinae	9.8 & 11.1	18.2	3.2
M. roeselii	Tettigoniinae	10.4	13.3	
M. saussuriana	Tettigoniinae	8.3		
Platycleis albopunctata	Tettigoniinae	5.6		
P. affinis	Tettigoniinae	6.4		
P. nigrosinata	Tettigoniinae	6.4		
P. sepium	Tettigoniinae	7.5		
Tettigonia cantans	Tettigoniinae	17.1		
T. viridissima	Tettigoniinae	14.5	15.6	7
Yersinella raymondi	Tettigoniinae	6.6		
Kawanaphila nartee	Zaprochilinae	21		>12

*Meal size includes both spermatophylax and ampulla because the female eats both. Information from Wedell (1993a, 1993b, 1994b), Belwood (1988), Bowen et al. (1984), Gwynne (1983b, 1988a, 1990b), Heller et al. (1998), Simmons and Gwynne (1991), Vahed and Gilbert (1996), Vahed (1997), and Busnel and Dumortier (1955). Note that data in parentheses represent an estimate (mean) for the genus. More than one entry indicates multiple estimates from different studies.

reviewed the ejaculate-protection (Table 6.1, hypothesis IV) and paternal-investment hypotheses (Table 6.1, hypothesis V), but other adaptive scenerios for the origin, evolution, and current maintenance of nuptial meals have also been proposed. I outline all of these hypotheses and rule out all as unlikely to account for the katydid spermatophylax, except for the paternal-investment hypothesis and two versions of the ejaculate-protection hypothesis (Gwynne 1997a).

One hypothesis for the nuptial offerings that are presented to females before copulation is that they are sexually selected devices that function to attract females and induce them to mate (Table 6.1, hypothesis I). However, this is not the function in katydids, because the male donation is not produced and eaten until after the female has been attracted and copulation with spermatophore transfer has occurred.

Another hypothesis that only applies to gifts held out to the female when she meets the male is that the meal is a naturally selected device that prevents the female from turning her mate into a meal (Table 6.1, hypothesis II). The few observations of sexual cannibalism in katydids indicate that this morbid meal takes place before the sperm ampulla and the putative life-saving spermatophylax are transferred (in *R. verticalis:* I. Dadour and L. Simmons, pers. comm. 1993). There is also one report of postcopulatory cannibalism. In this case, however, the spermatophylax meal had already been handed over to the female when she passed up the usual fare and ate the male instead (in *Saga natoliae*: Burr et al. 1923).

Snodgrass (1905) proposed that the spermatophylax "is a plug to close the bursa copulatrix" (Table 6.1, hypothesis IIIa). He was apparently referring to preventing sperm loss from the female. However, it is not the spermatophylax but the ampulla itself that blocks the female's gonopore (Boldyrev 1915). And it is unlikely that the ampulla itself prevents sperm loss, as sperm are stored in the female's spermatheca, the duct of which is probably kept closed by muscular control.

A second version of the "plug" hypothesis (hypothesis IIIb) views certain structures inserted by males into female insects as devices that block copulation attempts by rivals (Parker 1970) so that the male gains and the female loses because her mating (meal-acquiring) frequency is reduced. Male ensiferans do appear to have evolved devices that delay female remating, but rather than mechanical devices these appear to be chemical substances that turn off the mated female's receptivity, as suggested by the data of Simmons and Gwynne (1991) on a katydid and demonstrated in several gryllids (Loher 1984). More importantly, the days-long refractory period that follows mating in a diversity of female katydids (Wedell 1993b) (Table 6.3) renders redundant any structural "chastity belt" that the male attaches.

Finally, Eberhard (1996) proposed that male glandular donations that effect an increase in female fecundity may not be nutritional in nature at all but a sexually selected "hormonal" message from the male that turns on egg production. The finding that female insects use ejaculated chemicals as a cue to lay eggs has been known for some time (Leopold 1976; Gillott and Friedel 1977), but Eberhard's view on this is novel; he suggests that ejaculation chemicals are used as signals by females to discern high-quality males. The females favor high-quality males by increasing the rate of laying eggs fertilized by his sperm (or by allocating more of her own resources to individual eggs: Burley 1988). This is a form of cryptic or postcopulatory female choice, a topic that I discuss further in Chapter 7 (Table 6.1, hypothesis VI). It is important to distinguish the signal-for-cryptic-choice hypothesis from proposals that it is nutrition from the male donation that boosts fecundity after copulation (e.g., paternal investment) (Eberhard 1996). However, Eberhard's hypothesis is more applicable to insects (e.g., the acridid grasshoppers and butterflies mentioned earlier) in which the male contribution is introduced into the genital tract of the female. Oral meals such as the ensiferan spermatophylax are unlikely to function as chemical cues because such cues in the Ensifera are known to be passed in the ejaculate (Loher et al. 1981; Loher 1984). Furthermore, the positive effects of glandular gifts on female life span (Table 6.2; Kaitala and Wiklund 1994) are difficult to explain with Eberhard's hypothesis. Therefore, for the katydid spermatophylax, I consider all these alternative hypotheses unlikely. Let us now return to the ejaculate-protection and paternal-investment hypotheses—outlined earlier in the chapter.

As we have already seen, when the spermatophylax nurtures the offspring of the male (paternal-investment hypothesis), the sexes receive mutual benefits (Table 6.1, hypothesis V), but is there any conflict if this nuptial gift functions to protect the ejaculate? The answer, in theory, depends on what the female gets out of mating. Mutual benefits to the sexes will occur if the male protects his ejaculate with a highly nutritious meal (Table 6.1, hypothesis IVa) (e.g., Brown 1997a). A nutritious meal benefits the male because it reduces the necessity for the female to seek food by copulating with rivals. But, at the same time the investment by the male in a highly nutritious gift will reduce his opportunities for additional matings. Note that the difference between this hypothesis and the paternal-investment hypothesis is that the former does not assume that the nutritious contribution benefits the contributing male's own offspring.

However, the spermatophylax is known to be of little nutritive value in species such as the wart biter katydid (Wedell and Arak 1989) so that the female may gain little from mating. A low-quality meal might protect the ejaculate if its properties keep females busy during insemination (hypothesis IVb). One way the male could achieve this is by clogging the female's mouthparts with a sticky spermatophylax. In katydids the spermatophylax can be a gluey mass, as discovered by Gabriel Brunellius (1791), who found spermatophore glands to contain "a certain white material which adhered in great quantity, and which, when it came into contact with the air, immediately solidified and could not even, by any means, be washed away with water." However, the spermatophylax has to be more than just a mechanical annoyance because the female actually eats it all before eating the sperm ampulla (e.g., Wedell and Arak 1989). Why? One answer is that the male's strategy is equivalent to a "candy manufacturer"—make the less-wholesome offering taste good (e.g., using chemicals that stimulate female feeding; Warwick et al. 1998) so that the female does not discard it before insemination is complete (Gwynne 1997a). However, under these conditions, counteradaptations in females would be expected, such as remating quickly after receiving a poor gift to maximize the number of such meals (e.g., Simmons and Gwynne 1991).

A second possibility is that females eat a gift of low nutrition as a taste test of male genetic quality. This idea is similar to Eberhard's (Table 6.1, hypothesis VI) except that female preference for a high-quality male is not expressed as an increase in fecundity. Using gifts to assess male quality has been suggested as a reason for the precopulatory inedible silk balls offered by some male empidid flies (Cumming 1994) (other empidids use gifts of fresh prey: Kessel 1955; Cumming 1994) and the postcopulatory low-protein spermatophylax of the gryllid *Gryllodes supplicans* (Will and Sakaluk 1994). However, a postcopulatory gift would seem to be an unlikely way to assess male quality given the expectation that females would assess males as early as possible in the pairing-mating sequence.

The ample variation in protein content and size of different katydid spermatophylax gifts (Table 6.3) indicates that paternal investment or ejaculate protection with a nutritious spermatophylax occurs in some species, whereas ejaculate protection with a low-nutrient contribution occurs in others. However, additional information is required to separate the hypotheses. Let us now examine the evidence, first considering the origin and evolution of these mating meals.

6.3.ii. Origins of courtship feeding in katydids and their kin

As Gerhardt (1913, 1914) and Boldyrev (1915) first noted, the spermatophylax crops up as a nuptial meal in many katydids and in certain other ensiferan groups. Does the distribution of meals in the suborder represent multiple evolutionary origins or a trait that originated once in a common ancestor of these orthopterans? I provided a preliminary answer to this question by tracing the presence and absence of the spermatophylax on the phylogeny of ensiferan families (see Chapter 2, Fig. 2.3) expanded to include more detailed trees of Tettigoniidae (Fig. 2.12).

The results (Gwynne 1995, 1997a) reveal a striking pattern of convergence in the design of meals (Fig. 6.8); the spermatophylax evolved at least three times (four if we count the separate origin in megalopterans [Hayashi

140 Evolution of the Spermatophylax

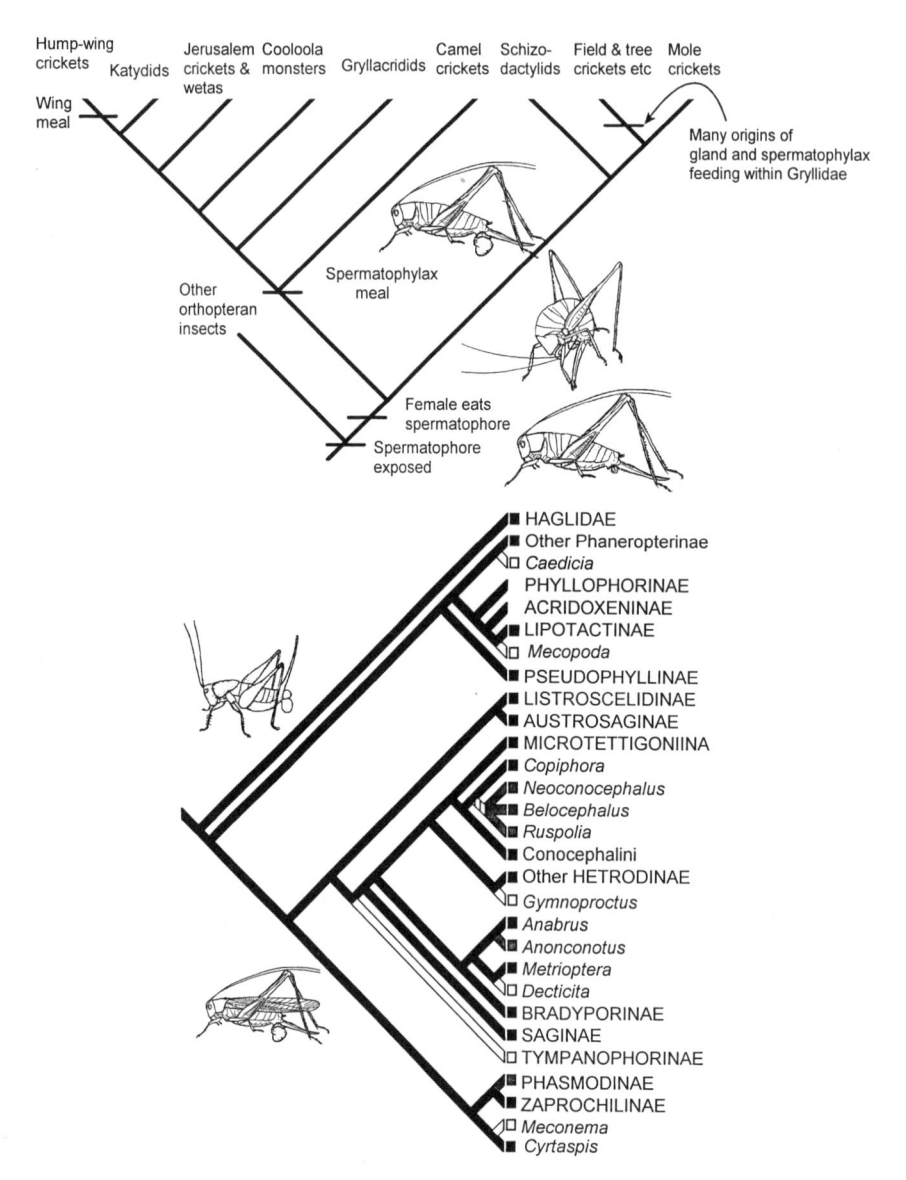

Figure 6.8. The origins and evolution of courtship feeding and related characters in katydids and their allies (suborder Ensifera). Top. A phylogeny of families in the suborder (from Fig. 2.3). The bars on the phylogeny indicate the most parsimonious explanation for origins of the spermatophore: being exposed externally on the female's abdomen, being eaten by the female, and acquiring a spermatophylax. The two bars at the top of the figure show the origin of other courtship meals in Ensifera. Katydids redrawn from Chopard (1965), after Boldyrev. Bottom. Evolutionary change in the spermatophylax meal within the Tettigoniidae (phylogeny from Fig. 2.12). Subfamilies (and the outgroup Haglidae) are uppercase names ending in *ae*. The phylogeny shows that the ancestral full spermatophylax meal (black parts) has been lost independently (white bars) or greatly reduced (stippled) in several groups (analysis using McClade 3.0) (lines indicate areas of the tree where the character state is equivocal). Katydids redrawn from a figure by D. Scott, in Clutton-Brock (1991) (top); and from Chopard (1965) after Boldyrev (bottom).

1992]). In Ensifera, one origin apparently occurred early in the diversification of this suborder in a camel cricket–like ancestor at the base of tettigonioid clade. Two other origins occurred within Gryllidae (Fig. 6.8). The tettigonioid origin was coincident with another change in the complexity of the spermatophore, the division of the sperm ampulla into two sperm cavities.[2]

Character tracing also shows a loss or great reduction in the spermatophylax in a few taxa within the katydid family (Fig. 6.8). Male *Tympanophora* (Tympanophorinae: D. T. Gwynne, unpublished data 1986), *Mecopoda elongata* (Mecopodinae: Vahed and Gilbert 1996), *Caedicia*, *Meconema* (Meconematinae: Vahed 1996), *Gymnoproctus sculpturatus* (Hetrodinae: Schmidt 1990), and apparently *Decticita* (Tettigoniinae: Rentz 1963) do not supply a spermatophylax, whereas the nuptial meal has been reduced to a mere hors d'oeuvre (less than 4% of the male's weight, see Table 6.3) in Phasmodinae (Bailey 1998b, D. T. Gwynne, unpublished data 1986), grassland cone-heads such as *Belocephalus*, *Neoconocephalus*, and *Ruspolia* species (Boldyrev 1915; Vahed and Gilbert 1996; Gwynne 1997a), some phaneropterines (e.g., *Acripeza reticulata*) (Wedell 1993b; Vahed and Gilbert 1996), *Anonconotus alpinus* (Vahed and Gilbert 1996), and *Coptaspis* species (Wedell 1998). See Table 6.3 for the size variation in spermatophore offerings.

A few katydid species have dispensed with spermatophores and mating completely. Warchalowska-Sliwa (1998) reported four cases of parthenogenesis that appeared to evolve independently, as they were found in widely different taxonomic groups. These comprise two Eurasian phaneropterines (*Poecilimon intermedius* and some populations of *Leptophyes punctatissima*), a European sagine (*Saga pedo*), and a Hawaiian meconematine (*Xiphidiopsis lita*). As well, the presence of all-female populations in a North American *Steiroxys* (Tettigoniinae) species points to parthenogenesis (D. T. Gwynne and P. Lorch, unpublished data 1999).

Boldyrev (1915) suggested that prolonged copulation after sperm transfer might function as an ejaculate-protection device in some katydids with small nuptial offerings. This hypothesis has some support in several species with small or no spermatophylax offering, because ejaculation in these katydids is not a postcopulatory event as in most. *Meconema meridionale* (Vahed 1996), *Uromenus rugosicollis* (Vahed 1997), some *Coptaspis* species (Wedell 1998), and *Pterophylla* species (Barrientos and Jaramillo 1998) are unusual among katydids in that the male's genitalia continue to clasp the female during sperm transfer from the spermatophore. However, there must be selection pressures other than ejaculate protection acting on

---

[2] The virtual ubiquity of the spermatophylax and the lack of a single sperm cavity in species in this clade containing the katydids are two of the characters that support the phylogenetic separation of this group from the other main group of ensiferans, the grylloids (Gwynne 1995).

copulation duration in katydids since long copulations also occur before sperm transfer in species with lengthy nuptial offerings. In particular, the 30- to 50-minute copulation of *Metrioptera roeselii* is followed by a meal of 6 to 12 hours (Kevan et al. 1962), and in *Acanthoplus speiseri*, the marathon mating, over 5 hours, is followed by a lengthy repast in which the female takes a further 8 hours to consume the spermatophylax (Mbata 1992a, 1992b).

As we have seen, the katydid spermatophylax appears to have originated in a camel cricket–like ancestor of the tettigonioid clade. We can now use this origin and other evolutionary events to investigate the original function of these mating meals. The Boldyrev-Gerhardt hypothesis predicts that each evolution of a "protective" device such as the spermatophylax always comes after the origin of an ampulla that is external, and the subsequent female exploitation of this vulnerable and presumably tasty sperm package.

As in any comparative analysis, it is important that we conduct a conservative test by using "origins" rather than taxa themselves, because taxa may not be independent data points; it is possible that a trait is shared by two taxa not because it was selected for in each lineage but because it was inherited from a common ancestor (Brooks and McLennan 1991; Harvey and Pagel 1991). To determine the independent origins of a discrete character, for example, "spermatophylax present or absent," we trace back down the tree from the state of the character in each living taxon at the tips of each tree branch.

The results (Fig. 6.8, top) show the sequence of evolutionary events predicted by the Boldyrev-Gerhart hypothesis: (1) the katydid spermatophylax evolved after female consumption of the ampulla first originated; and (2) female consumption of the ampulla first evolved at the same time or after the origin of a sperm ampulla positioned externally on the genitalia (Gwynne 1997a) (details in legend to Fig. 6.8). Because these ancestral females did not receive a nuptial meal, they probably found nutritional benefit by mating frequently and eating the proteinaceous cases of the sperm ampullae. A living analogue of this ancestor may be a species of *Arachnocephalus* (Gryllidae), a cricket that mates frequently and in which only a few ampullar sperm survive the gauntlet of the female's jaws to make it into her reproductive tract (Boldyrev 1915). Other field crickets can also mate multiply, perhaps for the nutrition contained in spermatophore cases; an experimental increase in the number of ampullae eaten by *Gryllus bimaculatus* females increases fecundity (Simmons 1988).

Although this sequence of three evolutionary events supports the Boldyrev-Gerhart hypothesis, it is a weak test because it is based on a sample size of a single origin of a "protective" meal. We can increase the power of the test somewhat by including the origins of other meals and other postcopulatory behaviors thought to function as protective devices within the Ensifera. These include the independent origins of spermatophylaxes in the clade of crickets and their kin, mating meals such as the thoracic gland secretion of tree crickets (Fulton 1915), and other behaviors that Boldyrev (1915) and Gerhardt (1913, 1914) proposed to function in ejaculate protection such as standing guard over the female after the ampulla has been transferred (see Sakaluk 1991 for alternative hypotheses to explain postcopulatory guarding). When traced onto our ensiferan phylogeny, all of these protective devices and behaviors evolved after the origin of an external ampulla and female consumption of this vulnerable sperm package (Gwynne 1997a) (Fig. 6.8). The power of this evolutionary test will be increased when information on the origins of a spermatophylax meal in neuropterans and megalopterans is available (Withycombe 1922; Hayashi 1992).

Finally, has our evolutionary test been able to distinguish the separate hypotheses for the original function of the spermatophylax? It clearly supports the sequence of evolutionary events predicted by the ejaculate-protection hypothesis but sheds no light on whether the spermatophylax originated as a highly nutritious or a low-quality meal. With regard to the paternal-investment hypothesis, although our sequence of evolutionary events is not specifically predicted by this hypothesis for the origin of the spermatophylax, it does not refute it (Gwynne 1997a).

After each origin of the spermatophylax, has there been any subsequent evolutionary change in function? One bit of evidence supporting change is that both postcopulatory mate feeding and mate guarding by males co-occur in some species. This indicates that one or the other behavior has undergone a change in function (Gwynne 1997a). Experiments with some true crickets showed that postcopulatory guarding does protect the ejaculate (e.g., Hockham and Vahed 1997), but with the spermatophylax-bearing *Gryllodes sigillatus*, it is this mating meal that appears to fill the ejaculate-protection role, with postcopulatory guarding functioning to keep rivals away from the mated female (Sakaluk 1984, 1991). What about katydids? Has there been any diversification in function of the spermatophylax since its origin in the stem group leading to this speciose family (Fig. 6.8)? Although postcopulatory mate guarding has never been observed in the family, results of comparative studies of katydids indicate that the function of the spermatophylax has changed in some species.

6.3.iii. Evolutionary change in the function of the
katydid spermatophylax

So little was available on the reproductive trait information necessary for a comparative analysis of katydid mating that Nina Wedell (who studied Australian and European taxa) and Karim Vahed (using mainly European species) collected almost all of the original data and subjected them to comparative analysis. One problem with comparing traits between species is that some variation in the traits will be due to between-species differences in body size. Larger species tend to have absolutely larger ejaculates (Wedell 1997) and spermatophylaxes. Body size variation is usually controlled for in comparative analyses by examining the size of the trait of interest relative to body size (typically by using as data points the "residuals" from a statistical regression of the trait of interest on body size).

The initial papers from the comparative studies of katydid species seemed to support ejaculate protection as a general function of the mating meal in katydids. First, in a test of the paternal-investment hypothesis, Wedell (1993b) predicted that taxa with relatively larger spermatophylaxes should have greater egg production "if all other things [are] equal." Her analysis refuted this prediction by finding no significant correlation (but see Wedell 1994a). A problem with the prediction in the first place, however, is that if the other variables are *not* equal, an absence of a positive correlation is also consistent with the paternal-investment hypothesis. For example, spermatophylax nutrients may have evolved "paternally" either as a replacement for a diet low in the nutrients necessary for egg production or to provide specialized nutrients to offspring.

We saw earlier that the spermatophylax protects the sperm ampulla by distracting the female with a meal. With this in mind, Wedell (1993b) predicted that species with larger sperm ejaculates should have a larger protective spermatophylax. Thus, an evolutionary increase in ejaculate size (ampulla size) and an important effect of an ejaculate chemical, the length of the female's refractory period, should correlate positively with increasing size of the protective spermatophylax. Both correlations were supported in an analysis of 16 genera of katydids (Wedell 1993b) (Table 6.3, Fig. 6.9).

By using genera as data points, Wedell attempted to reduce the problem of the lack of independence in comparative analyses when species are used as data points. However, analyses at the level of genus still lack much control for common phylogenetic history (Harvey and Pagel 1991). For example, katydid genera within the same subfamily tend to share traits such as spermatophylax size and length of the female's sexual refractory period (Gwynne 1994) (Fig. 6.9). To control for phylogenetic effects, Vahed and Gilbert (1996) used an "independent-contrasts test," a comparative method for the analysis of characters that show continuous variation (such as number of sperm in the ampulla) (Harvey and Pagel 1991). Their analysis supported Wedell's finding by showing that larger spermatophylaxes correlated positively with both ampulla size and number of sperm transferred (Fig. 6.10).

However, correlations between gift size and ejaculate size found by Wedell and Vahed are predicted by both the ejaculate-protection and paternal-investment hypotheses (Wedell 1994a; Vahed and Gilbert 1996);

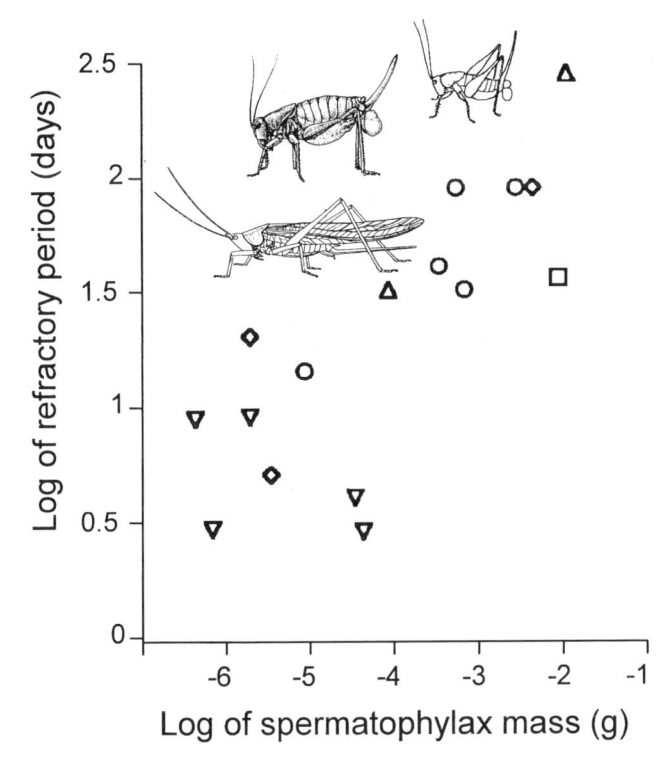

Figure 6.9. Relationship between duration of the female's refractory period (after mating) and mass of spermatophylax (correlation $r = 0.78$, $P < 0.001$) for 16 genera of katydids (original data from Wedell 1993b). The subfamilies of different species are represented by different symbols: diamonds for Conocephalinae, squares for Bradyporinae, triangles for Listroscelidinae, upside-down triangles for Phaneropterinae, and circles for Tettigoniinae. Refractory period also correlates with size of sperm ampulla (correlation $r = 0.76$, $P < 0.001$). Redrawn from Gwynne (1997a). Katydids redrawn from a figure by D. Scott, in Clutton-Brock (1991) (top); From artwork by K. Hansen McInnes, in Gullan and Cranson (1994) (middle); and from Alexander et al. (1972) (bottom).

if males nurture their offspring, whether it is through nuptial feeding or more directly by caring for progeny, they are expected to evolve paternity-assurance mechanisms to prevent cuckoldry—the costly investment in the offspring sired by rivals (Thornhill 1976b; Werren et al. 1980; Westneat and Sherman 1993).

A prediction that does appear to support the ejaculate-protection hypothesis is that as a protective spermatophylax increases in size to protect a larger ejaculate, its protein concentration (Table 6.3) is expected to decrease (assuming that the protein reserves of males are limited). Wedell (1994b) confirmed this pattern in an analysis of species. However,

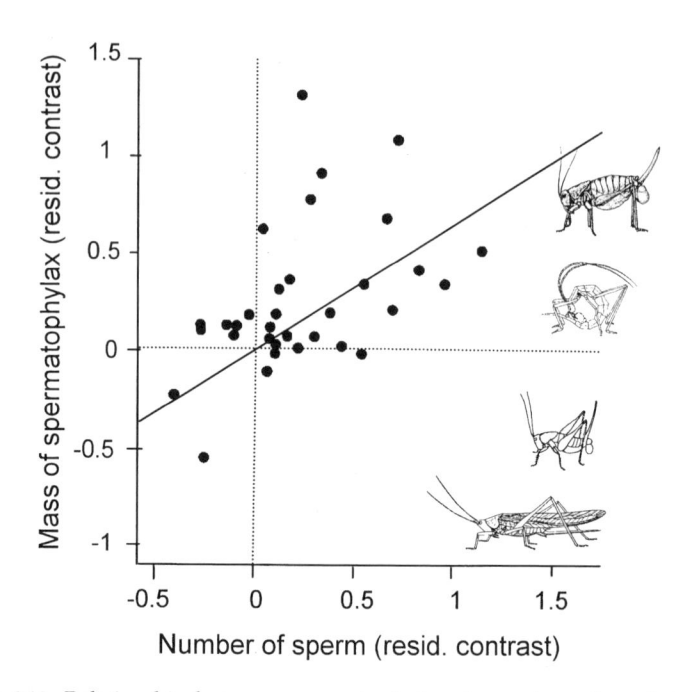

Figure 6.10. Relationship between spermatophylax size and number of ejaculated sperm in katydids. Data are independent contrasts ("residual" [resid.] controls for differences in body size referring to the residuals from a regression of body size versus spermatophylax size). Graph redrawn from Vahed and Gilbert (1996) (from Gwynne 1997a). Katydids redrawn from artwork by K. Hansen McInnes, in Gullan and Cranson (1994) (top); artwork by H. Proctor (second from top); a figure by D. Scott, in Clutton-Brock (1991) (second from bottom); and from Alexander et al. (1972) (bottom).

further research on the size and protein content of the spermatophylaxes suggested that these gifts function as paternal investment in some species and as ejaculate protection in others. In this study, Wedell (1994a) reasoned that if the function of male offerings in some katydids had changed from ejaculate protection with a low-quality spermatophylax (Table 6.1, hypothesis IVb) to paternal investment, then the design of the spermatophylax in different katydids should reflect the two functions: the former should be a smaller, less nutritious contribution that does little for female reproduction (fecundity), whereas the latter should be a nutritious and extra large contribution; recall the "larger-than-necessary" *Requena* spermatophylax discussed earlier in this chapter (Gwynne et al. 1984; Gwynne 1986). The test of the prediction used genera as well as species in an attempt to reduce some of the lack-of-independence problems.

The results with 19 species from 14 genera supported the predictions by revealing a cluster of several species on the graph with a larger-than-

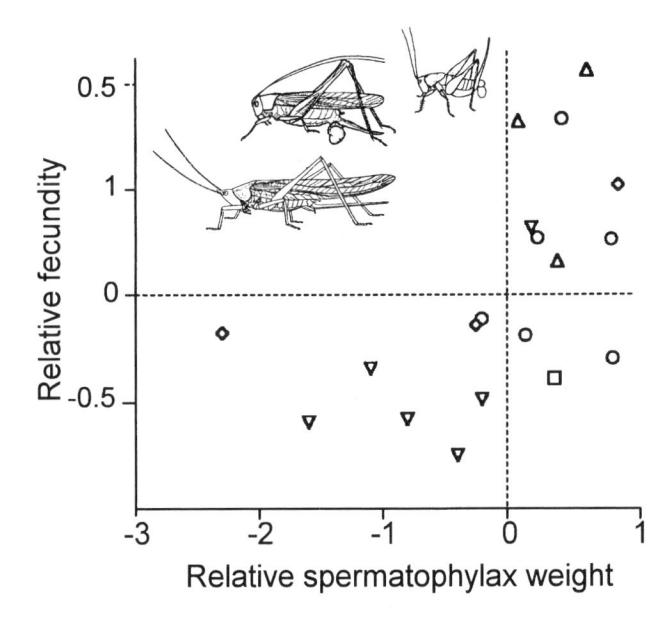

Figure 6.11. Relationship between female fecundity (mass of eggs laid) and spermatophylax size for 19 katydids from Europe and Australia (correlation $r = 0.78$, $P < 0.001$). Points above the horizontal line are fecundities larger than expected. Points to the right of the vertical line are spermatophylaxes heavier than expected. Symbols represent different subfamilies (see Fig. 6.9 legend). Original data from Wedell (1994a). Redrawn from Gwynne (1997a). Katydids redrawn from a figure by D. Scott, in Clutton-Brock (1991) (top); from Chopard (1965), after Boldyrev (middle); and from Alexander et al. (1972) (bottom).

average spermatophylax and greater fecundity than expected, perhaps indicative of "paternal" meals (Fig. 6.11). Furthermore, when the protein content of the spermatophylax offerings in this cluster of species was analyzed, the frequency distribution (Fig. 6.12) contained all taxa with the most nutritious mating meals, with one exception. In fact, the low-protein meals formed a separate distribution. Interestingly, males in these species actually paid more to produce the high-protein nuptial banquets by requiring more time to regenerate their "meal glands" (at least 2 days required to remate) than did species with less nutritious meals. It is important to note, however, that total protein content may not be a complete reflection of the quality of the mating meal, particularly for a manufactured meal such as the spermatophylax. Males could be blending into the meal specialized ingredients that are important to offspring fitness (see Gwynne 1988a; Heller et al. 1998) or providing other important products to their mates such as water (Ivy et al. 1999).

Wedell (1994a) suggested that taken together, these results point to two types of nuptial gift strategies in katydids: low-quality meals designed to

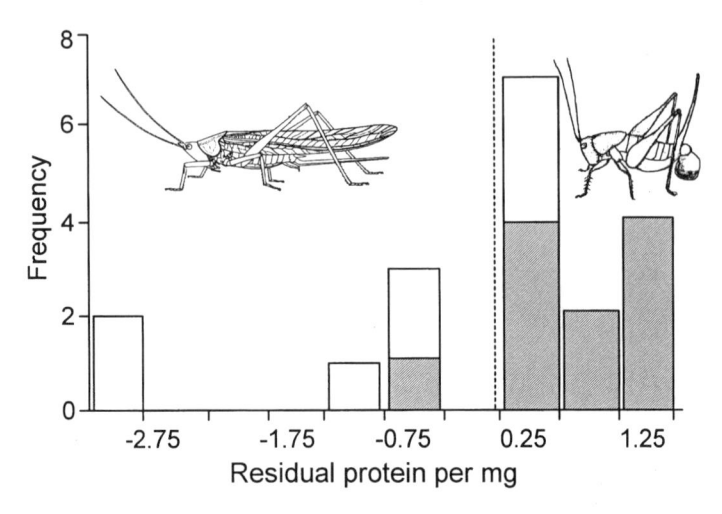

Figure 6.12. The frequency distribution of the amount of protein in the spermatophylax for 19 katydid species. Cross-hatched bars represent spermatophylaxes that may be larger than necessary to protect the ejaculate (from the right side of the dotted vertical line in Fig. 6.11). Redrawn from Wedell (1994a). Katydids redrawn from a figure by D. Scott, in Clutton-Brock (1991) (right); and from Alexander et al. (1972) (left).

enhance paternity (Table 6.1, hypothesis IVb) and high-quality meals that enhance the fitness of the male's offspring. But gifts that ensure effective insemination also can be nutritious, if a quality meal is what it takes for a female to be distracted so that full insemination can occur (hypothesis IVa). However, interspecific variation in the quality of spermatophylax gifts due to these different strategies of ejaculate protection does not predict the distribution of protein content found in Wedell's study.

Manipulative experiments with different species could tell us whether the spermatophylax has different *current* functions. A number of these sorts of experiments have now been conducted with spermatophylaxes of orthopterans.

6.3.iv. Current utility of spermatophylax feeding

Fortuitously, the species used in experimental studies on the current function of the male's gift span the phylogenetic range of katydids in that each is in a different subfamily (see Gorochov 1988 and Chapter 2). The listroscelidine *Requena verticalis* has a large bilobed spermatophylax (Fig. 6.4), while the wart biter *Decticus verrucivorus* is in a tettigoniine genus in which the male gift resembles "an opalescent bag similar in size and colour to a mistletoe berry" (Fabre 1917) (Fig. 6.13). *Uromenus (Steropleurus) stali* has the enormous spermatophore gift (40% of male weight) typical of

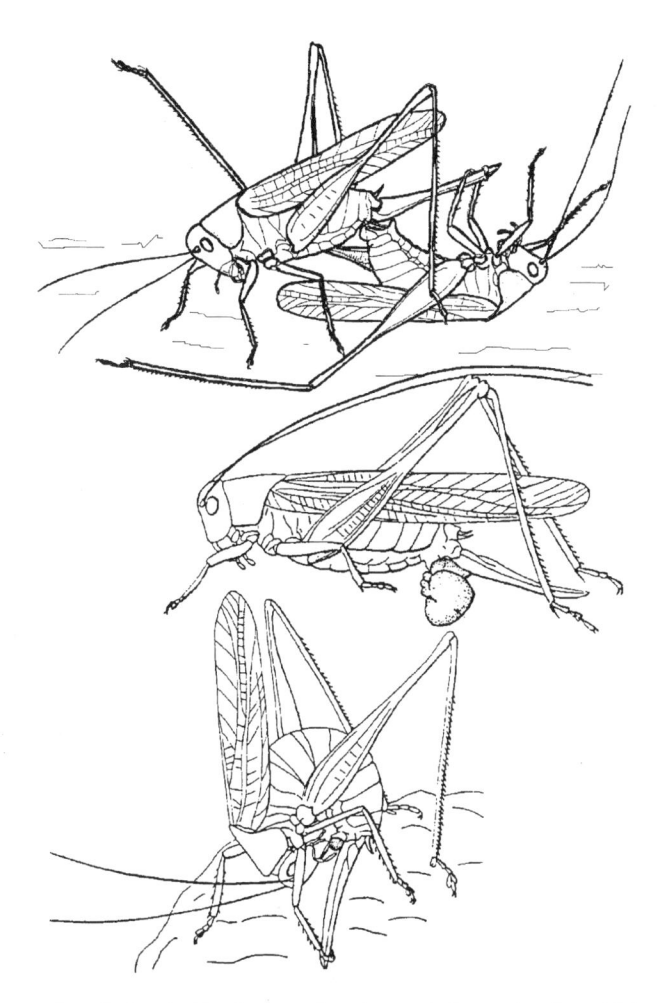

Figure 6.13. Copulation in *Decticus* species. A copulating pair of wart biters (*D. verrucivorus*) (top) and a female white-faced decticus (*D. albifrons*) with a spermatophylax (middle) that she eats (bottom). Redrawn from Chopard (1965), after Boldyrev.

many Ephippigerini (Fig. 6.14, Table 6.3), whose gifts were described by Fabre (1917) as "a sort of opalescent raspberry with large seeds." The phaneropterines *Poecilimon veluchianus* and *Leptophyes laticauda* are representatives of a subfamily with spermatophylax gifts that come in a variety of shapes and sizes (Table 6.3; Fig. 6.2, top; Figs. 2.13, 2.14, 6.15; Plate 3) including the "diaphanous, oval phial, . . . hanging from a crystal thread . . ." of *P. falcata* (Fabre 1917). The last species, *Kawanaphila nartee*, one of the pollen katydids (Zaprochilinae), has a translucent spermatophylax surrounding its more opaque sperm ampulla (Fig. 9.2. and Plate 23). For com-

Figure 6.14. A mated female *Uromenus* (*Steropleurus*) *stali* with her enormous spermatophylax. Photo by K. Vahed.

Figure 6.15. *Poecilimon affinis* female showing the large spermatophylax just after mating. Photo by K. Vahed.

parison, I include a species representing an independent origin of the spermatophylax, the decorated cricket (or stove cricket, "Zhao Ji"), *Gryllodes supplicans* (Gryllidae, Gryllinae: Fig. 6.16). Now that I have introduced the subjects of the studies, let us examine the predictions.

1. *The paternal-investment hypothesis* (Table 6.1. hypothesis V) *requires that*

Figure 6.16. Mating in *Gryllodes supplicans* showing the female (top) receiving the spermatophylax from the male. Photo by Scott Sakaluk.

male-derived nutrition increases offspring production over a period sufficiently long for the investing male to father the offspring he helps produce with his food donation (Thornhill 1976b; Wickler 1985; Gwynne 1986). This prediction has been refuted in studies of wart biters and the cricket *G. supplicans*. First, an experimental increase in the size of the wart biter's spermatophylax meal has no detectable influence on the size and number of eggs produced or on female survival (Wedell and Arak 1989). Moreover, the investing males' paternity in these progeny is not ensured (Wedell 1991) because females tend to remate before the nutrient contribution from the investing male can be incorporated into eggs (Wedell 1993a).

Female *G. supplicans* crickets also remate frequently, resulting in high levels of sperm competition that make it unlikely that males invest in their own offspring (Sakaluk 1986). For this species, Will and Sakaluk (1994) found a positive but nonsignificant effect of spermatophylax consumption on fecundity, and a later study by Kasuya and Sato (1998) showed a significant increase in egg production (with number of spermatophylaxes eaten) in the first few days after mating. There is also evidence that *Gryllodes* females might receive hydration benefits from water contained in the spermatophylax (Ivy et al. 1999).

Surprisingly, given its enormous size, the spermatophylax from *Uromenus* males does not appear to be a parental contribution. Bateman (1998b) found a positive but nonsignificant correlation between mating frequency (number of spermatophylaxes eaten) and both the size and the number of eggs a *U. stali* female produced after mating. A *Uromenus* female will remate after 3 to 7 days (mean 4.8 days) but is unlikely to invest in eggs during this period because a longer time is required for egg maturation. Thus, males are likely to be cuckolded by the high fertilization rate of eggs (some 95%) by the female's subsequent mates (Vahed 1998b).

For *L. laticauda*, Vahed and Gilbert (1997) also found no effect of a spermatophylax meal on egg number or weight, even when the insects were maintained on a poor diet. But for both *Uromenus* and *Leptophyes*, there may be significant effects on unmeasured aspects of offspring fitness (Vahed and Gilbert 1997) (see Table 6.2). For example, in another phaneropterine, *P. veluchianus*, the amount of spermatophylax eaten had no effect on the size or number of eggs that experimental females laid (Reinhold and Heller 1993) but had a significant effect on the relative dry weight of her offspring. In a subsequent paper, Reinhold (1999) suggested that the increase in relative dry weight might indicate that paternally nurtured larvae have higher energy reserves. Reinhold (1999) supported this hypothesis by showing that the amount of spermatophylax eaten by a female had a significant effect on the survival of starved larvae. Other evidence indicates that the *Poecilimon* male's nutrients benefit his own offspring; even though females mate more than once (Heller and Helversen 1991), and her most recent mate fertilizes about 90 percent of her eggs (Achmann et al. 1992), field evidence indicates that the common interval between matings (4 or 5 days: Heller and Helversen 1991) provides sufficient time for males to sire offspring benefiting from their own courtship gifts (Reinhold 1999).

Results with *R. verticalis* and *K. nartee* also support the paternal-investment hypothesis. Our first result with *R. verticalis* showed no fecundity effects of courtship feeding (Gwynne et al. 1984), probably because of the large variation in fecundity from the use of old, "eggbound" virgin females. However, two subsequent studies (Gwynne 1984a, 1988a) using a more natural age at first mating for the females showed positive effects on egg size and subsequent overwintering survival of the eggs, as well as growth rate and possibly adult size of sons (Gwynne 1988a) (Fig. 6.17).

Finally, *Requena* males invested directly in their own offspring because male-derived nutrients are incorporated into the eggs (Bowen et al. 1984) that are matured and laid before females remate in nature (Gwynne 1988b; Gwynne and Snedden 1995). Moreover, the first male to mate fertilizes most of the subsequent offspring (Gwynne and Snedden 1995). For the *Kawanaphila* species, nutrition from the spermatophylax has rapid post-mating effects on the production of eggs (both size and number) that are probably fertilized by the mating male's sperm (Simmons 1990). *K. nartee* may have less specialized spermatophylax nutrients than *R. verticalis*: both the pollen diet and spermatophylax nutrients appear to have similar influences on fecundity in *K. nartee* females (Simmons and Bailey 1990), whereas only the size of the spermatophylax meal, and no other component of the diet, affects egg size in *R. verticalis* (Gwynne 1988a).

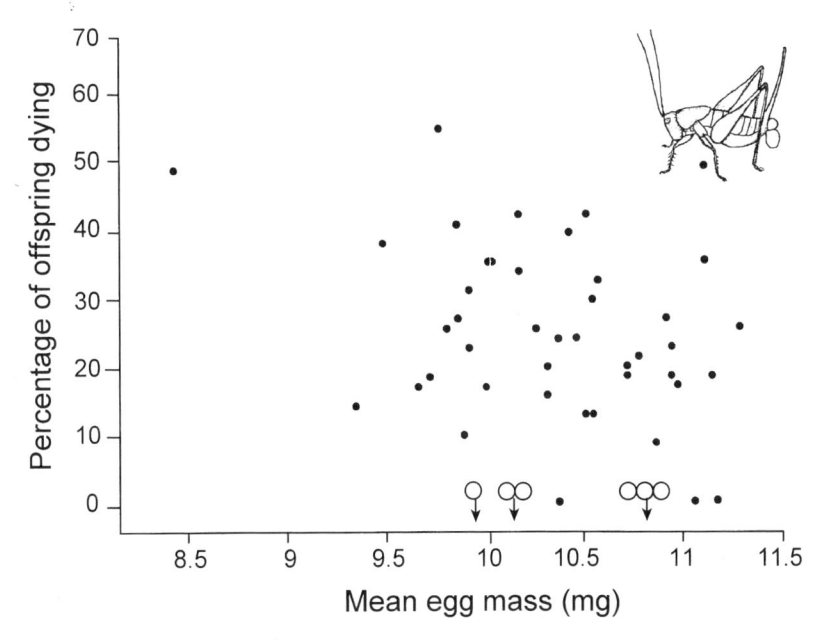

Figure 6.17. Spermatophylax feeding and the fitness of female katydids, *Requena verticalis*. The bilobed spermatophylax was experimentally manipulated to provide three meal sizes of one, two, or three lobes. Increasing the size of the spermatophylax eaten by females resulted in increased egg size (weight of five eggs), as shown by the three open-circle symbols on the x-axis (F = 3.9, P = 0.03, covariates such as body size held constant). Egg size appears to be important to offspring fitness, as larger eggs showed a significantly higher rate of overwintering survival (filled circles, correlation r = –0.32, P = 0.02). Redrawn from Gwynne (1988a). Katydid redrawn from a figure by D. Scott, in Clutton-Brock (1991).

2. *The ejaculate-protection hypothesis* (Table 6.1, hypothesis IV) *requires that the time taken for the female to eat the spermatophylax is just long enough for the sperm and other ejaculatory substances to be transferred* (because males should invest no more than what is necessary to complete ejaculation). This prediction appears to be met by all species studied except *R. verticalis*, as summarized in Figure 6.18. For four of the studied species, no evidence exists that the spermatophylax is any larger than necessary (on the horizontal axis of Fig. 6.18, an ampulla-attachment duration of 100%) for the transfer of a full complement of sperm or inducement of a complete refractory period. This prediction even holds for the two subspecies of *P. veluchianus*. Figure 6.18 shows the *P. v. veluchianus* data; in the other subspecies, *P. v. minor*, sperm is transferred more rapidly than in *P. v. veluchianus* but in a time period that matches the duration of the smaller spermatophylax meal of this subspecies (Heller and Reinhold 1994).

In the exceptional *R. verticalis*, the spermatophylax meal is more than

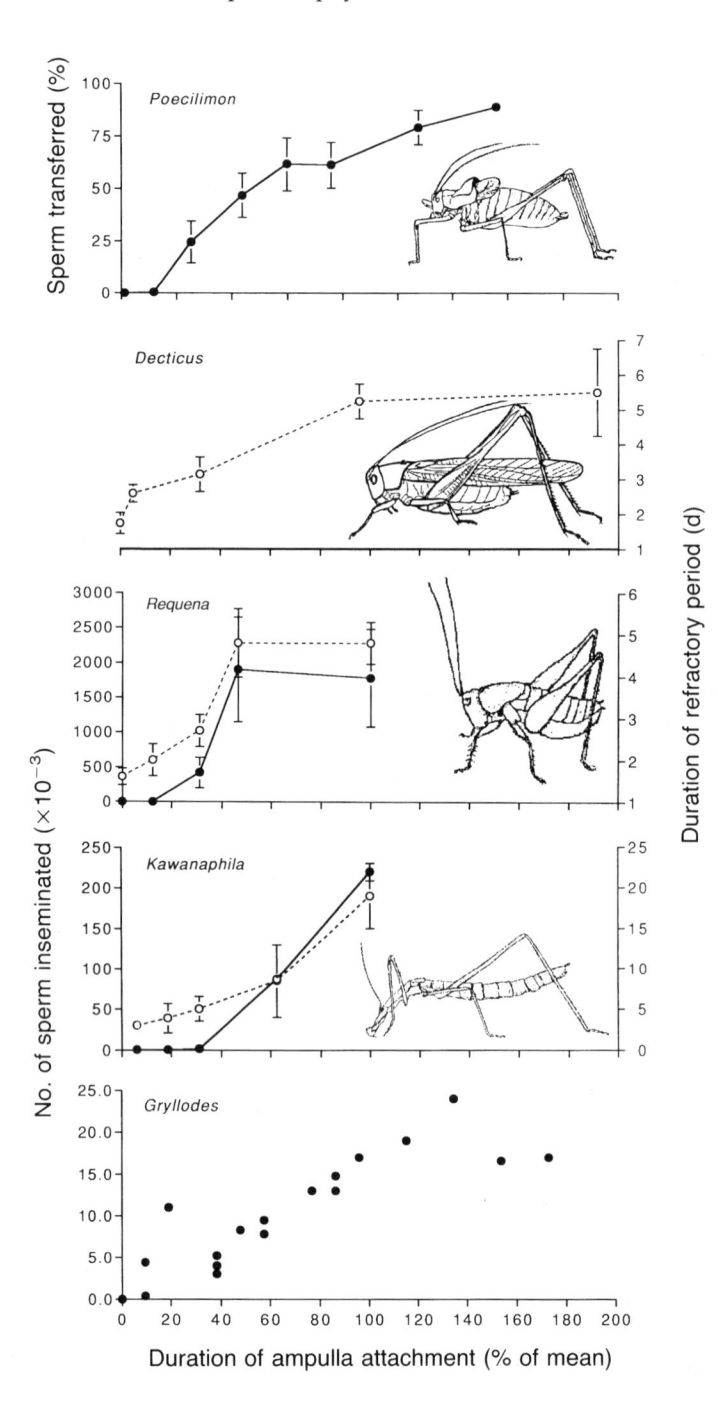

154 Evolution of the Spermatophylax

twice the size necessary to delay ampulla removal sufficiently for complete insemination (Fig. 6.18). This conclusion is supported by three studies of unmanipulated matings (Gwynne et al. 1984; Gwynne 1986; Simmons et al. 1999) (Fig. 6.18). Although one experiment by Simmons at al. (1999) demonstrated a meal consumption time (mean) of just under 4 hours (more on the significance of this variation of meal size later), this time was still much longer than the approximately 2-hour time period necessary for full insemination. Thus, even though there was large variation in ampulla-attachment times, all meal sizes were associated with full insemination (see also Simmons 1995a), so even the smallest offerings cannot be considered to function merely as ejaculate protectors. This contrasts with work on wart biters (Wedell and Arak 1989) and *Poecilimon* (Reinhold and Heller 1993) and *Gryllodes* (Sakaluk 1984) species in which males that produced smaller meals compromised their ability to transfer a full ejaculate.

Variation in the size of the parental meal supplied by *R. verticalis* males appears to be related to variation in expected paternity returns "in accord with sperm competition theory" (Simmons et al. 1993). Just as male birds decrease parental care (feeding of chicks) when reproductive returns from the mating are likely to be low (e.g., Davies 1992), *Requena* males reduce the size of their courtship meal when increased competition with rival ejaculates is likely. In particular, when males mate with old and thus nonvirgin females, they reduce the spermatophylax size and also increase the number of sperm inseminated (Simmons et al. 1993).

Requena and *Poecilimon* males also reduce the size and protein content of spermatophylax meals when the time between copulations is experimentally reduced (Heller and Helversen 1991; Simmons 1995b; Simmons et al.

Figure 6.18. The relationship between size of spermatophylax meal and ejaculate transfer in a true cricket, *Gryllodes supplicans* (Gryllidae) (from Sakaluk 1984) (bottom) and four katydids (Tettigoniidae): *Poecilimon veluchianus* (Phaneropterinae) (from Reinhold and Heller 1993), *Decticus verrucivorus* (Tettigoniidae) (from Wedell and Arak 1989), *Requena verticalis* (Listroscelidinae) (from Gwynne 1986), and *Kawanaphila nartee* (Zaprochilinae) (from Simmons and Gwynne 1991). Simmons and Gwynne presented refractory period data for several treatments. To be consistent with other studies, the data here are for females on an ad libitum diet that ate their spermatophylax. In order to compare taxa, the duration of ampulla attachment (x-axis) is expressed as a percentage of the mean time for the female to remove the ampulla. Therefore, values in excess of 100% are mainly those in which the spermatophylax meal was experimentally increased. Sperm inseminated (number except for *P. veluchianus*, which was reported as a percentage of sperm in the ampulla transferred) (filled circles and lines) is on the left y-axis (note the large interspecific differences in the number of sperm inseminated). Length of refractory period (open circles and dashed lines) is on the right y-axis. All data are shown as mean ± standard error (except for *Gryllodes* in which original data points are shown). Graphs redrawn from Gwynne (1997a). Katydids (from top) redrawn from Harz (1957); Chopard (1965), after Boldyrev; D. Scott, in Clutton-Brock (1991); and Rentz (1993).

1999; Fig. 7.7). Simmons (1995b) with his colleagues (1999) interpreted this result as evidence that males use a high copulation frequency (rather than mate-encounter frequency) as a mechanism for determining the presence of many receptive females and thus adaptively decreasing the size of the meal to achieve more copulations. An alternative hypothesis, however, is that the increase in meal size with increasing copulation interval is simply a nonadaptive effect of the increased material stored in the spermatophylax (rough accessory) glands (Simmons 1995b). This hypothesis is not ruled out by the fact that experimentally abstinent males still retained material in glands when they eventually mated (there may be constraints on the upper size of a spermatophore that a male can produce). Moreover, given that a high encounter rate with receptive females is used as a cue to increase male choosiness in katydids (see Chapter 9) and to reduce gift size in tree crickets (Fig. 6.5) (Bussière, submitted manuscript), it is puzzling that the same mechanism is not seen in *R. verticalis*; there was no effect of the encounter rate with females on spermatophylax size in the studies by Simmons (1995b) and Gwynne (1990b). The latter examined repeated matings by the same males and found that the full-size spermatophylax meal transferred at the first mating was not reduced in size in the subsequent four matings during which males had daily encounters with virgin females. Interestingly, even though these males had maximum opportunities to mate with many females, there was not a significant reduction in spermatophore size in these subsequent matings (see Fig. 9.14, closed points). Finally, I should note that the only estimate of spermatophylax size of *Requena* males in nature (males brought into the lab and mated immediately) showed the spermatophylax meals to be every bit as large as those produced by well-fed lab-raised virgins (Gwynne 1986). This contrasts with *P. veluchianus*, of which lab-raised males produced significantly larger spermatophylaxes than did males from the field (Heller and Helversen 1991).

3. A *central prediction of the ejaculate-protection hypothesis is that the relative size of the nuptial meal should be directly proportional to the donating male's paternity*. This prediction received support in a study of the wart biter in which two males were mated to the same female. Wedell (1991) found a positive correlation between relative meal size and paternity (Fig. 6.19). Such a pattern may also apply to the cricket *G. supplicans* (Sakaluk 1986). However, this relationship is not expected for the parental contribution of *R. verticalis* because all sizes of spermatophylax seem sufficient to fully inseminate the female. Initial studies with *R. verticalis* by Gwynne and Snedden (1995) failed to turn up such a correlation, although in contrast to the other studies we used the size of the whole spermatophore meal (spermatophylax plus ampulla), not just the spermatophylax, as an estimate of the male's contribution.

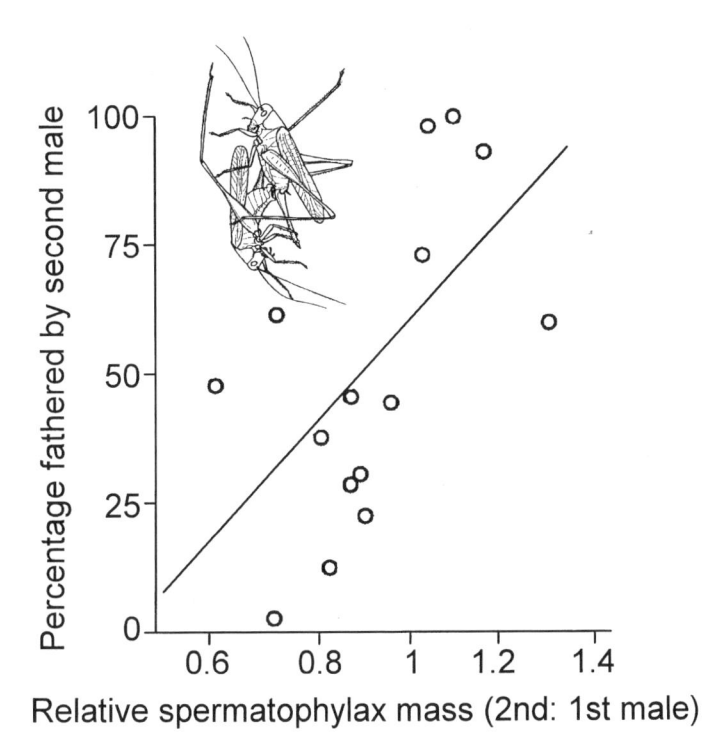

Figure 6.19. Paternity and relative spermatophylax size in an experiment in which two male wart-biter katydids, *Decticus verrucivorus*, were mated to a single female. The plot shows percentage of eggs fertilized by the second of the two males versus the mass of his spermatophylax relative to that of the first male ($r = 0.58$, $P < 0.025$). Redrawn from Wedell (1991). Mating katydids redrawn from Chopard (1965), after Boldyrev.

4. As we have seen, *spermatophylaxes serving an ejaculate-protection role* (Table 6.1, hypothesis IVb) *should be less costly for males than when the nutritional investment is in offspring*. Wedell's (1994a) comparisons of gift protein and male remating time suggest that there are two different types of katydid matings, with different costs that correspond to the ejaculate-protection and paternal-investment functions. In which of the two groups do the spermatophylax gifts of the five experimental species fall? First, I examine the male remating interval. The conclusions about spermatophylax function from predictions 1, 2, and 3 suggest that it should be longer than 2 days (Wedell's costly meal group) for paternally investing katydids such as *P. veluchianus*, *K. nartee*, as well as *R. verticalis* where a male gives up 20 percent of his body mass (some 24% of his daily energy budget: Simmons et al. 1992). This turns out to be the case with estimated male

remating intervals of 4, 3, and 5 days, respectively, for the three species (Davies and Dadour 1989; Gwynne 1990b; Simmons 1990; Reinhold and Helversen 1997).

In contrast, males from katydid species with apparently low-cost, ejaculate-protecting gifts are expected to remate in less than 2 days. This prediction is supported by *D. verrucivorus*, which has a 1-day remating interval (Wedell and Arak 1989), but not for *U. stali*, which has a relatively long remating interval of 5 days or longer (Bateman 1997; Vahed 1997). For the cricket *G. supplicans*, a mere half a day passes before the investing male is again ready to seek mates (Burpee and Sakaluk 1993).

The only data available on gift quality for the species studied are consistent with the prediction that paternal males serve higher-quality meals. At 10 and 13.2 percent protein contents (wet mass), respectively (Table 6.3), the *P. veluchianus* and *R. verticalis* meals are relatively rich repasts compared to the wart biter offering of 4.2 percent protein (Wedell 1994b).

6.4. Discussion and Conclusions

A clear conclusion from these studies is that there is no universal function for the katydid spermatophylax, as I suggested might be the case (Gwynne 1990a). Instead, comparisons between katydid taxa and the experimental work on individual species strongly suggest that the function varies, as argued by Wedell (1993a, 1994a). Although the spermatophylax has mainly a paternal-investment function in *R. verticalis* (hypothesis V in Table 6.1), for most katydids this male gift appears to retain its ancestral ejaculate-protection function. This appears to be the sole function in the low-nutrition offering of *D. verrucivorus*, and possibly *U. stali* and *L. laticauda* (hypothesis IVb) as well as the cricket *G. supplicans* (hypothesis IVa). In contrast, experimental evidence for the katydids *P. veluchianus* and *K. nartee* suggests that the spermatophylax gift is maintained by both natural selection (paternal investment) and sexual selection (ejaculate protection).

The mating behavior of a couple of other katydids suggests that there may be other, as yet unknown functions of the spermatophylax. Males of the cone-head *Ruspolia nitidula* produce a very small spermatophylax yet are able to inseminate the female without the lengthy "guarding" copulation during insemination that we saw for other katydids with small meals (Boldyrev 1915). Boldyrev (1915) noted that although *R. nitidula* females eat the spermatophylax quite quickly, their jaws do not destroy the ampulla for some time after. He concluded that females have a "biological instinct of abstention" from eating the ampulla, and this conclusion may be reasonable; females are expected to refrain from interfering with sperm transfer if they copulate only once. This may be the case for *R. nitidula*;

Vahed (pers. comm. 2000) found that two thirds of the 31 females mating in the lab did not remate when they were given access to other males over subsequent days. Females of another conocephaline, *Orchelimum*, also rarely remate (Morris et al. 1975c; Feaver 1977, 1983).

For katydids in which the spermatophylax functions only in ejaculate protection, it is clear how the male benefits, via increased paternity, but the benefits to females of consuming the meal while being inseminated are less clear. The comparative data indicate that these "protective" gifts are lower in quality (protein content and size: Table 6.3) than are paternal ones. However, the data are not sufficient to determine whether females consume their entire mating meal for the nutrition they do obtain (Table 6.1, hypothesis IVa) or whether these donations provide virtually no material benefits to females. If females gain little of value, the "male-deceit" or "assessment-of-male-quality" hypotheses for why females eat the whole contribution would have some support. The spermatophylax appears to contain little material value in species where there is no influence of spermatophylax nutrients on female fitness (Wedell and Arak 1989; Bateman 1998a). However, for some of these species more powerful tests (larger sample sizes) may be required to reveal whether nonsignificant trends in the influence of spermatophylax nutrients on fecundity (e.g., Bateman 1998a) are real or not.

Regarding the origin and evolution of mating meals in katydids and related orthopterans, however, we can conclude that the contributions originated from Darwinian sexual selection among males to inseminate females and fertilize eggs. Similar selective histories have apparently occurred in other arthropods in which males feed their mates. For example, the capture and donation of prey by male bittacid scorpionflies (Thornhill 1976a), empidid flies (Svensson et al. 1990), and a pisaurid spider (Austad and Thornhill 1986; see also Lang 1996) are maintained by sexual selection; these gifts can serve both to attract females and to prolong copulation until ejaculation is complete.

In the above-mentioned arthropods, males typically provide no more prey material than what is necessary to allow time for the safe transfer of sperm. A similar function appears to be the case for a more dramatic meal in the Australian redback spider, *Latrodectus hasselti*, where the male himself often becomes prey when he places his body in the jaws of his much larger mate (Forster 1992). As the redback male's somatic sacrifice is his only donation, it was originally suggested that the male's complicity in his own demise functions in nurturing offspring (e.g., Buskirk et al. 1984). As it turns out, however, the self-sacrifice functions like most other mating meals: it distracts the female and enhances the male's paternity (Andrade 1996).

In light of the usual function of nuptial gifts, the spermatophylax meals of three katydids are all the more surprising. The evidence suggests that in *P. veluchianus*, *R. verticalis*, and *K. nartee*, the meal at least in part is maintained by selection on males to invest directly in their offspring. There is perhaps one other insect system in which a male donation might have a paternal role. In a number of beetles the male ejaculates the chemical cantharidin, better known as "Spanish Fly", into his mate (Snead and Alcock 1985; Eisner 1988). In one of these species, a fire-colored beetle, *Neopyrochroa flabellata*, cantharidin is an important antipredator component of eggs (Eisner et al. 1996a). The chemical's legendary role in human mating[3] is paralleled in these beetles. *Neopyrochroa* males eat the chemical and store it in specialized glands in the head and abdomen. During courtship, females place their mouthparts on the cephalic gland and appear to be stimulated to copulate if they taste cantharidin (Eisner et al. 1996b). The fact that males then ejaculate most of their chemical store into the female (for later incorporation in eggs), rather than using it to obtain additional mates, suggests that the chemical may function in the protection of the mating male's future offspring (Gwynne 1997b).

The elaboration of the mating meal via insemination competition among males has produced a nutritious item in some katydids so that females might even prefer males that supply higher-quality meals. "Material benefits" and other advantages of mate choice are dealt with in the next chapter. The elaboration of the meal presents something of a paradox in other species, however. The gift apparently has become so large and valuable that patterns of sexual competition have become reversed; females take on the aggressive, fighting role and compete for what the males offer. The link between the offering, katydid hunger, and sexual differences in competitive behavior are the subject of Chapter 9.

---

[3] Meloid beetles (genus *Lytta*) were the source of this aphrodisiac ("Spanish Fly").

7

The Nature of Sexual Selection: Females Choosing and Males Competing

> *Motionless, as though turned to stone, with their foreheads almost touching, the two exchange caresses with their long antennae, fine as hairs. . . . The male seems somewhat preoccupied. . . . Yet one would think that this was the very moment at which to make the most of his strong points. . . . The interview, a mere exchange of greetings between friends of different sexes, does not last long. What do they say to each other, forehead to forehead?*
>
> J. H. Fabre (1917), *The Life of the Grasshopper*

7.1. The Struggle for Reproduction in a Michigan Marsh

"Not much, apparently . . ." was Fabre's (1917) answer to his question about the courting activities of these katydids—the white-faced decticus (Fig. 7.1). But research since Fabre's time has revealed that courting male animals can "say" quite a bit to females. And there is convincing evidence that this information can be used by females to assess a male's "strong points" and thus to make decisions about whether or not to copulate with him (Andersson 1994). A female katydid in particular could gain substantially by choosing among males if she gets the best mating gift, the spermatophylax that can boost her fitness, as we saw in Chapter 6. In fact, katydids, especially meadow katydids, are exemplar systems for understanding female choice, particularly for the information about the male conveyed in his sexual signal (for a recent essay, see Dugatkin and Godin 1998). In meadow katydids, the male signal, his song, also mediates the other main component of Darwinian sexual selection—sexual competition between males.

Sexual selection and mating in one meadow katydid were studied in detail by Marianne Feaver (now at North Carolina State University). Her research on individually marked black-legged meadow katydids over three seasons in Michigan's E. S. George Nature Reserve (Feaver 1977, 1983) remains one of the most detailed field studies of insect mating behavior ever conducted. Like the black-sided meadow katydids we first met in Chapter 3, black-legged meadow katydids, *Orchelimum nigripes* (Plate 21), can be found in riparian areas, particularly in mosaic habitats of floating and emergent aquatic vegetation. The association of this katydid with different vegetation types is a key factor determining the nature of mate choice as well as direct competition among males.

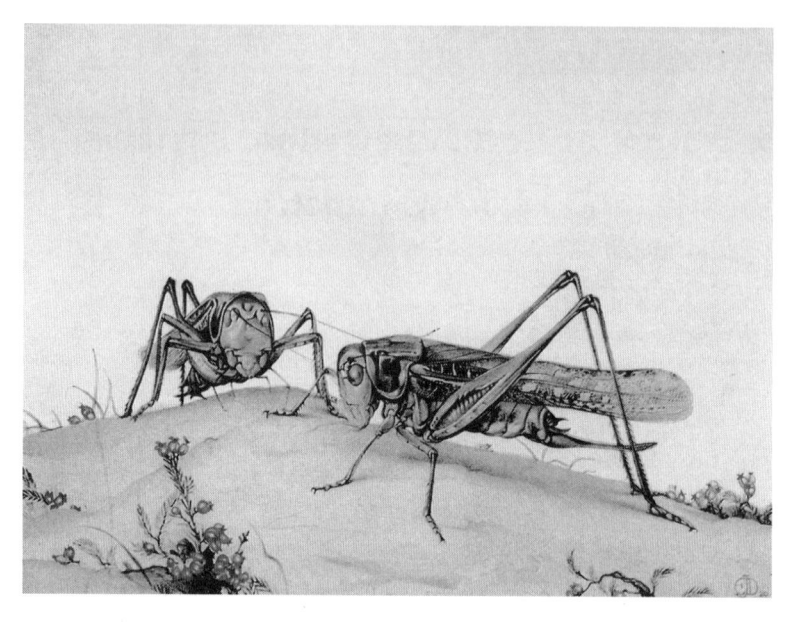

Figure 7.1. A white-faced decticus male fences antennae with a female. Illustration by E. J. Detmold from *Fabre's Book of Insects* (Stawell 1921).

The story of the black-legged katydid begins with the early instars. These youngsters can be seen in June and early July leaving the cover of grass clumps near the water's edge and hopping across lily pads to feed on the flowers. However, habitat preferences change about a week after the larvae molt to adulthood. Then, adults of both sexes converge on tall stands of cattails (*Typha*) and sedge, which are the main arenas for reproductive activities. The first to migrate are the full-grown females, who after many days feasting in grass and water-lily areas are replete with mature eggs. The females are seeking out their main oviposition sites in the thick, pithy stems of cattails. In contrast, males are less attracted by the plants per se and respond more to the movements of the fecund virgins. For a male black-legged katydid, a copulation virtually guarantees reproductive success because unlike most insects (Ridley 1988, 1989), his ejaculate does not compete with rival stored sperm for fertilization of the female's eggs; this species and other *Orchelimum* species are unusual katydids in that females appear to mate only once (Morris et al. 1975c; Feaver 1977).

The valuable virgins are tracked by the males in different ways. Young adult males actually follow females into the hedges of cattail and sedge that tower above the water lilies, whereas the older males contact females by calling from territories staked out in the high vegetation. Feaver noted that the exact location of male territories in sedge and cattail shifted almost daily to match the changing directions of female migration out of the larval habitats (Fig. 7.2). It is on the boundaries between the two habitats that vociferous border disputes between singing territory holders herald the first stages of sexual competition, a competition that can be very injurious indeed.

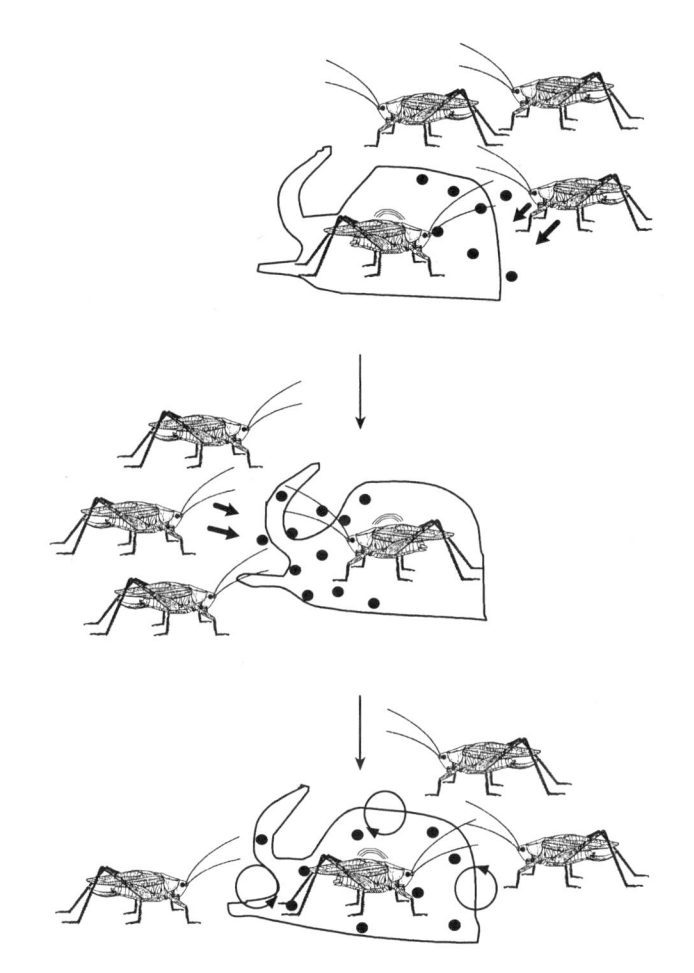

Figure 7.2. Territorial areas of black-legged katydids in a cattail-sedge area: the changing patterns of male calling territories (black dots) and the direction of movement of females from the surrounding water lily areas (thick arrows) into the cattails (outline area). In the bottom figure, the spread of calling males around all the borders correlates with the migration of females into cattails from all directions and out again to feed. Females laid their eggs in the stems of cattails and sedges in the centers of cattail clumps. Redrawn from Feaver (1983). Katydids redrawn from Alexander et al. (1972).

Interactions between callers, the older males that settle into territories in cattail and sedge, either can be settled acoustically or can escalate to fighting, when the males rake each other with tarsal claws and bite one another. During acoustical interactions, particularly dominant males will move around and synchronize their song elements with each of several neighbors in turn. Losing males will move away from these more aggressive callers. The costs involved in not backing down from a stronger rival can be great; about a third of all fights result in torn wings and lost limbs. In general, it is the larger, older males who win the fights, and as a consequence, these heavyweights hold the prime central territories where most of the migrating females are encountered. The result is that lighter males mate significantly less frequently than do heavy, dominant individuals.

The young males tagging along behind migrating females also tend to be losers in the mating game. The failure of these youngsters is not for want of trying, as they often hang around as "satellites" within territories and will even accost females that respond to local callers. During interactions with females, an upstart satellite will occasionally call, but he is quickly ousted by any resident male within earshot. However, residents will put up with a silent satellite, even on occasion allowing him to share the same plant stem as a perch. Silent youngsters are tolerated because they pose no cuckoldry threat; females will not mate with any male that does not begin his courtship routine with a song. Some satellites will even appear to bluff the hard-to-get females by raising and vibrating their tegmina but producing no sounds. Although nearby females occasionally move toward these furtive signals, they quickly turn away to continue on their paths toward the calls of the dominant males. The advantage to a young male of remaining in a caller's territory is that he can be in line to take over the territory when the resident departs. Feaver established this experimentally by removing the original caller and noting that a nearby satellite suddenly becomes vocal and takes over the area. A resident will decamp from his territory in nature after he mates and thus drops about 10 percent in body mass due to spermatophore production (see Chapter 6).

Female black-legged katydids not only spurn silent satellites but also show considerable coyness in their interactions with territorial callers. First, they rarely respond to lone callers in the outlying areas of cattail and sedge, spending most of their time in areas where there are groups of singers. A second level of choosiness occurs once females contact a caller. They often withdraw several times before facing him to initiate courtship. During courtship "the two exchange caresses with their long antennae," but this activity goes on much longer than Fabre observed for the white-faced decticus. While courting, the pair of *O. nigripes* can break up or proceed to mating; copulation occurs when they slide past each other in the vegetation, and the male curls his abdomen, engaging his genitalia and transferring the spermatophore and the spermatophylax meal (see Chapter 6). Although males can leave before the spermatophore is transferred, most breakups are caused by the departure of the female. Females spend an average of 1 hour in a suitor's territory and many hours moving between and interacting with adjacent territorial callers. In fact, females can take days "doing the rounds" before finally submitting to what appears to be their sole copulation; they first respond to calling males a few days after the adult molt but do not mate until they are, on average, 12-day-old adults. These movements by females appear to function in mate assessment.

The mating habits of black-legged katydids, in which males compete for access to territories and females choose among males, is typical for many animals, including most katydids. There can be a reversal in these mating roles, as we saw in Chapter 1 for Mormon crickets (we will revisit role-reversed katydids in Chapter 9), but, as also noted in Chapter 1, even some Mormon crickets show similar mating behavior to the black-sided katydids. Studies of male competition and female choice in katydids, and the role of male song in these processes, have been extensive, beginning with Darwin's (1874) illustration of the singing apparatus as an example of a sexually selected trait (Fig. 7.3).

I begin the discussion of the nature of sexual selection in katydids with female choice, because choice has become the main focus of research in

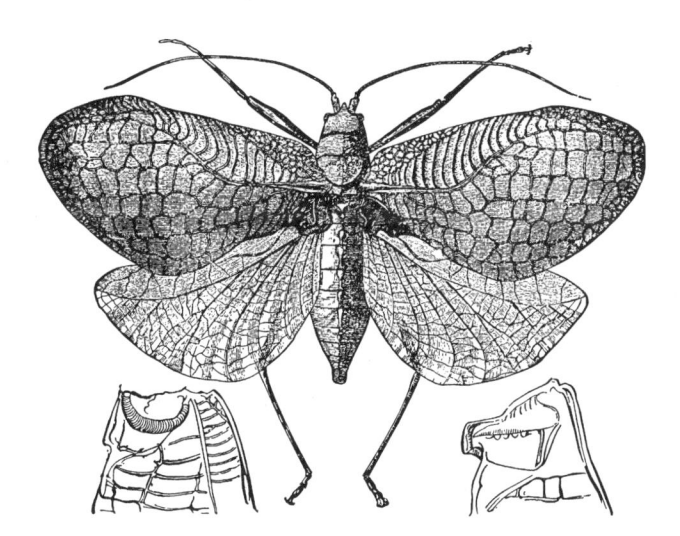

Figure 7.3. The male pseudophylline katydid, *Thliboscelus hypericifolius* ("from Bates"), illustrated in Darwin (1874) and showing the file (below left) and scraper (below right) on the male's forewings.

these insects and in other animals (in part because of the controversial nature of mate choice; see Cronin 1991).

7.2. Mate Choice

Coyness or choosiness by female black-legged katydids is first seen in the rejection of noncallers. The rule of thumb to never mate with silent types may benefit females who are more likely to get the best-quality mates by inciting competition between males (see Cox and LeBoeuf 1977; Burk 1983), particularly between resident callers and the younger males that are testing out their acoustic apparatus for the first time. Singing males are certainly tried and tested individuals. Feaver (1977) found that 49 percent of young adult males died before getting a chance to sing (a large source of mortality being infection by horsehair worms: see Chapter 1).

More obvious examples of choosiness come from observations of female black-legs spending more time near groups of signaling males than around solitary callers. These females appear to "window shop" among callers before finally picking one as a mate.[1] These observations raise a number of questions about preferences, which include the basic "ultimate" (evolutionary) query about fitness, "How do females benefit from being choosy?" (Table 7.1), as well as a "proximate" question, "What cues do females use when choosing?" Katydids have proved to be good subjects for answering these questions, because like certain crickets (Brown 1999), females are likely to choose among males that vary in their abilities to supply food at mating (see Chapter 6). There is also the potential for cryptic female discrimination (Lloyd 1979a; Eberhard 1996) (Table 7.1) if, for example, a female discriminates against a male by removing the externally placed spermatophore before ejaculation is complete (see Chapter 6). Most of the interest in mate choice by katydids, to answer both proximate and ultimate questions, has stemmed from a focus on female selective responsiveness to male calling songs (Gerhardt 1994b). Demonstrating that females prefer certain songs or other signals requires experimental evidence that they respond preferentially to certain signals over others in the absence of non-acoustical cues (Searcy and Andersson 1986). Finally, demonstrations of adaptive mate choice need to consider other explanations for the differential mating success of males (Table 7.1), such as whether females are being "passively attracted" to a signal that is simply more conspicuous to females because it contains more energy per unit of time (i.e., has a larger *duty cycle* [more signal per unit of time] or is louder)

---

[1] There is good evidence (for a gryllid cricket) that females can recall locations of previously encountered males, returning to them even after they have stopped calling (Wiegmann 1999).

Table 7.1. Why females might prefer to mate with certain males over others

1. a. Active choice (Parker 1983). An adaptive choice of mates to obtain either (i) immediate benefits for themselves or their offspring (e.g., goods and services such as spermatophore food gifts, or a low-cost copulation such as mating with a nearby male) or (ii) genetic benefits ("good survival genes" that are passed on to the offspring) or solely because the female's sons will gain an advantage because they inherit the attractive trait ("runaway" sexual selection). Good-gene and runaway models of female choice are difficult to separate empirically, although good-gene models alone predict beneficial effects of choice on offspring quality (Kirkpatrick 1987; Brown 1999).

   Thus, females mate with certain males or favor the fertilization success of these males through "cryptic" or postmating choice (Thornhill 1983; Eberhard 1996), after assessing traits such as size or colorful displays. Females also may have evolved to behave in such a manner to indirectly favor certain males (e.g., by inciting male competition).
   b. Another type of active choice involves female mating preferences that evolved to avoid potentially costly pairing or hybridization mistakes with other species ("species recognition").
2. Females did not evolve to make mating choices; i.e., no genes that cause females to show adaptive mate choice for certain signals or displays have been directly selected. Thus, males are favored indirectly by:
   a. Being the male who produces the greatest amount of song energy per unit of time, i.e., one that produces more signal per unit of time or a louder signal ("passive attraction": Parker 1983).
   b. A female rule of thumb to not discriminate. The sorting among males of different qualities is done by intermale competition (but see Qvarnstrom and Forsgren 1998).
   c. Being favored indirectly due to a specific trait or resource; e.g., a hungry female might feed on a male's mating meal while he (or his spermatophore) inseminates her, but she ends insemination as soon as the meal is finished, thus terminating copulation or ejaculation and indirectly favoring males able to supply the largest gifts (see Chapter 6).
   d. Their ability to locate a particular place and time where receptive females are found; i.e., females mate with any male but favor males that are able to best use the environmental cues to increase the probability of encountering these females.
3. Females did not evolve to make mating choices, but ancestral genes that evolved in another context, e.g., preferences for food (Proctor 1991), select for male signals that exploit female sensory biases such as female preferences for "leading" males in choruses of singing males (Greenfield 1993; see also Ryan 1990, 1998).

Adapted mainly from Parker (1983) and Alexander et al. (1997).

(Table 7.1). Note, however, that even passive attraction is expected to impose sexual selection on song loudness or the song's duty cycle or correlated characters such as body size.

Male songs are likely to be useful signals for choosy females because the calls are variable and costly and so should be reliable indicators of the quality of potential mates (Brown et al. 1996). Moreover, calls can be assessed from a distance; at this level mate rejection has a relatively low cost for females because it can be done without any direct struggle with the male (Brown et al. 1997). After the male makes contact with the female, he usually stops stridulating, in contrast to many crickets and grasshoppers (Brown and Gwynne 1997; Greenfield 1997). At this point there are nonacoustical cues that male katydids might use to advertise their quality, including, for example, vibration displays such as tremulation (see Chapter 5).

7.2.i. Female preference for singing groups

Female preferences for a chorus of territorial males might be adaptive if females gain by saving the typical energetic costs and risks that come with travel between widely separated individuals (Alexander 1975). For katydids, the obvious cue that would allow a female to assess the density of males in an area is the calling song. Female black-sided katydids, *Conocephalus nigropleurum* (Fig. 7.4), are attracted to the tape-recorded sounds of a chorus, both a "real" chorus of 10 males and a song model, which included the chorus-like broad band of frequencies of this species' song and a lack of an obvious song pattern (Morris and Fullard 1983). More importantly, however, female black-sided katydids prefer the

Figure 7.4. A female black-sided meadow katydid, *Conocephalus nigropleurum*, eats her spermatophylax meal.

sounds of a male chorus (a recording of two callers) over a solitary singer (Morris et al. 1978). However, these data do not rule out the possibility that females were passively attracted to the louder sound of two males' combined calling (Table 7.1, item 2a). In fact, when the acoustic power of the two-male call was lowered to a level comparable to that of the one-male call, females reversed their preference. Perhaps the reversal is caused by mate choice in which the loud male is perceived to be of higher quality than the two softer-calling males in the chorus (Fig. 7.5). An alternative hypothesis, that the single male is preferred because its call is not "masked" by another call, seems less likely given that important song elements are masked only by the overlap of many conspecific calls (e.g., Morris and Fullard 1983).

Although there is some evidence of a preference for choruses in singing animals (Doolan 1981; Ryan et al. 1981; Shelly and Greenfield 1985), studies of other singing animals have found no support for such a preference (e.g., Tejedo 1993). In fact, in one katydid species, solitary callers were more likely to attract females. Arak et al. (1990) attracted female great-green grasshoppers, *Tettigonia viridissima*, to caged callers, a method first used by Brunellius in 1791 for the same katydid (see Chapter 5). Trials conducted by Arak et al. in daytime showed that virgin females preferentially oriented to caged solitary callers over caged groups. This result may have been due to an apparent lack of choosiness by these virgins; 95 percent of all released females hopped immediately to the closest caller. Because virgins had been isolated from males, perhaps they were so responsive that they were no longer choosy (Butlin and Hewitt 1986). Alternatively, mating with the closest individual may be adaptive if daytime movement exposes the female to predation (Table 7.1), as seen in an Australian frog in which a high risk of predation by snakes causes females to prefer close callers (Grafe 1997). Predation risk has been proposed as an explanation for the preferences by female katydids for the higher frequency or louder songs typical of closer males (Table 7.2). Thus, female great-green grasshoppers may become more choosy after dark when most mating takes place (and risk from visually hunting predators is lower). In the related upland-green grasshopper, *Tettigonia cantans*, copulation is common at night when callers form small groups. Dadour (1990) suggested that females might prefer groups of callers. Once darkness falls, sexual activity, and thus mate rejection, appear to be more common wherein males appear to shift from a more dispersed (Schatral et al. 1984) to a clumped distribution (Dadour 1990).

7.2.ii. Female preference for song leaders

As a female upland-green grasshopper or black-legged katydid approaches a chorus of males, what differences among the cacophony of calls might cause her to select one male over the others? If a call contains

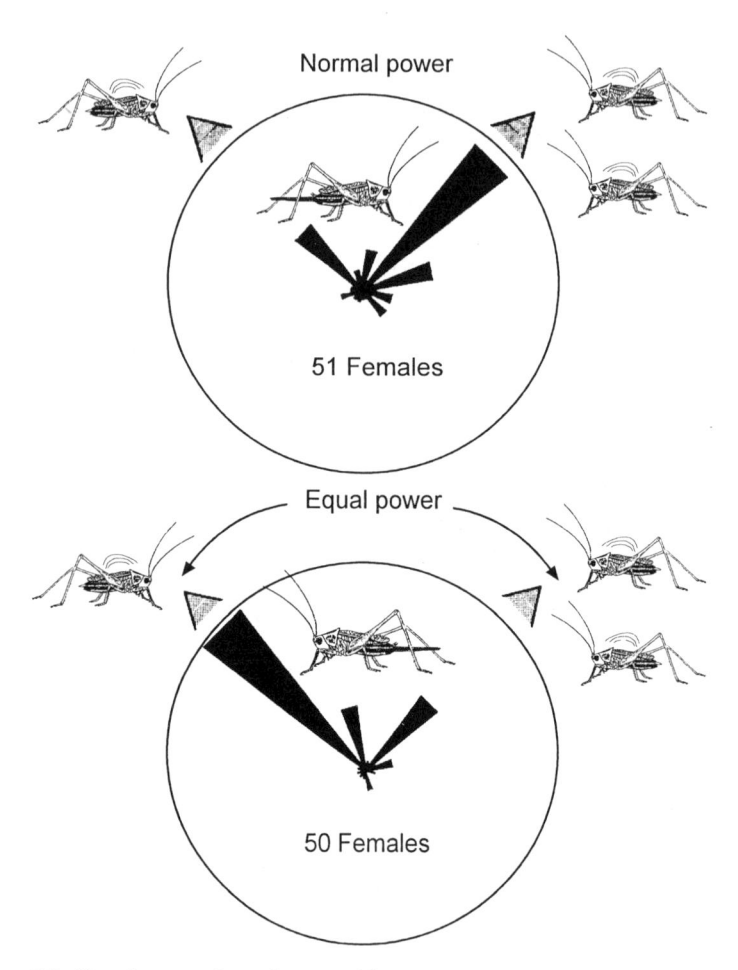

Figure. 7.5. The phonotactic preferences (directions where the moving female exited the circular arena) of black-sided katydid females for paired song models in an arena (outer circle) with two speakers (shaded triangles). The heights of the black triangles are proportional to the number of exits occurring from that sector of the arena. The figure on the top shows a clear preference for the song of two males. However, this preference reverses when the two-male model is reduced in power to the level of the one-call model (bottom figure). Redrawn from Morris et al. (1978). Katydids redrawn from Alexander et al. (1972).

honest indicators of the male's quality, such as his ability to provide a meal or good genes, females should respond preferentially to the best males (Table 7.1). However, one aspect of the calls that strongly influences the phonotactic decisions of a female may have little to do with male quality, as it results from a sensory (perceptual) bias for the call chirps that she hears first. The result from this sensory bias is that the male constantly

7.1 The Struggle for Reproduction 171

Table 7.2. Female choice of songs indicating size or other aspects of a signaling male

Species (a katydid unless otherwise stated)	Preferred song or male (usually in two-choice trial)	Proportion of variation in male weight explained by song cue (r^2)	Adaptive significance of preference suggested by authors	Remarks
Neoconocephalus spiza (Conocephalinae)	Song with the leading chirp (Greenfield 1993; Snedden and Greenfield 1998)		None suggested; preference is due to perceptual bias from a neural precedence effect (Greenfield 1993)	
Tettigonia cantans (Tettigoniinae)	Both lower mean song carrier frequency (pitch) (Latimer and Sippel 1987; Schul et al. 1998) and song with the louder high-frequency component (also louder male: Latimer and Sippel 1987)	0.18 (NS) with mean pitch (negative correlation), but see Remarks (Latimer and Sippel 1987)	Low-frequency preference: (i) for male status, but not size (but see Remarks [Latimer and Sippel 1987]) or (ii) given Schul's (1998) data, to avoid mismatings with *T. viridissima*, whose carrier frequency is close to Latimer and Sippel's high-frequency model	A very small sample size for weight × mean pitch correlation
Scudderia curvicauda (Phaneropterinae)	Greater syllable repetition rate in song (Tuckerman et al. 1992)	0.71	For male genetic quality (only 3% of variance in meal size explained by male weight)	Passive attraction to male with highest duty cycle not excluded
Oecanthus nigricornis (Gryllidae)	Lower calling song carrier frequency (Brown et al. 1996)	0.17 (negative correlation)	For male genetic quality or resources	
Conocephalus nigropleurum (Conocephalinae)	Greater tremulation pulse interval (De Luca and Morris 1998)	0.86	For spermatophylax meal (41% of variance in meal size explained: Gwynne 1982)	

Table 7.2. (continued)

Species	Preference	Value	Notes
Amblycorypha parvipennis (Phaneropterinae)	Larger, louder male and males with longer, leading songs (Galliart and Shaw 1991)	0.12 (with song intensity)	For meal (21% of variance explained by male weight) or the most vigorous male (Galliart and Shaw 1992)	Passive attraction to louder signal not ruled out
Requena new species no. 5 (Listroscelidinae)	Larger male with lower mean calling song carrier frequency (Wedell and Sandberg 1995)	0.3	For meal (44% of variance explained by male weight)	Experiment based on preference for a live calling male
R. verticalis (Listroscelidinae)	(i) Mating advantage in field for heavier male with lower mean call carrier frequency (Bailey and Yeoh 1988; Schatral 1990) but not in speaker trials (see Remarks) (Bailey et al. 1990); (ii) for song with louder high-frequency component (also louder male) (Bailey et al. 1990)	0.35	For closest male or passive choice (Bailey et al. 1990). Possibly for meal (Schatral 1990) (40% of variance explained: Gwynne et al. 1984)	Bailey et al. (1990) report small sample size (n = 7) of speaker trials (see text).
Kawanaphila nartee (Zaprochilinae)	For highest carrier frequency (Gwynne and Bailey 1988)	0.02 (NS) (late-season field sample) to 0.16 (early sample)	For closest male; large-male-preference hypothesis refuted in phonotaxis trials and field mating samples	Passive attraction for more easily located frequency not ruled out
Ephippiger ephippiger (Bradyporinae)	For songs without missing pulses caused by missing file teeth (Ritchie et al. 1995)		For younger male	Passive attraction for lower duty cycle not ruled out

NS = not significant.

adjusts his call's timing in attempts to be the leading singer (Greenfield 1993) so that collectively, call chirps appear to be synchronized or alternated (Jones 1966; Shaw 1968). Such unison signaling has been noted in a number of animals (Alexander 1975; Greenfield et al. 1997),[2] including the black-legged katydids studied by Feaver (1977), who suggested that call interaction was a sort of song one-upmanship in which aggressive males, the ones most likely to win fights, dominated their neighbors. Both Feaver's hypothesis and the one proposed by Greenfield and Roizen (1993) differ from most previous suggestions that unison singing was a result of "cooperation among competitors" to maintain species-specific signaling rates or to maximize the peak intensity of displays, thus attracting more females to the chorus as a whole than to the same number of separate singleton callers (Walker 1969; Alexander 1975; Otte 1977). In contrast to Feaver, however, Greenfield and Roizen argued that song interactions among males are driven by female preferences, not male-male competition.

Greenfield and Roizen (1993) developed their hypothesis in a study of a neotropical cone-head, *Neoconocephlus spiza*, for which playback experiments in the field showed that call synchrony resulted from individual males competing (although not all species of *Neoconocephalus* do this: Meixner and Shaw 1986). The clinching experiment was conducted in the laboratory and showed that females preferred the leading chirp of two playbacks even when the following call is longer (Greenfield 1993) or louder by 2 to 4 dB (Snedden and Greenfield 1998), thus refuting the possibility of passive attraction to the male that signals more (Table 7.1). Subsequent experiments showed similar leading-call preferences in other katydids, *Ephippiger ephippiger* (Bradyporinae) and *Sphyrometopa femorata* (Conocephalinae), as well as in some crickets (Gryllidae), grasshoppers (Acrididae), and frogs that show call synchrony or alternation (Greenfield et al. 1997, and references therein). This broad taxonomic sensory bias in females appears to be a result of an acoustic "precedence effect" (also known for humans), a perceptual bias evolved in other unknown adaptive contexts[3] in which the first of two sounds heard suppresses temporarily the animal's neural response to subsequent calls (see references in Snedden and Greenfield 1998). Additional work with *N. spiza* (Snedden and Greenfield 1998) supported this hypothesis over alternative ones, such as the hypothesis that the following call is physically masked by the leading call. This work also showed that preferences for the leading call remained when subject females were presented with a large "chorus" of four taped signals

---

[2] One side effect of song interactions is that males of different katydid species can affect each other's calls (reviewed by Latimer and Broughton 1984).
[3] The precedence effect might allow individuals to more easily locate predator and prey sounds or suppress responses to echoes by reduced response to following sounds (Snedden and Greenfield 1998).

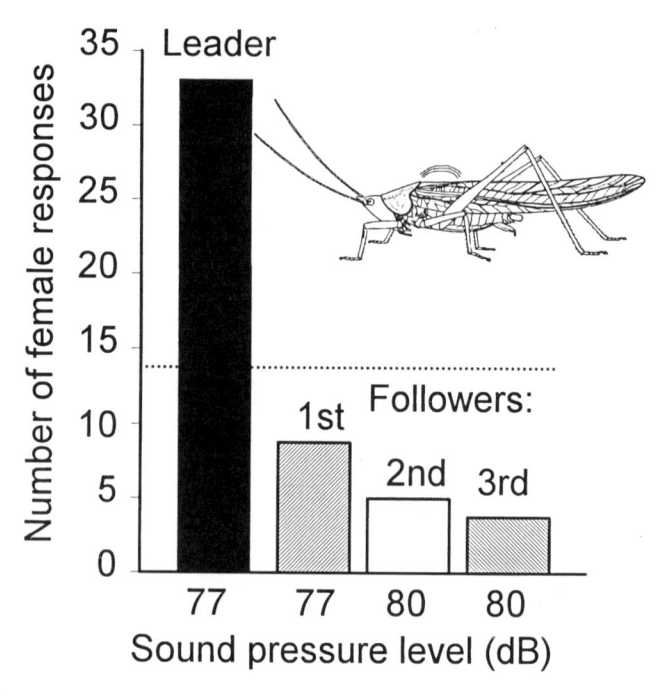

Figure 7.6. Orientation of female cone-heads, *Neoconocephalus spiza*, to each of four speakers playing call chirps. The dashed line represents the expected responses to each speaker if females orient randomly. The female response that to the leading call was significantly higher than that to the other calls ($P < 0.001$). Redrawn from Snedden and Greenfield (1998). Katydid redrawn from Alexander et al. (1972).

(Fig. 7.6). However, as Snedden and Greenfield pointed out, because a following call 5 dB or more above the leading call will cause the female to reverse her preference and orient to the louder call, a female in nature will respond preferentially to the call leader only in a local neighborhood of males (i.e., not when the leading call is distant from her and faint). Nevertheless, as males respond to their closest rivals, these local neighborhoods would be linked acoustically and thus could cause the "massive synchronies" observed not only in singing by many katydids, crickets, and cicadas (Alexander 1975) but also in the Christmas tree–like synchronized flash displays of certain tropical fireflies (Lloyd 1973a).

7.2.iii. Female preferences for older males and heavyweights: how and why?

A key unanswered question in the research on call one-upmanship between neighbors is whether female sensory biases impose sexual selection for call leading as a trait. For sexual selection to occur, there must be inherited variation among chorusing males in the ability to be the chorus leader. A male with a high calling rate might be expected to lead the chorus more often because he produces more chirps or because he calls more than his rivals (Snedden and Greenfield 1998). If differences in the amount of time calling respond to selection (e.g., Hedrick 1988), this song attribute, in theory, could evolve as a signal of a male's vigor and thus of his quality (Greenfield and Shaw 1983) or via "runaway" intersexual selection (details in Table 7.1). For example, Galliart and Shaw (1992) suggested choice for vigorous *Amblycorypha parvipennis* males, after showing that females preferred call leaders and louder males with longer phrases, and even initiated copulation more with the louder, heavier individuals (Galliart and Shaw 1991).

There is evidence that rates of calling and other signals in katydids can vary among males and can also signal male body weight and other aspects of the phenotype correlated with the ability to supply direct or indirect (genetic) benefits (Table 7.2). Recent research into the indirect benefits of mate choice in animals suggested that an important indicator of male genetic quality to females is the degree of developmental asymmetry in body parts; asymmetry indicates low quality, as the developing individual was not well buffered from forces such as disease that tend to affect the phenotype asymmetrically (Watson and Thornhill 1994; Leung and Forbes 1996; Møller and Thornhill 1997). Little research has been conducted into this topic using katydids and other calling animals. We do know that the degree of developmental symmetry can be signaled by song in at least one austrosagine katydid (*Sciarasaga quadrata*) (Hunt and Allen 1998) (see Chapter 8), but it is unknown whether these song cues are used by females.

As we saw earlier, mating male black-legged katydids were heavier and older, and Feaver (1977) suggested that female choice of calls may explain their mating success. Most experimental work on female choice in katydids has examined body size, although some studies have examined age. It is often argued that older male animals should be preferred for genetic benefits. Old black-legged males, after all, have proved their ability to survive the high mortality of young adults (the sort of life history in which older-male preferences should be adaptive: Beck and Powell 2000). Although Zuk (1987) found preferences for older male song in her study of a cricket (*Gryllus*), the only two experimental studies on katydids showed the opposite: females prefer the songs of young males (Ritchie et al. 1995; Bateman and Marquez, manuscript). Ritchie et al. (1995) found that *Ephippiger* females discriminate against signals indicating wear and tear, such as those produced with missing stridulatory-file teeth. This kind of discrimination may also be due to passive attraction to songs with more song energy. On the other hand, females might actively prefer younger males, if these katydids have different patterns of mortality than do *O. nigripes* (Beck and Powell 2000), especially if young-male preference results in a better spermatophore meal or higher-quality offspring (the latter was shown by Price and Hansen [1998] in fruit flies).

With regard to the size of mates, females are expected to prefer heavyweights either because they are genetically superior (e.g., vigorous or parasite free: Hamilton and Zuk 1982) individuals that pass on these fitness attributes to their progeny or because large males provide more direct benefits to females in the form of high-quality goods or services (Andersson 1994) such as large spermatophylax meals (Table 7.3). In calling animals, song is likely to be a reliable indicator of size because the acoustic spectrum of the song (frequency) is determined by the size of the sound-producing structures (Sales and Pye 1974); small males are physically incapable of sounding like large rivals (Brown and Gwynne 1997). There are other signals that might also convey honest information about body size, especially tremulation (a vibratory display; see Chapter 5). Heavyweights should be able to shake the substrate more vigorously than lighter rivals (Table 7.2).

Table 7.3. Female preferences for larger male animals: some key studies in which preferences for larger males have been linked to increased fitness of females

Species	Benefits of preference for large males
	Genetic benefits from males
Crickets (Gryllidae)	Offspring of large males are likely to be competitively superior (Simmons 1986; Brown et al. 1996).
Seaweed fly (*Coelopa frigida*)	Large males sire offspring of greater fitness (Crocker and Day 1987; Day et al. 1996).
Certain toads	Large male *Scaphiopus multiplicatus* (Woodward 1986), but not *Bufo americanus* (Howard et al. 1994), sire offspring of greater fitness.
Linyphiid spider (*Neriene litigiosa*)	Large males sire larger offspring with more rapid larval growth (Watson 1998).
	Goods and services from males
Katydid (*Requena*)	Larger males provide larger spermatophylax meals, and these nutrients increase offspring fitness (Gwynne 1988a; Schatral 1990; Wedell and Sandberg 1995).
Seed beetle	More ejaculate nutrients increase female fecundity (Savalli 1998).
Darkling beetle (*Parastizopus armaticeps*)	In this biparental species, larger males dig deeper nest burrows, allowing greater survival of offspring (Rasa et al. 1998).
Salamander	A large male's territory provides more food (Mathis 1991).
Fish (red-lipped blenny)	Larger males guard offspring longer (Côte and Hunte 1989).

7.2.iv. Heavyweights as genetically superior individuals

Black-legged females mating with heavy, territorial males may acquire genetic benefits that would be inherited by their offspring (Feaver 1983). However, genetic benefits that result from active female choice for honest signals reflecting traits such as body size have not yet been shown directly for any katydid species. We do know that body mass is an inherited trait in *Poecilimon veluchianus*. Interestingly, body mass in this species is apparently determined by X-chromosomal genes,[4] as are characteristics in other insects thought to be important in sexual selection (Reinhold 1994).

For another phaneropterine, *Scudderia curvicauda*, Tuckerman et al. (1992) argued that female preference for more vigorous (energetically expensive) calls (Table 7.2) might represent good-genes choice. This study showed both that male body mass correlated positively with the number of chirps in the song, a consistent (repeatable) call parameter within males, and that females preferred to answer multiple-chirp song models representative of large males. Other hypotheses to explain these female song preferences were ruled out, including "species recognition" (Table 7.1) and the material-benefit hypothesis (the correlation between male size and meal size was low and not significant). However, the data did not exclude the hypothesis that *Scudderia* females were attracted passively (Table 7.1) to songs containing more sound energy. Finally, although Tuckerman et al. favored the good-genes model, the runaway model of female choice (Table 7.1) cannot be ruled out because there is no direct evidence that mate choice based on song correlates with increased offspring fitness, as has been reported in other calling animals (Welch et al. 1998). It also should be pointed out that few studies reporting positive correlations between sire display and offspring fitness (Table 7.3) were able to determine whether there is a direct effect of male genes or whether females favor high-quality males by allocating more resources to eggs fertilized by these males (Burley 1988). Exceptional studies separating these possibilities have involved taxa with external fertilization, such as anurans, where sperm from males of differing phenotypes can be assigned experimentally to eggs (Welch et al. 1998).

A detailed investigation of female choice for the songs of high-quality, large males comes from research on a cricket (Gryllidae) that is similar to katydids in that males call from vegetation and offer a mating meal (secretions from the dorsal thoracic glands [Fig. 6.5]: Brown et al. 1996; Brown 1997a, 1997b). Females of the black-horned tree cricket, *Oecanthus nigricornis*, move preferentially toward the low-frequency tape-recorded songs of large males (Table 7.2). In this cricket, size appears to be an indicator of male quality because large males win fights and females mating with these

---

[4] Sex determination in most katydids is XX for female and XO for male. There are, however, some cases of XX/XY (4% of species), and two species have a multiple sex chromosome mechanism (Warchalowska-Sliwa 1998).

males lay more eggs. This beneficial effect of mating with a large male is not a result of a larger meal because meal size did not affect the female's egg-laying rate (although a large meal does increase the fitness of a female by increasing her lifespan). Rather, heavyweight males may be able to supply secretions that mimic high-quality nutrients. Alternatively, an increase in egg-laying after mating with a heavyweight may be due to a "cryptic" female bias (Table 7.1) for the sperm of genetically-superior males that along with premating choice for large male songs, really boosts the success of large male tree crickets (Brown 1999).

The mating biology of tree crickets and katydids brings up an important point; material-benefits choice and good-genes choice do not have to be in opposition as models of sexual selection often assume (Kirkpatrick 1996). A female may evolve to use premating choice to acquire the best spermatophylax or glandular meal, then "cryptically" choose the male's sperm if he is perceived to be of high genetic quality (Bussière, submitted manuscript). In fact, there may be a complex relationship between male genetic quality and meal size where high-quality males can get away with delivering fewer goods and services to females (see Møller and Thornhill 1998; Simmons et al. 1999). I return to this topic later in the chapter.

7.2.v. Heavyweights as good providers

As we saw in Chapter 6, females of many katydids receive a spermatophylax meal from each copulation. An increase in these direct benefits does not have to occur by direct choice. In black-legged katydids, the replacement of mated males by "satellite" youngsters would ensure that a female mating with a replacement male gets a decent-sized mating meal because recently mated *Orchelimum* males produce small spermatophores (Feaver 1977) (Fig. 7.7). Large males of many katydid species also produce large spermatophore meals. These include both *Orchelimum* (Fig. 7.7) and black-sided katydids, *C. nigropleurum* (Gwynne 1982), for which some 40 percent or more of the variation in the size of the spermatophore meal is explained by male body weight (Table 7.2). If song is a useful predictor of male size, females are expected to prefer the larger callers. I first examined this question by presenting female black-sided katydids with two males calling at equal distances from her. Females always mated with the larger male (Fig. 7.8; Gwynne 1982). Furthermore, when black-sided females in the laboratory interacted with several males over several days, like black-legged females in nature (Feaver 1983), they preferred to mate with the heavier of the males encountered (D. T. Gwynne, unpublished data 1981).

There has been no subsequent work on black-sided katydids to show whether the overwhelming success of heavyweights was actually due to female song preferences alone or whether the winning large males were able to "acoustically outcompete" their rivals by, for example, being the

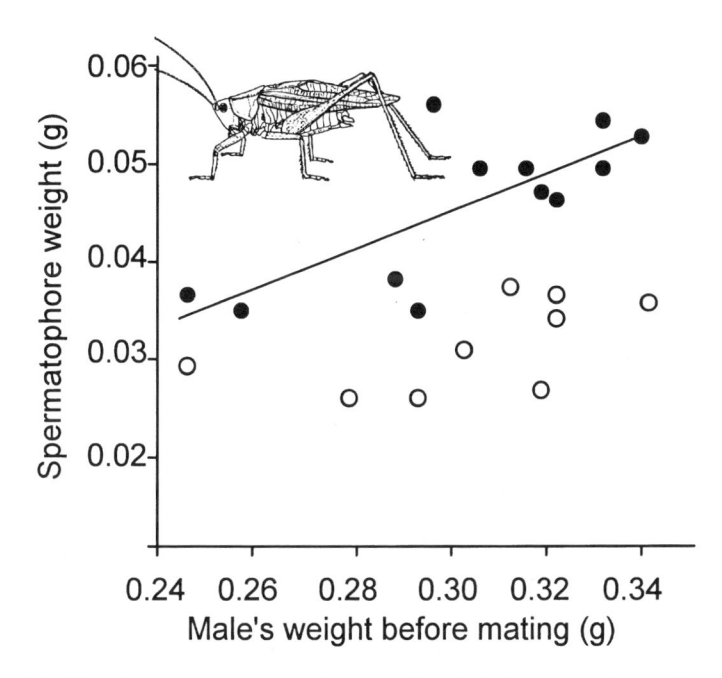

Figure 7.7. The size of the spermatophore meal in *Orchelimum* species is affected not only by male body size (positive correlation for closed points) but also by his mating status. Recently mated males (open points) produce significantly smaller spermatophores than do unmated males (closed points). These data are for *Orchelimum delicatum* (D. T. Gwynne, unpublished data). Katydid redrawn from Alexander et al. (1972).

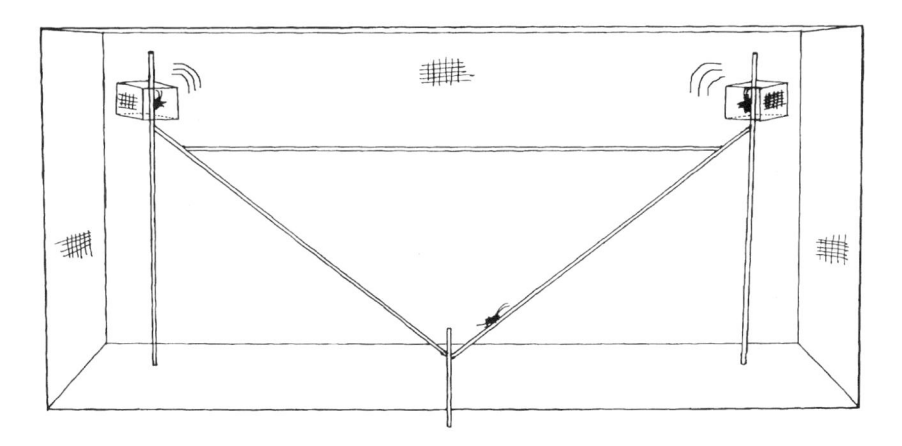

Figure 7.8. The apparatus used to test the response of black-sided katydid females to two live calling males. The Y-maze was constructed of thin wooden dowels to simulate plant stems and to resemble as much as possible the natural situation in which males call from plant tops. Each male called from within an open-fronted small cage mounted on a dried stem (1.2m apart, the mean distance between callers in nature: Morris 1967). Redrawn from Gwynne (1982).

song leader so that their calls were perceived much more strongly by the orienting female (Greenfield et al. 1997) (see above). There is, however, good experimental evidence that black-sided females can use a nonacoustical cue to select large males. One component of the male's tremulation courtship turns out to be a virtually foolproof predictor of body weight. The pulse rate produced by shaking (tremulating) black-sided males explains a very high 86 percent of the variation in male body weight (De Luca and Morris 1998) (Table 7.2, Fig. 7.9). Black-sided males tremulate after they make direct contact with a female (see Chapter 5), and this signal may have been involved in mate rejection in three of my original laboratory trials with this species where females were first courted by the small caller before they moved across to and mated with his larger rival (in all other trials females moved toward and mated with the large caller first: Gwynne 1982).

De Luca and Morris (1998) went on to investigate the role of tremulation in mate choice by black-sided females. They used an ingenious T-maze that allowed each female to ascend an artificial plant stem, a wood dowel, to encounter a choice of two horizontal dowels through which computer-

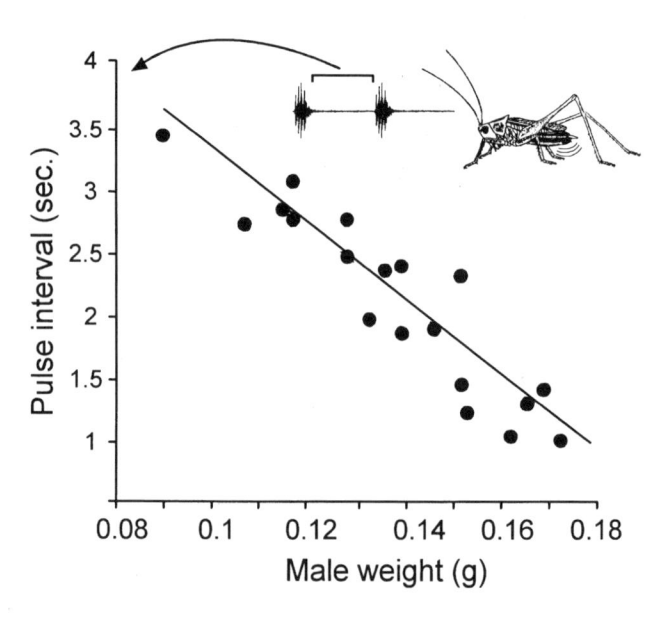

Figure 7.9. Scatterplot showing the close relationship between the pulse interval in the tremulation signal of the male black-sided katydid (top of figure) and his body weight. From Figs. 3 and 6 of De Luca and Morris (1998). Katydid redrawn from Alexander et al. (1972).

generated vibratory stimuli could be transmitted. Significantly more females were attracted to the dowel with conspecific tremulation[5] and more importantly, showed a significant preference for vibrations of a large male over those of a small male (17 of 24 females). These results may indicate passive choice by females for the more intense signal, although another of the experiments showed that females (14 of 18) appeared to actively choose a signal by preferring conspecific tremulation over an artificial tremulation of equal duration and intensity. The preference by females for tremulation with a pulse rate shorter than the species mean (average) also suggests that preference by females is not simply species recognition (Table 7.1).

Little is known about the mating success of large black-sided males in nature. In fact, other than Feaver's work with black-legged katydids, the only field studies examining the mating success of male katydids were on pollen katydids, *Kawanaphila nartee* (showing no large-male advantage—for a possible explanation, see Chapter 9; Gwynne and Bailey 1988), and the Australian listroscelidine, *Requena verticalis* (Schatral 1990). In both species the male spermatophylax meal has significant influences on female fitness by increasing the size of the eggs produced (see Chapter 6). Female preference for heavyweights is expected in *Requena* species because heavier males donate larger meals; 40 to 44 percent of the variation in the mass of the highly nutritious spermatophore of two species in this genus is explained by male body weight (Gwynne et al. 1984; Wedell and Sandberg 1995). Size (pronotum length) also predicts spermatophylax mass in *R. verticalis*, although it is unclear at this point whether size per se is a predictor of the benefit of the meal, because there is no significant correlation between male size and the absolute protein content of the spermatophlyax (Simmons et al. 1999). Paradoxically, Simmons et al. (1999) also found no correlation between male size and percentage of protein in the meal; a negative correlation is expected if the heavy spermatophylaxes of large males contain no more protein than the small meals transferred by small males. Furthermore, it is possible that overall protein in the courtship meal may be less important to mating females than specialized components of this meal; egg size is increased only by an increase in spermatophylax meal size, not by an increase in protein content in other components of the mating female's diet (Gwynne 1988a; see Chapter 6).

---

[5] These results concern only flightless females, by far the most common wing phenotype in nature. However, a separate experiment found that a significant number of long-winged females (14 of 17) showed negative "vibrotaxis" to male tremulation. De Luca and Morris (1998) suggested that long-winged youngsters in these trials may have been adaptively avoiding mating with males as part of a strategy to delay reproduction until after they had migrated (Ando and Hartley 1982).

Schatral (1990) supported the prediction that heavy *R. verticalis* males have a mating advantage. She compared the weights of males in mating pairs with those of each closest caller to the pairs. The analysis revealed that body weight and song frequency (a negative correlate of size in this and other katydids: Table 7.2) were "equally important determinants" of male mating success.

Wedell and Sandberg (1995) took the research one step further using an undescribed species of *Requena* (species no. 5). They showed that when given a choice of two callers, females preferred the heavier male with the lower song frequency. But does this result reflect female choice for the songs of heavyweights, or intermale acoustical interaction (competition), or even choice for some other attribute of the male? To determine whether female choice for a male song cue did indeed occur, experimental manipulation of the song cue is necessary, as shown in work already discussed on female preference for the songs of large male tree crickets and *Scudderia* katydids (see Table 7.2 for a summary of these and other studies).

It was Bailey et al. (1990) who examined preferences for different components of the calls of *R. verticalis*. In the first part of this experiment with live males, females showed no preference for the larger of two callers (note that this study examined size, not weight). Next, when tape-recorded songs were played to females, they also showed no preference for the song of the larger male or even for the song of the lowest frequency, although these results were based only on a sample size of seven females. However, this small sample was sufficient to show a significant (possibly adaptive) preference for the song that would be produced by the closest male, that is, the loudest call with the greater power in the high-frequency part of the broad-spectrum song of this species (high frequencies of the song attenuate with increasing distance from the caller; see Chapter 5). Trials using a larger sample size may reveal a significant preference for lower-pitched (large-male) calls as reported for the katydid *T. cantans*. In this species females not only oriented to a close-sounding song (with relatively louder high-frequency components of the call) but also when this distance cue was controlled, preferred calls with a lower average frequency, a possible indicator of male quality, perhaps size or species recognition (details in Table 7.2) (Latimer and Sippel 1987). The finding by Bailey et al. that *R. verticalis* females prefer louder songs, even when the difference is as small as 2-dB intensity (Römer et al. 1998), supports a passive-attraction hypothesis. In fact, a mix of song frequencies may make the song more easily localizable by the orienting female (Table 7.1).

So what can we conclude from tests of the hypothesis (Gwynne 1982) that female katydids prefer males that supply the largest spermatophylax meals? In other insects with courtship meals, females are known to prefer males that supply the most nutrients (Thornhill and Alcock 1983; Vahed

1998a), but these are systems in which nuptial meals, such as prey carried by the male, can be assessed directly by the female (e.g., Thornhill 1976a). In orthopterans such as katydids, the female cannot directly assess the meal and so must rely on indirect cues from the male's display or his body size. This indirect nature of assessing potential meal size makes demonstrations of adaptive female choice for large gifts more difficult; field observations of male mating success must be followed up with laboratory manipulation of the male traits hypothesized to contain cues about meal size (see Searcy and Andersson 1986). Unfortunately, complete support for the hypothesis about preference for large spermatophylaxes has yet to come from any study system. The work with black-sided katydids has pointed to a nonacoustical signal, courtship tremulation, as a highly reliable indicator of male size used by discriminating females. However, further information on both the fitness benefits of the nuptial meal and the mating success of large males in nature is needed for this species.

In *Requena*, large males in nature seem to be the most successful in transferring their nutritious meals to females, but a demonstration that females actually choose these males remains elusive. Ironically, the best evidence that female katydids respond preferentially to the song of large males indicates that female behavior is less for direct benefits than a form of passive attraction or a preference for males with good genes (e.g., *Scudderia*: Table 7.1), both of which can impose sexual selection on song and correlated traits. The good-genes hypothesis may hold particular promise in future studies of *R. verticalis*, because there is recent evidence that compared to symmetrical males, asymmetrical males, likely of lower genetic quality (see above), provide spermatophylaxes containing more protein (Simmons et al. 1999) (Fig. 7.10). Could males of higher genetic quality get away with investing less in their mating meals because they are highly attractive to females (see Møller and Thornhill 1998; Simmons et al. 1999)?

7.2.vi. Female preferences: getting the right species

Given the strong selection on female animals to choose the signals and displays of the best conspecific males (Dugatkin and Godin 1998), it seems almost paradoxical to find that females can be attracted to and even mate with males from completely different species. Such reproductive interactions can be costly to females, either in terms of movement toward the wrong signals (Gwynne and Morris 1986) or if the species are closely related, in terms of producing fewer offspring or hybrid offspring with reduced fitness (e.g., Gerhardt 1987; Howard et al. 1993; Shapiro 1999). In the Michigan marsh, which opened this chapter, there is little chance that black-legged females (*O. nigripes*) make this sort of pairing or mating mistake. Even though these katydids share the marshland habitat with two other *Orchelimum* species, *O. gladiator* and *O. vulgare* (Feaver 1977), the

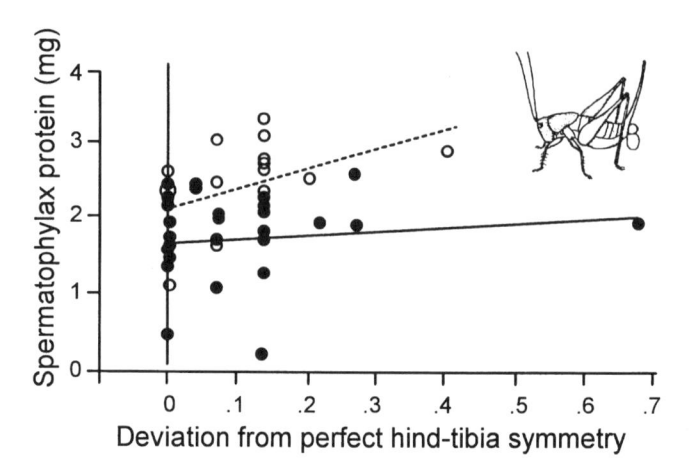

Figure 7.10. In *Requena verticalis* there may be an indicator of male genetic quality in the male's spermatophylax meal. The greater the deviation of the male's hind-tibia length from perfect symmetry (= 0), the greater the overall protein content in his spermatophylax meal. This is true, however, only when males are allowed a long time period (12 days) between matings (the dotted line through the open symbols: Spearman-rank $r_s = 0.49$, $P = 0.17$). There is no such pattern if males have only 4 days between matings (continuous line through the filled points: correlation $r_s = 0.14$, $P = 0.48$). Redrawn from Simmons et al. (1999). Katydid redrawn from D. Scott's figure in Clutton-Brock (1981).

three species have very different song patterns (G. K. Morris and T. J. Walker, unpublished data 1975; Fig. 7.11), making it unlikely that a female will respond to the incorrect song "code." This is certainly true for the *O. vulgare* song, which in laboratory trials is completely ignored by female black-legged katydids (*O. nigripes*) (Shapiro 1996). However, the song code of a fourth meadow katydid is a potential cause for confusion for black-legged females in two locations quite distant from Michigan (Shapiro 1998). In these locations, black-legged katydids coexist with the "handsome" meadow katydid, *O. pulchellum* (Plate 22). Although *O. pulchellum* is genetically and morphologically distinct from *O. nigripes*, it has a virtually identical calling song that in laboratory trials using taped calls, can attract responsive *O. nigripes* females (Shapiro 1996).

One of the areas of species overlap (sympatry), along the Potomac River near Washington, D.C., is a result of black-legged katydids recently (probably in the last century) crossing the Appalachian Mountain barrier that separates most of the *O. pulchellum* distribution from that of *O. nigripes*. A second area is found in the Deep South of the United States, where the two species meet around the "bottom end" of the mountain barrier (Fig. 7.11). Here the two species interbreed freely across a broad and apparently old hybrid zone; linkage disequilibrium between the genes of the two species

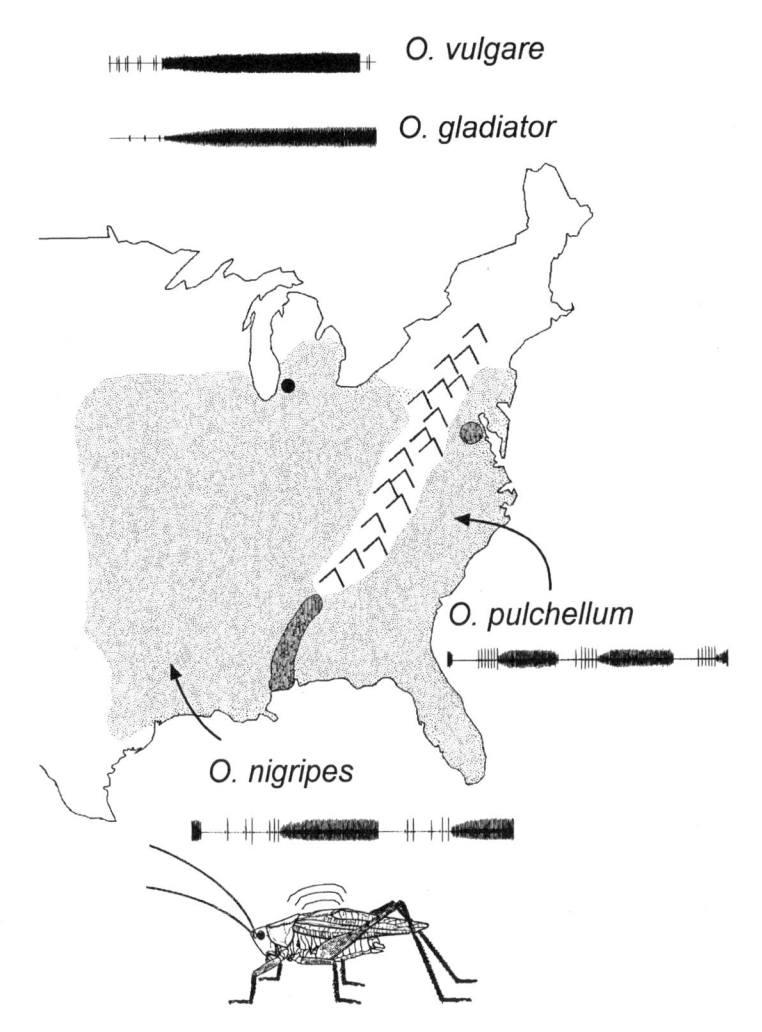

Figure 7.11. The black-legged katydid (*Orchelimum nigripes*) at Marianne Feaver's Michign study site (black dot) shares its habitat with two other meadow katydids, *O. vulgare* and *O. gladiator*, two species with song patterns quite different from the *O. nigripes* song (compare the top two traces with the bottom trace). However, the *O. nigripes* song is very similar to that of the handsome katydid, *O. pulchellum* (compare bottom and right traces), and these two species co-occur and hybridize (heavy shading) in the deep south at the bottom of the Appalachian mountains and on the Potomac River, east of the mountains. Five-second song traces of each species were from recordings made between 23 and 28°C. Map redrawn from Shapiro (1998). Katydid redrawn from Alexander et al. (1972). Song traces courtesy of G. K. Morris: G. K. Morris and T. J. Walker, unpublished data 1975.

indicates no behavioral or postzygotic barriers to gene flow, thus no evidence of a mismating cost through lower fitness of the hybrid offspring. Indeed, in this region the two species may be slowly fusing. By contrast, in the area of sympatry along the Potomac River there is strong linkage disequilibrium (Shapiro 1998) that may result in part from discrimination by females against heterospecifics after pair formation (Shapiro 1996), to avoid the mismating costs of reduced fecundity (Shapiro 1999) and possibly sterile sons (Cabrero et al. 1999).

Discrimination by Potomac females was first demonstrated when they directly contacted a male of each species in lab cages and rejected the heterospecifics, possibly using visual (note differences between Plates 21 and 22) or olfactory cues. This result was especially strong for black-legged females who rejected the handsome katydid male in each of 16 trials. However, this mating barrier could be breached if a heterospecific male was the only available mate. If mismating occurred, there was another impediment to hybridization (Shapiro 1999): compared to matings with conspecifics, mismated females laid far fewer eggs, which Shapiro argued may be caused by heterospecific males passing ineffective or insufficient chemical substances in the ejaculate, substances that are known to stimulate oviposition in orthopterans (Loher and Edson 1973) (ejaculatory chemicals in insects are known to evolve rapidly within species as a result of sexual selection: Rice 1996). Shapiro (1996) even suggested that the lack of a full oviposition response to heterospecific ejaculatory chemicals might be a form of cryptic choice (Table 7.1) whereby females favored fertilization by conspecifics. Cryptic fertilization biases against heterospecific sperm are known for several insects (reviewed by Gwynne 1998) and are expected to evolve if there is a cost of producing hybrid offspring. Although studies of black-legged × handsome katydid crosses have yet to reveal any decreased hatching success or survival of hybrid offspring (Shapiro 1999), there is evidence that hybrid sons may be sterile (Cabrero et al. 1999).

Any lower fitness of black-legged × handsome katydid hybrids is expected to select for a cost saving in premating discrimination against heterospecific calls, the "reproductive character displacement" of mate choice predicted by theory (Littlejohn 1981, but see Paterson 1982; Spencer et al. 1987). Even if hybrids do not have lower fitness, character displacement in mate choice might be expected to evolve: why pay the costs and risks of moving to the song of a heterospecific male only to discriminate against him at some later stage in the mating-oviposition sequence? One answer to this question is that there has not been time for character displacement to evolve because the origin of the Potomac hybrid zone is very recent (Shapiro [1998] estimates 50–70 years ago). It is also possible that the two species eventually may begin to fuse into one, as appears to be the case in the Deep South hybrid zone (Shapiro 1996). One selection pressure in some katydids that actually may accelerate such fusion (and explain why females do not discriminate among songs) is that any costs of sexual interactions with heterospecific males may be offset by the spermatophylax meal the female receives (Shapiro 1999). Even a courtship meal received from a male of the wrong species is nutritious! Nutritional benefits should enhance the selection on females suggested by Shapiro (1996): to mate with a heterospecific but then show a fertilization discrimination against his sperm until a conspecific male is available.

Is there any evidence for katydids that mispairings or mismatings between species have resulted in character displacement in song preferences? Studies, including work on katydids, showing that orthopteran females prefer their own species' song have often implied that rejection of a heterospecific song in favor of the conspecific signal (e.g., Bailey and Robinson 1971; Heller and Helversen 1986; Duijm 1990; Hartley 1993) is the "ghost" of past interspecies interactions. However, a demonstration of conspecific song preferences does not only support the species-recognition hypothesis. Rejection of the "inappropriate" calls of other species might simply be an effect of selection on females to choose high-quality males within their own species (Table 7.1) (Gwynne and Morris 1986; Ryan and Rand 1993). Some evidence that a conspecific call might function in species recognition is if females prefer the mean of a call parameter that varies interspecifically (e.g., Butlin et al. 1985). The best evidence for species recognition, however, is a demonstration that females from areas of sympatry (areas of species co-occurrence) are more discriminating of conspecific song elements than are allopatric females (e.g., see work on treefrogs by Gerhardt 1994a). An alternative is finding that females recognize and ignore (or actively avoid) the call of a sympatric "problem" heterospecific (Gwynne and Morris 1986), although heterospecific recognition would be a costly mechanism in areas where many species share similar communication signals (e.g., the tropics) (H. C. Gerhardt, pers. comm. 1999).

In a location with few species with similar signals, Morris and Fullard (1983) examined heterospecific recognition in the response of female black-sided meadow katydids, *C. nigropleurum*, to tape-recorded calls. Although females were remarkably nondiscriminating in their response (positive phonotaxis) to song models, including both a "random-noise" model containing the same broad range of frequencies found in the songs of many *Conocephalus* species[6] and the call of the lance-tailed meadow katydid, *C. attenuatus*, a species with which they are never found, they showed no response to the call of the short-winged meadow katydid, *C. brevipennis*, a species that commonly coexists with black-sided katydids.

---

[6] As noted earlier in the chapter (section 7.2.i), females probably perceive white noise as the featureless sound of a chorus of conspecific callers.

The cue to heterospecific recognition appears to be the "ticks" in the "buzz-tick" song found in short-winged katydids but not black-sided ones. This cue remained when the calls of five short-wing males were overlapped to produce a song model that was still ignored by black-sided females. However, when the calls of 10 short-winged males were overlapped, the ticks were finally obscured, and black-sided females showed an equal preference for this playback and the equivalent 10-male conspecific song model (Morris and Fullard 1983).

Morris and Fullard (1983) concluded that black-sided females recognized the short-winged call, so as to avoid the time and risk costs of movement toward the wrong song (see Gwynne and Morris 1986). It is unknown whether black-sided and short-winged katydids hybridize in nature, but this seems unlikely since the only available phylogeny of the North American *Conocephalus* genus places each species in a different monophyletic species group (Rehn and Hebard 1915).

Two species of European *Tettigonia*, *T. viridissima*, the great-green grasshopper, and *T. cantans*, the upland-green grasshopper, do produce hybrids in nature that survive at least until adulthood, although the relative fitness of these hybrids is unknown (Schul 1994). Detailed experiments by Schul (1998) showed convincing evidence that females recognize the calls of the other *Tettigonia* species; the paths (measured as the mean vectors through the points where females exited from the experimental arena) taken by female great-green grasshoppers toward a speaker playing a conspecific song are deflected by some 15° away from another speaker playing the song of the sympatric upland-green grasshopper (compare A and C in Fig. 7.12). This result was in striking contrast to the reverse experiment using upland-green grasshoppers. Upland-green females showed evidence of heterospecific recognition in a comparison of the mean phonotaxis vector for conspecific song alone (Fig. 7.12B) versus when the great-green song was played from a second speaker (Fig. 7.12D). This time, however, the significant vector deviation (18°) was *toward* the speaker playing the other species's song. Moreover, whereas great-green females did not respond positively to the other species's song (Fig. 7.12E), upland-green females moved toward the great-green song when it was the only call present (Fig. 7.12F). One hypothesis for the great-green grasshopper alone avoiding heterospecific song is that females of this species are more at risk of a mismating than are upland-green grasshoppers because great-green grasshoppers are more prone to migrating and thus settling in a population with only male upland-green grasshoppers available as mates (Ingrisch 1981, cited in Schul 1998). The main hypothesis discussed by Schul, however, is that one of the two song rates in the complex great-green song (Fig. 7.12) is similar to the upland-green rate and so attracts upland-green females, whereas the upland-green song contains no elements attractive to females of the other species (see also Schul 1998). To be more than

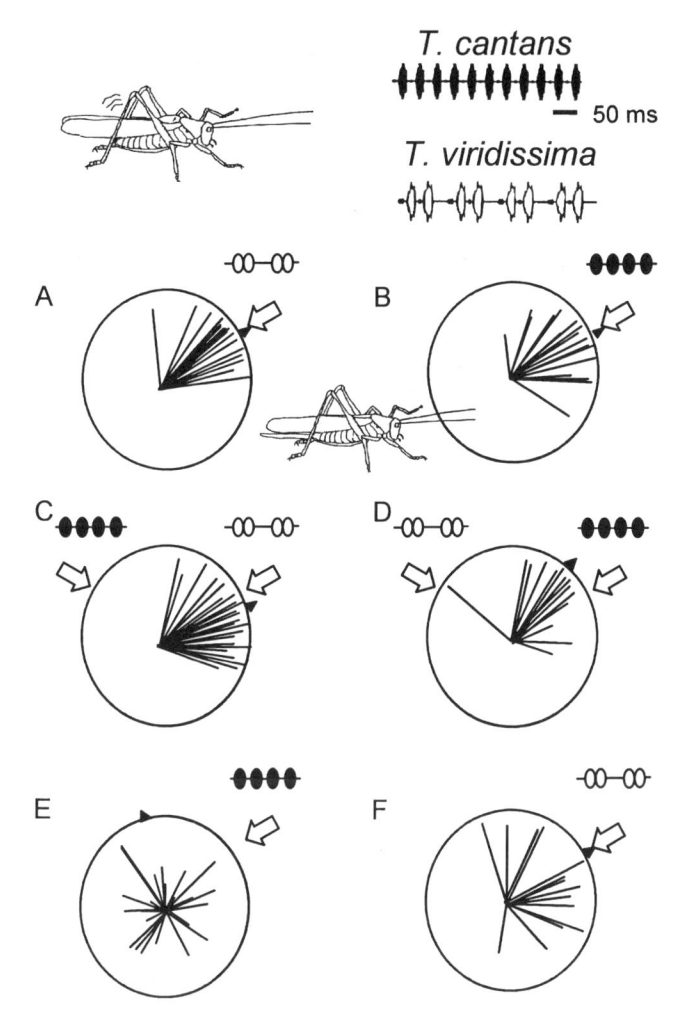

Figure 7.12. The callings songs and phonotactic responses (directions where the moving female exited the circular arena) of *Tettigonia cantans* and *T. viridissima* females to conspecific and heterospecific songs. Top: Temporal patterns of the calling songs of the two species. Bottom: Circular outlines representing experimental arenas, showing mean exit vectors (triangles) and speaker positions (arrows). Responses of *T. viridissima* females are on the left (A, C, E) and of *T. cantans* females on the right (B, D, F). A and B. Conspecific song alone. C and D. Simultaneous playback of conspecific and heterospecific songs. E and F. Heterospecific song alone. Redrawn from Schul et al. (1998).

just a proximate explanation for an adaptive change—such as upland-green males have evolved a song structure to avoid attracting great-green females—Schul's hypothesis must assume that upland-green grasshoppers are somehow constrained phylogenetically by the original contexts in which their song patterns and song preferences evolved.

To fully understand the adaptive significance of conspecific or heterospecific recognition in *Conocephalus* and *Tettigonia* species, we require more field research on how species interactions affect the fitness of individual males and females. We know very little about the costs, most likely to be hybridization in *Tettigonia* species (Schul 1994), and the risks of phonotactic movements to the wrong species by *Conocephalus* katydids (Gwynne and Morris 1986). Moreover, there is the intriguing possibility that the costs of interspecific reproductive interactions for a female katydid might be offset if she obtains a nutritious spermatophylax meal from a heterospecific mating. The expense of this costly courtship gift also raises the question of whether males might have been selected to discriminate against females of other species. Males of certain *Pterophylla* species that mate in dense populations (see Chapter 3) show a strong preference to court and mate with females from their own populations (Barrientos 1998), although interestingly there is no obvious spermatophylax gift in these katydids (but there is evidence of feeding by females on the male's dorsal gland: Barrientos 1998).

7.2.vii. Female preferences: concluding comments

What are the pressing questions with regard to choice of males by female katydids? As stressed at the beginning of this chapter, most research on this topic has centered on selective female phonotaxis for calling songs. At one time, calling was thought to signal species information alone. However, it is now clear that the calls of orthopterans and other singing animals communicate far more than this (Gerhardt 1994b), and that except for few species (see above), heterospecific songs may rarely be a problem, owing to the tight conspecific signal preferences that are expected to result from sexual selection on male calls. What we currently lack is an understanding of how various attributes of the male, as revealed in his song, interact to influence the response of the female. For example, can she overcome a sensory-biased response to the leading call (Greenfield et al. 1997) if one or more of the subsequent calls is from a higher-quality male? And, if the risk of phonotactic movement toward a better song is high, will females choose another male? These questions can be answered only with a combination of solid information about mating in nature and playback experiments in which there are manipulations of male song as well as manipulation of perceived risk by the female who seems to be less choosy when the risk of predation is high (for a study of crickets see Hedrick and

Dill 1993). Finally, there are the other signals produced by the male. For example, perhaps females of some species do not use song to assess male quality and rely instead on his courtship by assessing tremulations (De Luca and Morris 1998) or antennal "caresses" (Fabre 1917).

7.3. Competition for Mates

As we have seen, females of a variety of katydid species respond to the songs of males. Female response to song must be the rule for Tettigoniidae, because females of most species have fully functional ears that are tuned to conspecific calls (see Chapter 5). However, a look at the anatomy of males also reveals that with few exceptions (e.g., Bailey and Römer 1991; Fig. 9.17), their ears are every bit as elaborate as those of females. Therefore, some sort of eavesdropping by males on the songs of rivals must be the rule throughout the family. The most detailed information we have on listening to rivals comes from territorial katydids, mainly meadow katydids, cone-heads, and tettigoniines, that live in field or heathland vegetation. For example, territorial male black-legged katydids not only appear to listen for courtship attempts by newly matured males (who may be accosting nearby females) but also detect and interact with neighboring males. Other important stages of male-male competition in katydids include physical fights and finally the cryptic competition that takes place inside the female genital tract, competition between the ejaculates of rivals.

7.3.i. Calling and sexual competition

Competition between calling katydids is for access to sexually responsive females. As we saw earlier in the chapter, the most intense battles among black-legged katydids occurred in pathways through which females migrate into tall vegetation (Feaver 1983). In these areas, territories were packed in so tightly that distances between callers were reduced to a mere 2 m (Fig. 7.13), and about half of all male-male interactions ended in physical fights. Other studies reported aggregations of calling male katydids around patches of vegetation important to the biology of the species involved, either the oviposition and food plants attractive to females (Spooner 1968b; Shaw et al. 1981; Gwynne 1984c; Gwynne and Bailey 1988) or the best calling perches, usually tall plants from which callers are able to broadcast their signals over long distances (Schatral et al. 1984; Arak and Eiriksson 1992). Arak and Eiriksson (1992) reported an almost eightfold increase in the call's maximum detection range when male great-green grasshoppers broke from the cover of dense, low-lying vegetation to sing from high bushes.

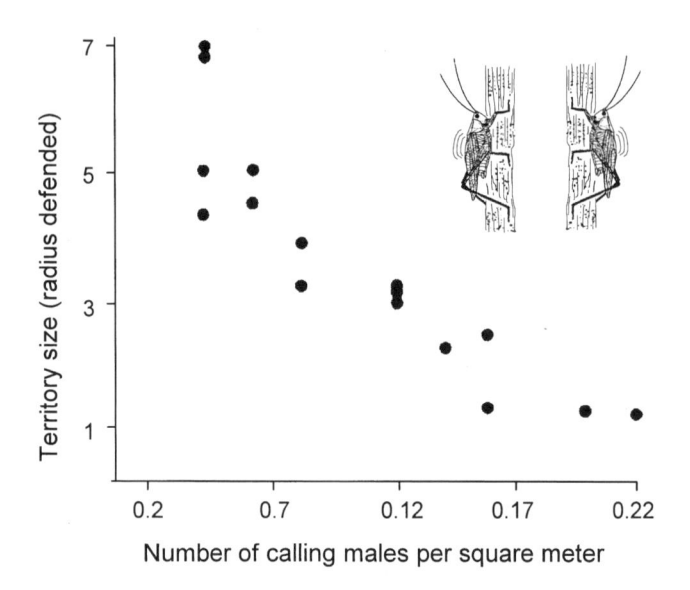

Figure 7.13. For black-legged katydids, *Orchelimum nigripes*, there is a highly significant positive correlation between the density of callers and male territory size (correlation $r = 0.89$, $P < 0.001$). Similar relationships (and correlation coefficients) were found at two other sites. Redrawn from Feaver (1977). Katydid redrawn from Alexander et al. (1972).

Many other territorial katydids show far less site-dependence because they inhabit more homogeneous areas of vegetation. For example, although Feaver (1977) reported aggregations of calling sword-bearing katydids (*O. gladiator*) in plants near female migration areas, Morris (1967) found territories of the same species spread throughout a homogeneous area of sweet flag (*Acorus calamnus*) and sedge (*Carex* species). In areas such as this, males have far less opportunity to monopolize sites attractive to females. Instead, the regular spacing of callers, as revealed by a significant statistical analysis of the distribution of calling locations, appears to result from callers' maximizing mating success with (the presumed) randomly distributed females by spacing out as far as possible from their rivals (Arak et al. 1990).

Sizes of territories, as estimated by the mean distance between calling male katydids that show a regular or "spaced" distribution, vary from 1 to 20 m (*estimated* is an appropriate word, as these distances are not necessarily defended areas) (Morris 1967; Bailey and Thiele 1983). Within species, territory size is known to decrease with increasing density of callers (Fig. 7.13) (Feaver 1983);[7] with decreasing vegetation density, owing

---

[7] Unlike certain field crickets (Alexander 1961), there is no evidence that territoriality in katydids breaks down at high population densities.

to the degradation and attenuation of sounds (Römer and Bailey 1986, see Chapter 5; Arak and Eiriksson 1992); and with increasing male size (independent of vegetation density: Bailey and Thiele 1983). There is also a between-species positive correlation between male size and distance between calling males, although this drops below statistical significance when the phylogenetic relationships between species are accounted for (Fig. 7.14). The simplest explanation for this relationship is that because "little [animals] are much more common than the big" (Colinvaux 1978), an increase in the body size of a species necessitates an increase in the size of the territory or there will be a decrease in the encounter rate with females; larger males should control on average a larger "acoustic space" because they produce louder (Table 7.2; Bailey and Thiele 1983) and lower-frequency calls. Bailey and Thiele (1983) demonstrated the direct effect of song intensity on territory size; *Mygalopsis marki*, whose call intensities were experimentally reduced, had smaller territories compared to unmanipulated callers.

The spacing of male territories is a dynamic process resulting from constant acoustic interactions between males in which losers retreat and more aggressive males expand the areas they control following their phonotactic forays, sometimes ending in battle, into neighboring territories (Morris 1972; Bailey and Thiele 1983; Brush et al. 1985). The role of song in the repulsion of competitors was shown in experimental releases of male *M. marki*; intact callers spread out from the release point whereas experimentally deafened males did not (Bailey and Thiele 1983). Repulsion of *M. marki* males by rival songs is also seen when males move to a lower position in vegetation in response to an increase in intensity of the playback of taped the conspecific calls directed at them (Dadour and Bailey 1985). In another cone-head, *Neoconocephalus affinis*, males in nature were attracted to the taped rival song. Brush et al. (1985) randomized the order of playback of a range of intensities of the conspecific song. Ten of 18 males moved to one of the song intensity treatments, with half of the responsive males actually jumping onto the loudspeaker (no males responded to the control calls of congeneric *N. maxillosus*). There was no evidence, however, that the phonotactic, presumably more aggressive, males were larger individuals. The probable reason why Dadour and Bailey observed no male phonotaxis in *M. marki* is that the intensity of song in each trial was increased steadily so that all experimental males backed down when they perceived the increasingly loud playback as a larger and possibly more aggressive intruder moving toward them.

Male song in some katydid species probably attracts more than just aggressive calling neighbors. Recall that Feaver observed silent "satellite" black-legged males moving into the territories of dominant callers. Because the dominant males tolerated satellites, as long as these newcomers did not sing, Feaver suggested that the intruders were youngsters waiting for

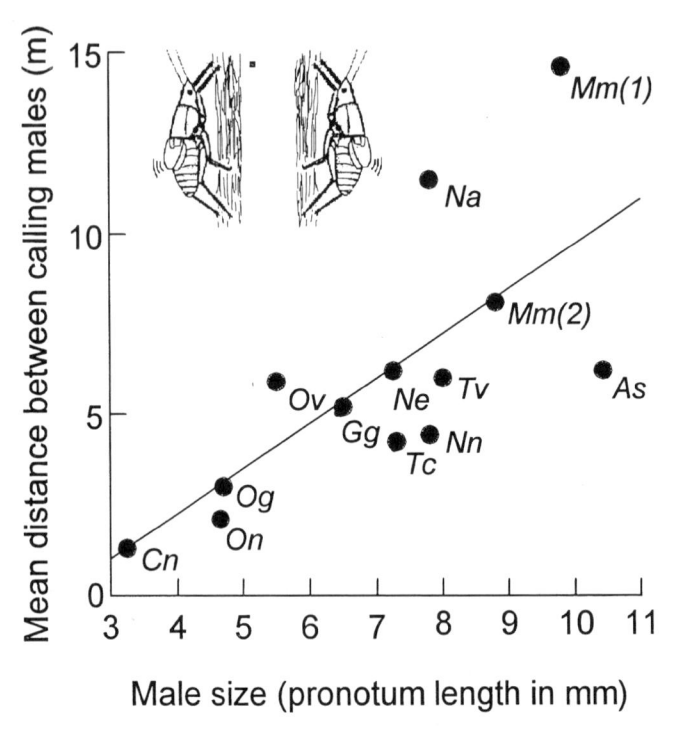

Figure 7.14. The relationship between male body size and the distance between calling males for 12 species of old-field and heathland katydids from Europe, North America, and Australia (Conocephalinae and Tettigoniinae). For one species there were data for two different populations. The pattern of actual species points (letters next to each point identify each species, see below) showed a significant correlation (Spearman-rank $r_s = 0.80$, $P < 0.01$, $n = 13$). However, on an independent contrasts test (Burt 1989), which controls for the tendency of closely related species to share characters through common descent, the positive correlation drops to nonsignificance (Spearman-rank $r_s = 0.37$, $P < 0.1$, $n = 13$). I inferred the relationships between the 12 species from a phylogeny of katydid subfamilies (Chapter 2) and phylogenies of the subfamily Tettigoniinae (Rentz and Colless 1990). For other subgroups I used taxonomic arrangements as proxies for phylogenies. I held within-species influences on territory size constant by calculating overall mean distances to the nearest neighbor when there was more than one estimate for a particular study. Data for *Anabrus simplex* (As) from Gwynne (1984c); *Conocephalus nigropleurum* (Cn) from Morris (1967); *Orchelimum gladiator*, *nigripes*, and *vulgare* (Og, On, Ov) from Feaver (1977) and Morris (1967); *Neoconocephalus affinis* (Na) from Greenfield (1983) (size data from Walker and Greenfield 1983); *N. ensiger* (Ne) from Shaw et al. (1982), *N. nebrascensis* (Nn) from Meixner and Shaw (1979); *Mygalopsis marki* (Mm) from Bailey and Thiele (1983) (size data from Bailey 1979); *Gampsocleis glabra* (Gg) from Latimer (1980); *Tettigonia viridissima* (Tv) from Schatral et al. (1984); and *T. cantans* (Tc) from Arak and Eiriksson (1992) (size data from Latimer and Sippel 1987). Other size data from Blatchley (1920) (North American Conocephalinae).

an opportunity to replace the resident after he was depleted by mating. In other katydids, however, silent males are more than mere pretenders to territorial positions. Mbata (1992b) frequently observed phonotactic male armored "ground crickets" (Hetrodinae), *Acanthoplus speiseri*, stopping about a meter away from a caller and mating with some of the females responding to the caller. Such sneaky male behavior is better known in field crickets such as *Gryllus integer*, a species in which satellite sneakers appear to have low mating rates relative to callers but live longer by avoiding the risk of lethal parasitism from acoustically orienting tachinid flies (see Chapters 4 and 8) (Cade 1979).

A particularly exploitable system for satellite males is one in which female responses to signaling males can alert any potential interlopers in the vicinity. As we saw in Chapter 5, many phaneropterines have a call-answer system of pair formation, and for *Elephantodeta nobilis*, Bailey and Field (2000) recently showed that silent satellites are attracted when females stridulate (click) in response to a male call (Fig. 7.15). Females time their answer clicks to occur some 570ms after the final component of the male's complex call, and a satellite male's response is to insert a "brief volley" of clicks about 200ms *before* the caller's final call component. The acoustic interruption by the satellite can allow him to take over the dialogue with the female by causing her to reply earlier. Although the relative mating success of satellites has not been shown, lab experiments support the contention that this alternative male strategy of acquiring mates might be successful in some cases. In two-speaker arena trials, females were attracted mainly to the main singer's call. However, they were attracted to the interloper's call 20 to 30 percent of the time even when it was more faint (distant) than the main call. Similar deceit tactics of interloping males are known in the call-answer flash communication of certain fireflies (Lloyd 1979b).

More typical song interactions of rival katydids occur when two males interact acoustically, before competitive interactions escalate to a level where two males can be attracted to each other, fight, or retreat. We have already seen (in section 7.2.ii) how males monitor the songs of rival neighbors in the constant competition to vary chirp rate so as to remain attractive to females (Greenfield et al. 1997). But there are a number of other ways in which males compete acoustically. Although katydids do not have the specialized aggressive songs noted in other Orthoptera (e.g., grylline crickets: Alexander 1961), elements of the calling song in certain species do appear to function in a competitive context, such as when interacting males increase their call rate (Simmons and Bailey 1992). One of the clearest examples are the "ticks" in the "tick-buzz" song of many *Orchelimum* and *Conocephalus* meadow katydids (Fig. 7.11) (Alexander 1960). When neighboring black-legged males interact, the ticking part of the song of both males is extended in 20 percent of all observations. In two other

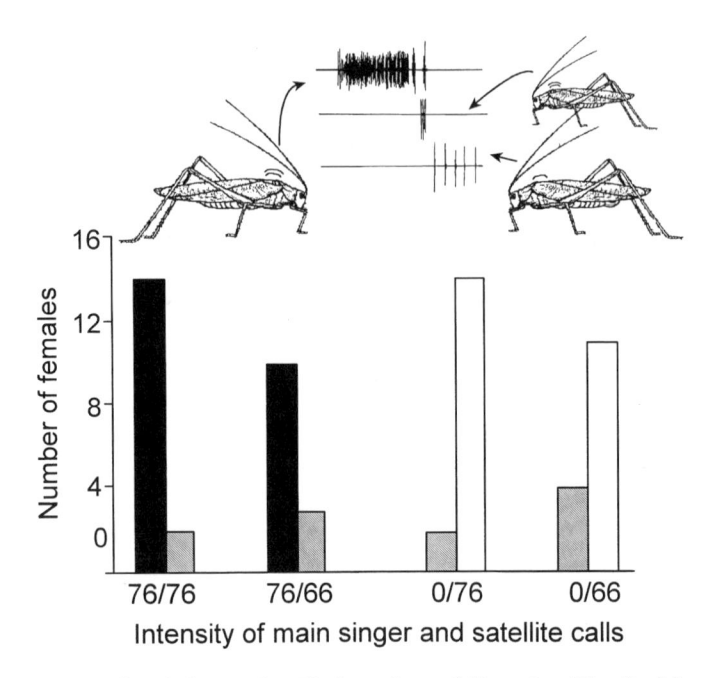

Figure 7.15. Satellite behavior by *Elephantodeta nobilis* males. The final burst of a calling male's (left katydid) four-part song (top trace) is answered by a female (right foreground) with a series of clicks (lower trace). However, an eavesdropping satellite male (right background) can disrupt this dialogue between a pair by producing a volley of clicks just before the caller's last burst of sound (middle trace) (all traces = 6.5 seconds). Such song disruption can likely result in the female being attracted to the satellite male, as revealed in the results of trials in which the phonotactic response of females was examined (histogram). The two pairs of bars on the left show that females responded mainly to the song of the main caller (black bars) but would occasionally move to the satellite's song (hatched bars), even when the latter song was at a lower intensity (66 dB). Females rarely responded to the satellite's song when it was the only song present (two right pairs of bars: compare the number of females responding [hatched bars] with the number of females showing no response [white bars]). Figures redrawn from Bailey and Field (2000).

Orchelimum species, the common and sword-bearing meadow katydids, ticking duels are even more frequent, being observed in 85 percent of interactions (Feaver 1977). The function of these duels is unknown but may be a way in which males assess the strength or body size of opponents before deciding whether to engage them in a fight. Ticks may be less localizable sounds that are useful in advertising a male's presence to another male while avoiding being tracked down by this rival.

Further support for the role of ticking in aggression comes from detailed descriptions of fights between males of the "high-ticking" meadow katydids, the sword-bearers and common meadow katydids (Morris 1971). An

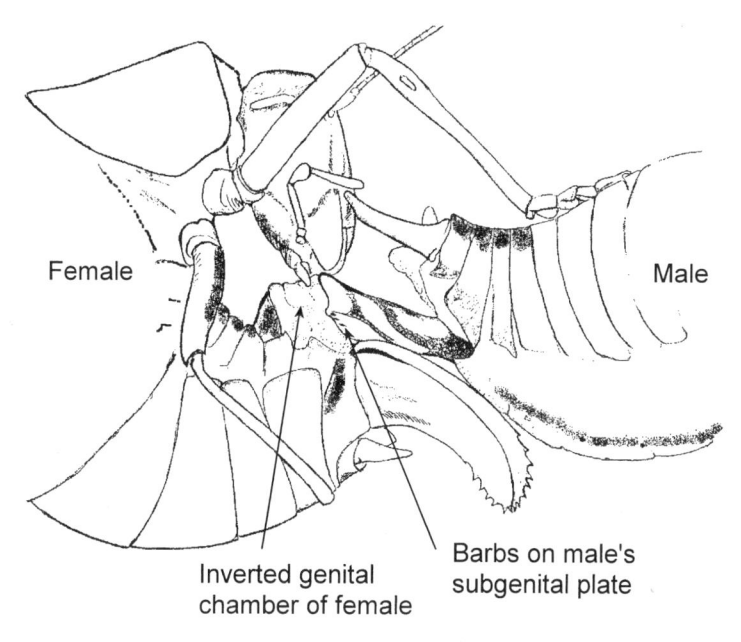

Figure 7.16. Removal of rival sperm by male *Metaplastes ornatus*. Barbs on the keel of the male's subgenital plate appear to catch the inside wall of the female's genital chamber and turn it inside out. The subgenital plate may simulate an egg in the genital chamber, thus causing the female to release stored sperm (of rivals) from the spermatheca. The Helversons supported this hypothesis by artificially moving an egg in the genital chamber and finding it covered with a large quantity of sperm. From Fig. 4 in Helversen and Helverson (1991).

intruding male was for the most part silent as he approached his calling neighbor, but occasionally he stridulated using only the tick sequence from the calling song. Tick sequences were especially common from both males as they closed in on each other. During fights the combatants grasped the underside of the opponent in a grapple that lasted up to 30 seconds. Males bit the exposed parts of the opponent's underside and often fell from the vegetation while fighting. Fighting and grappling between males has also been observed in other meadow katydids, cone-heads, and shield-backs (Morris 1971; Feaver 1977; Gwynne 1984c; Brush et al. 1985).

7.3.ii. Competition between ejaculates

After the acoustic interactions and the fights have ended, there is one final opportunity for competition between rival males to occur, but only in species in which females copulate with more than one partner. In these species the potential for competition between the stored sperm of rivals

(Parker 1970; Simmons and Siva-Jothy 1998) produces "cryptic sexual selection" on any trait that enhances a male's success in fertilizing ova (Eberhard 1996; Birkhead and Møller 1998). Chapter 6 described an intersexual component of cryptic sexual selection (see Table 7.1) whereby a more long-lasting spermatophylax meal allows a male to more fully inseminate his mate, and thus to achieve a higher chance of paternity (Wedell 1991). The classic example of cryptic intrasexual selection, the male dragonfly's penis "brush" that removes rival sperm (Waage 1979), is mirrored in at least one katydid, *Metaplastes ornatus* (Helversen and Helversen 1991). Before he transfers his spermatophore, the male of this European phaneropterine periodically inserts a specialized subgenital plate into the female's genital chamber , and moves the plate back-and-forth (Fig. 7.16). These genital movements are correlated with an approximate 85 percent reduction in the number of rival sperm in the spermatheca. The Helversens suggested that the movements of the male's subgenital plate remove rival sperm by stimulating the release of sperm from the spermatheca in the same manner as an egg passing through the genital chamber during fertilization.

Copulating pairs are known to attract predators (Gwynne 1989), so there is probably a cost to any additional time that the sperm removal movements of *M. ornatus* add to copulation. Earlier stages in the mating game are also risky endeavors. In this chapter we have seen that a large number of black-legged katydid males die during the first few days of competition. Does this mortality directly result from the competitive behavior of males? The next chapter addresses this question by examining the costs and risks of sexual competition.

8

The Hazards and Costs of the Mating Game

> *A nocturnal journey like this, at some twenty paces' distance, is a serious undertaking for the Cricket.... He dies a wretched death, forming a savoury mouthful for the Toad on his night rounds. His visit to the lady Cricket has cost him his home and his life. What does he care! He has done his duty as a cricket.*
>
> J. H. Fabre (1917), *The Life of the Grasshopper*
>
> *You're nobody 'till somebody loves you . . .*
>
> Dean Martin

8.1. Eavesdropping Flies

Fabre's take on the life and death of a male field cricket (*Gryllus* species) and the song lyrics represent the concept of fitness in evolutionary biology in which individual reproduction is the currency of success (fitness) even if survival is severely threatened. In the case of Fabre's cricket, the male pays the price in the necessary risks of seeking out mates because reproduction, not survival, is the focus of Darwinian selection. Activities and structures, usually in males, that function to secure mates often incidentally increase the risk of encounters with natural enemies (Darwin 1874; Thornhill and Alcock 1983). In the premating activities of the sexes, the typically competitive male usually takes the more risky role. For example, in most moths the female attracts the male using chemical signals, because pheromones appear to be difficult signals for most predators to detect (but see Eberhard 1977). But when the more risky acoustic or visual[1] medium is used to attract mates, as in most Orthoptera (and the occasional moth: Gwynne and Edwards 1986), males usually assume the signaling role and females move silently to them (Alexander and Borgia 1979; Thornhill and Alcock 1983; Sakaluk 1990; Zuk and Kolluru 1998). Mate-seeking movements are rarely risk free, as Fabre noted, but in singing insects calling is probably more hazardous given the array of predators and parasitoids that are known to eavesdrop on prey sounds (Burk 1982; Sakaluk 1990) (see Chapter 4). As inferred by Fabre, even the male crickets that opt for the silent, less risky way of contacting mates (see Cade 1979) can still become a predator's meal. In this chapter, I examine sexual differences in the risks

---

[1] Light emitters such as fireflies can also attract predators, including frogs. Lloyd's (1973b) review recalls two particularly interesting reports (one by Charles Darwin's grandfather Erasmus) of frogs hunting for glowing fireflies seen to gobble up live coals.

and costs involved in the mating of katydids, particularly the expectation that the sex competing for mates will carry the burden of these costs (Darwin 1874; Trivers 1972).

Studies of crickets and katydids have shown that singing carries a death risk (Burk 1982; Sakaluk 1990). Ormiine flies are one of the best-known eavesdropping killers of these nocturnal orthopterans, as we saw in Chapter 4. These tachinid parasitoids deposit their lethal larvae on or near singing males, and the larvae burrow into the host's tissues. The flies are almost exclusively a male problem, although a female may occasionally suffer collateral damage when both she and a larva-dropping fly encounter a caller at about the same time (Robert and Hoy 1994). One ormiine, *Homotrixa alleni*, has devastating effects on the populations of male austrosagines, *Sciarasaga quadrata*, studied by Geoff Allen (University of Tasmania). The katydid host has the short wings, fat abdomen, and large head reminiscent of a large field cricket (Fig. 8.1) and is found in a very small geographical area of southwestern Australia, on the broad peninsula between Cape Leeuwin and Cape Naturaliste that juts out into the Indian Ocean (Fig. 2.35) (Rentz 1993). Within the peninsula, the katydid's strange toadlike call is a familiar evening sound in late spring along the wind-blown coastal heathland and in the open woodland gardens of nearby holiday homes.

The first *Sciarasaga* songs are heard in late October, the middle of spring in this part of the world, and the evening songs continue until the last of the year's cohort of males falls silent by February (Allen 1995b). Few males are silenced by senescence, however; the fate of some 90 percent is death from the activities of up to 16 *Homotrixa* maggots (Allen 1998). Burk (1982) reported similar kill rates of the American *Neoconocephalus triops* by *Ormia lineifrons*.[2] The trauma of the mature maggots bursting from the male's body (Fig. 8.2) kills him after about 2 weeks of infection, whereas unparasitized males can survive for many more weeks. Males are free of fly attack during the first 2 weeks of the singing season (Fig. 8.3), probably because the young flies need this time to complete their first pregnancies, involving internal egg-hatching and development to the first-instar stage, which is ready to be larviposited near a songster. As Allen (1995a) pointed out, the timing of this early respite from attack with respect to when the katydids mate determines the impact of the parasitoids on the host population. If most of the male katydids (in Fabre's words) "have done their duty" as mates during the 2-week respite period, *Homotrixa*-caused death would have a negligible effect on the fitness of the hosts. The data available suggest, however, that katydid mating goes on well after this 2-week period. Indeed 75 percent of the females that Allen sampled 4 weeks into

---

[2] The number of maggots hosted by a katydid can be determined by the respiratory hole that each parasite punches through its host's abdominal integument (Allen 1995).

Figure 8.1. A male *Sciarasaga quadrata* from the top (**a**) and front (**b**).

the season were still virgins, so there appeared to be plenty of opportunities for parasitoid flies to eliminate certain males before the katydids have had a chance to mate.

Which of the callers is more likely to be eliminated, or to put the question another way, how can a caller avoid these predators before mating? One tactic is to become a moving target, perhaps to outmaneuver the larvipositing flies or to escape contact with any nearby larvae, which will die within 2 hours if they do not contact a host (Allen et al. 1999). Interestingly, in this heavily parasitized katydid, calling males do move frequently and as a consequence, do not show the regular spacing of ter-

Figure 8.2. Two ormiine fly (*Homotrixa alleni*) larvae (just under the right hind leg at the top of the figure) emerging from the silenced corpse of a male *Sciarasaga quadrata*. Photo by G. Allen.

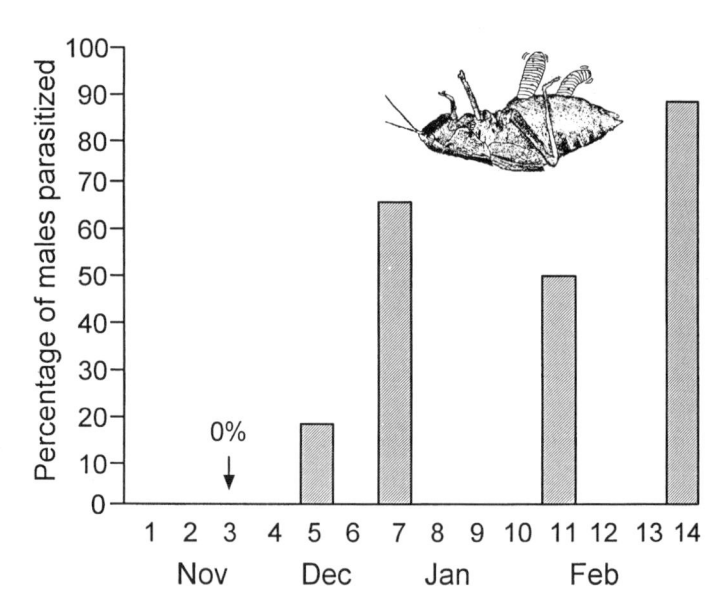

Figure 8.3. The proportion of *Sciarasaga quadrata* males dying from ormiine fly attacks increases over the season. Redrawn from Allen (1995a).

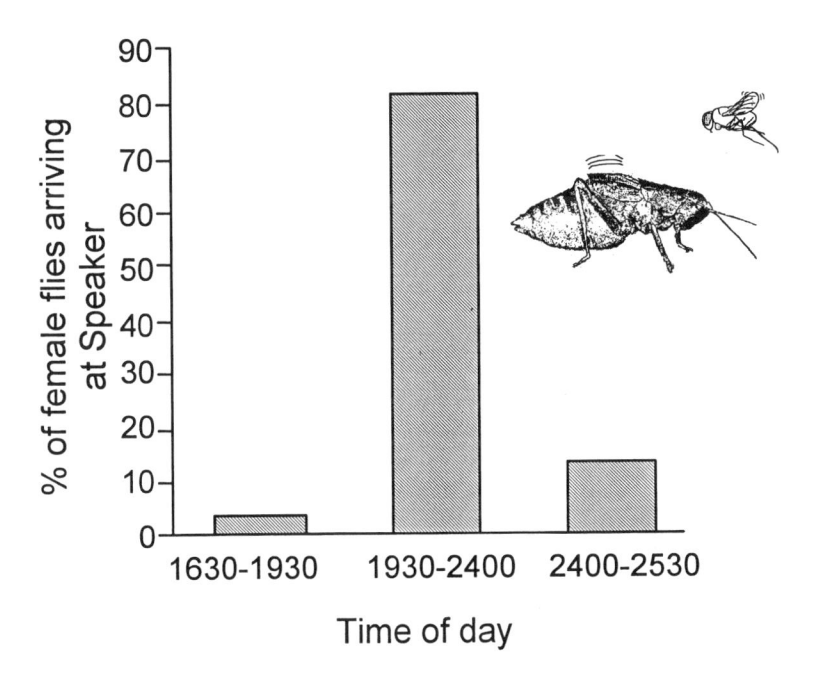

Figure 8.4. Ormiine fly predators are active at the same times that their host, *Sciarasaga quadrata*, sings: after sunset (approximately 1900h). Redrawn from Allen (1998).

ritorial callers that is common in other tettigoniids (see Chapter 7) (Allen 1995b).

Sciarasaga males might also reduce the risk of attracting flies via adaptive modifications of their singing. First, let's consider when they sing. Males might call when flies are less active or might reduce each evening's bout of calling (see Zuk et al. 1993). There is some evidence that these primarily nocturnal katydids may have adaptively shifted the start of their daily singing to several hours before the parasitoids begin flying (Fig. 8.4). However, although there are consistent, repeatable differences between males in their daily calling patterns, Allen (1998) could find little support for the prediction that callers who were still maggot free late in the season were those that put in shorter calling shifts each evening. One explanation for this result is that males putting in long shifts reduce their risk in the long term by logging less singing hours over the whole season (Allen 1998). This seemingly paradoxical result could occur if long-shift callers achieve a high copulation rate and each mating is followed by a several-day silent refractory period during which spermatophore glands are recharged, which is typical of katydids, such as *Sciarasaga* species, with large spermatophores (see Chapter 6).

In addition to how long they call, katydids might reduce the risk of attracting flies by making their calls less conspicuous to these natural enemies. Römer and Bailey (1998) suggested that the low-frequency song of *Sciarasaga* is an adaptation to avoid the sound frequencies to which the fly's ear is most sensitive (Fig. 8.5). The song is pitched well below that noted for other katydids (including other austrosagines; see Fig. 5.9). In fact, the song's frequency is much lower than the most sensitive part of the "tuning curve" of the katydid's own ear, necessitating a mechanism that partially occludes the auditory spiracle (see Chapter 5) and thus allows the ear to be more sensitive to conspecific song (Fig. 8.5) (Römer and Bailey 1998).

In addition to song frequency, *Sciarasaga* males might avoid flies by making the structure of their songs less detectable by these eavesdroppers.

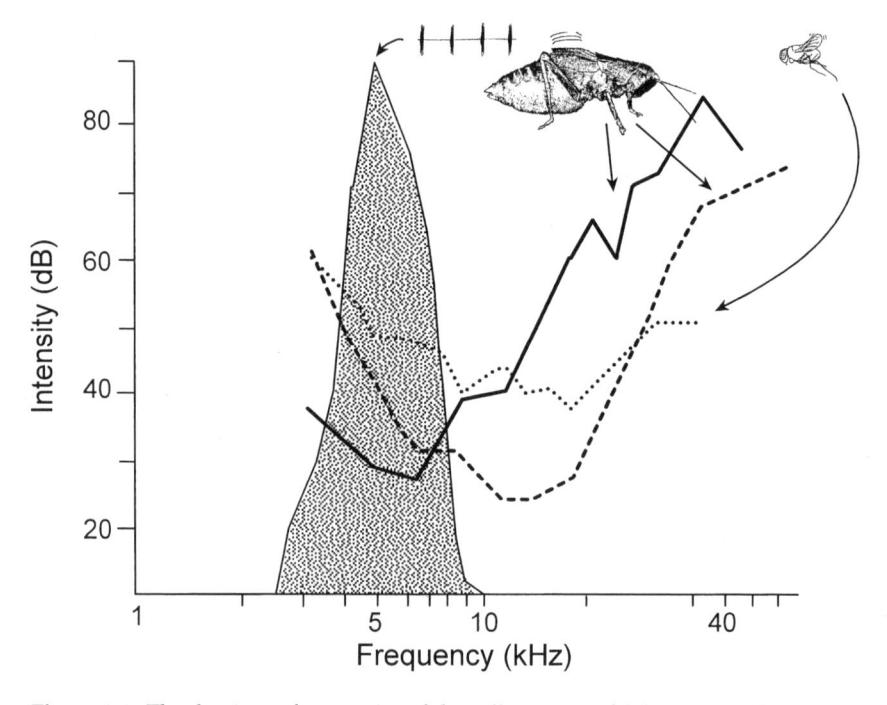

Figure 8.5. The dominant frequencies of the calling song of *Sciarasaga quadrata* (stippled area) are much lower than that for other katydids, owing to the unusually fleshy forewings. As a result, song frequencies are shifted below the most sensitive range (lowest parts of each curve) in the ears of both listening parasitoid flies (*Homotrixa alleni*) (dotted line: from Lakes-Harlan et al. 1995) and other *Sciarasaga* katydids (when the katydids have open auditory spiracles: dashed line). *S. quadrata*, however, can increase its sensitivity to conspecific song by partially closing its auditory spiracles, as shown by the shift in the "tuning curve" for katydids with open spiracles (dashed line) to those with experimental occlusions of the auditory spiracles (continuous line). Redrawn from Römer and Bailey (1998).

Because males surviving late in the calling season produced significantly longer chirps, Hunt and Allen (1998) suggested that shorter-chirping males may have attracted flies and thus been killed off. Interestingly, there are other signs that males with shorter chirps are generally less fit; short chirpers are more asymmetrical in hind leg length (Fig. 8.6), and asymmetry is usually argued to be an indication of low fitness (see Chapter 7). Field samples of males over three seasons and within each season showed that asymmetrical males were more heavily parasitized (insect morphological development is complete by the adult stage so the adult-infecting parasitoids were not the cause of the asymmetry) (Hunt and Allen 1998). But the conclusion that symmetrical *Sciarasaga* males of apparently high fitness produce songs that are least attractive to flies appears to conflict with theory, since conspicuous, easily located males should often suffer more costs when advertising for mates (Darwin 1874; Burk 1982; Sakaluk 1990). Furthermore, another study of ormiine parasitism involving the fly *Ormia ochracea* showed that male field crickets (*Gryllus lineaticeps*) able to produce longer-chirp songs were more attractive to both females and flies (Wagner 1996). However, Allen (2000) suggested that the opposite may be

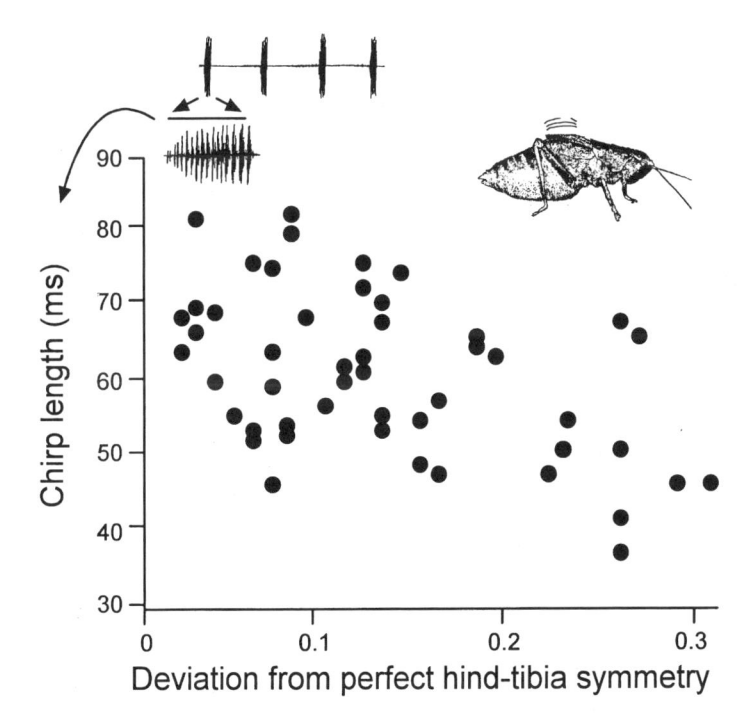

Figure 8.6. More asymmetrical *Sciarasaga quadrata* males tend to sing shorter-chirp songs (perfect symmetry of hind tibia length = 0). The trace of the song in the top part of the figure shows chirp length. Redrawn from Hunt and Allen (1998).

true for *Scarasaga*'s parasitoid *H. alleni*; these flies may have a sensory bias for short chirps. Alternatively, the explanation for the survival of *Sciarasaga* males with long-chirp songs may lie in the accumulated time-out from singing due to the large spermatophore costs for this katydid, a cost that is not present in *Gryllus*. Perhaps the old, symmetrical surviving *Sciarasaga* males with significantly longer chirps are not less attractive to flies but have simply outlived their rivals because, as noted earlier, they sang less over the season, owing to an accumulated recovery time from frequent matings.

Testing hypotheses about the relative attractiveness of males of differing fitness to hunting flies requires some sort of experimental manipulation of the system. Lehmann and Heller (manuscript) conducted a series of experiments on *Poecilimon thessalicus* to examine whether males attractive to females also attracted more flies. Like *Sciarasaga* males, a male *P. thessalicus* puts up with both the costs of producing spermatophores (each one weighing up to 30% of male weight) and the attraction of ormiine flies. Lehmann and Heller ranked the sexual attractiveness of males by seeing how many virgin females responded to each caller. Males of differing rank were then exposed to attacks by ormiines, *Therobia leonidei*, after release into a large field enclosure from which all female katydids had been removed, to eliminate mating as a confounding variable. Males that scored as highly attractive to females also attracted significantly more flies (Fig. 8.7). It is possible that the more attractive males were those with longer song elements; indeed, another field experiment by Lehmann and Heller (1998) showed that a *Poecilimon* species with multiple-syllable songs was

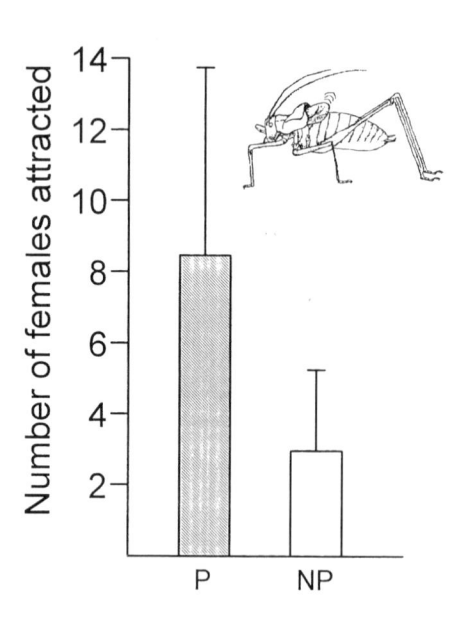

Figure 8.7. *Poecilimon* males that attracted parasitoid (P) ormiines, *Therobia leonidei* (hatched bars: mean ± standard devation), had been shown previously to attract significantly more females than unparasitized males (NP). From Lehmann and Heller (manuscript). *Poecilimon* redrawn from Harz (1957).

more heavily parasitized by *T. leonidei* than was a species from the same habitat with monosyllabic songs (Table 8.2). However, as the researchers did not directly manipulate the male signals in either study, it remains possible, although unlikely, that mates and enemies were attracted to some nonacoustical aspect of male phenotype (note, however, that male body size was not involved because the experiments did control for this variable) (Lehmann and Heller, manuscript).

The menace of ormiine flies to calling male katydids appears to occur worldwide, because these flies have been reared from a number of different host species from Western Australia, eastern and western Europe (Lakes-Harlan 1992), and North America (Burk 1982; Shapiro 1995) (see Chapter 4). Future studies will undoubtedly find katydid victims of this parasitoid on other continents, but even the katydids free of these enemies are expected to suffer risks and costs that silent females avoid. Energetic costs of calling include increased metabolic rates, which have been estimated to be up to 1.6 times the resting rate in a listroscelidine, *Requena verticalis* (Bailey et al. 1993), and a massive 15 times the resting rate in a cone-head, *Neoconocephalus robustus* (Stevens and Josephson 1977). These direct calling costs might come not only from use of the male's energy reserves but also from any reduction in his capacity to avoid predation, parasitism, or disease that is caused by the energy loss.

8.2. Eavesdropping Bats

In addition to ormiine flies, there are a host of other listening predators that use song to locate prey (Burk 1982; Zuk and Kolluru 1998). Nocturnally hunting vertebrates with a well-developed sense of hearing are a particular threat, as revealed in the attraction of Spanish lizards (Table 8.1) to speakers playing the calls of prey bradyporines, *Uromenus stali* (Bateman 1995), and in observations of bats homing in on katydid calls. Katydid-hunting bats tend to be "gleaners" that cruise over vegetation while listening for prey sounds rather than echolocating and capturing insects on the wing (Fig. 8.8).

Calling male katydids exposed on territorial perches should be easy pickings for these gleaners, as shown by Spangler (1984), who saw desert bats swooping in toward *Insara covillae* males calling from exposed perches atop creosote bushes. In response to these swoops (probably when katydids heard the bat echolocation squeaks), the vulnerable katydids quickly fell silent. Experimental examples of bats homing in on katydid sounds come from several studies. The Panamanian bat *Tonatia silvicola* (Fig. 8.8) was attracted to the calls of caged katydid callers but ignored the calls of a frog, whereas the sympatric frog-eating bat *Trachops cirrhosus* showed the expected reversal in these prey-sound preferences (Tuttle et al. 1985). In

Table 8.1. Call differences in Panamanian katydids: the low-risk, sporadic singing in six neotropical false-leaf katydids (Pseudophyllinae) compared to the continuous signals of two *Neoconocephalus* cone-heads (Conocephalinae) found in forest clearings

Species	Song length (sec)	Song frequency (kHz)	No. of calls/5 min
Cocconotus wheeleri	0.4	26	6.5
Docidocercus gigliotosi	0.1	22.4	21.6
Xestoptera cornea	0.5	27.5	4.4
Unknown species "a"	0.5	17.4	5.1
Unknown species "b"	2.7	17.4	18.8
Idiarthron major	0.1	23–27	2.1
Neoconocephalus affinis	Continuous	18	Continuous
Neoconocephalus punctipes	Continuous	16	Continuous

Adapted from Fig. 1 in Belwood and Morris (1987) and Table 4.6 in Belwood (1988).

Figure 8.8. A neotropical bat, *Tonatia silvicola*, preying on a katydid. Photo by Merlin D. Tuttle, Bat Conservation International.

the Northern Hemisphere, big brown bats, *Eptesicus fuscus*, appear to be less discriminating, as they can be attracted to tape-recorded sounds of both true katydids, *Pterophylla camellifolia*, and cricket frogs, *Acris crepitans* (Buchler and Childs 1981). In Europe, the lesser mouse-eared bats, *Myotis blythii*, prey more heavily on singing *Pholidoptera griseoaptera* males (Plate 7) than on experimentally muted males (Heller 1997). Similarly, back at the Panama study site, Belwood and Morris (1987) attracted several species of insectivorous bats into nets baited with the caged singing katydids, *Ischnomela pulchripennis* and *Scopiorinus fragilis*. The predators were attracted to calling males in cages but not to caged females. Neotropical gleaning bats are incapable of detecting motionless females but can home in on incidental sounds produced by any katydid as it moves through leaves or rustles dry vegetation on the forest floor (Belwood 1988). In fact, movement may turn out to be highly risky for the pseudophylline *Xestoptera cornea* female, as female prey made up about half of the hundreds of the remains of katydids that Belwood (1988) collected beneath the roosts of katydid-hunting bats, *Micronycteris hirsuta*. This finding refutes the prediction that bats should prey more on *X. cornea* calling males than on females.

The factors responsible for the large number of female katydids in this bat's diet are unknown (Belwood 1988). However, the chance of a *X. cornea* male ending up in a bat's jaws is clearly decreased by his extremely reduced calling schedule compared to most katydids; *X. cornea* males broadcast only four half-second songs every 5 minutes (Table 8.1). Such secretive signaling turns out to be typical for neotropical forest callers, as first noted by Rentz (1975), who hypothesized that a very low "duty cycle" (proportion of time spent calling: Table 8.1) was an adaptation to prevent bats from locating the katydid's call. In fact, males of many neotropical katydids appear to replace much of their sound signaling with the more elusive tremulation signals (shaking of the vegetation: see Chapter 5), which can elicit "tremulation answers" from females (Belwood 1988).

The antibat hypothesis for low-duty cycles has received some support in findings that katydids in the relatively bat-free environments of both paleotropical forest (Heller 1995) and neotropical forest clearings (Belwood and Morris 1987) sing much more frequently than do neotropical forest singers (43 species: see Table 8.1 for examples). The four singing species from clearings had virtually continuous songs. More importantly, however, there was direct evidence that reducing duty cycles does make singing a safer endeavor; Belwood and Morris (1987) showed that bats took over half an hour to locate the low-duty cycle calls typical of neotropical forest singers and only about half a minute to locate the frequent-calling song produced by katydids in forest clearings.

Differences in singing behavior found by Belwood and Morris (1987) may be confounded by phylogenetic relationships, because the lists of forest and clearing species differed in taxonomic representation. All four of the katydids from clearings were grassland cone-heads (three *Neoconocephalus* species plus *Bucrates capitatus*), whereas the 43 forest species were all pseudophyllines except for two copiphorines (*Copiphora*). Importantly, mating differences between the two groups raise an alternative explanation for between-habitat differences in the duty cycle of songs and in the response of bats to the songs: perhaps males of forest species sing less and decrease their risk because of coyness associated with a large investment in mating (see Chapters 1 and 9). Relative to the very small spermatophore meals typically produced by clearing species (about 2% of male weight: Table 6.3 and Fig. 6.7), the forest species studied by Belwood (1988) produced quite large contributions: four pseudophylline species and one *Copiphora* (see also Morris 1980) showed a postmating drop in weight averaging 17 percent of male body weight (ranging from 9.5% to 25.3%: Table 6.3). The low-cost matings typical of cone-heads in forest clearings are usually associated with male-male competition, which is expected to result in longer bouts of calling for hard-to-get females (see Chapter 7). In contrast, large nutritious spermatophores may mean an abundance of responsive females seeking these nuptial meals. In such situations males can afford a low-risk, coy signal by drastically reducing their call duty cycles (Gwynne and Bailey 1999). I briefly return to the topic of reduced song output as a form of male coyness at the end of this chapter and explore it in detail in Chapter 9.

8.3. Factors That Affect Mating Risk

Along with making sounds, the other "dead giveaway in nature" is moving (Ammons 1992). Calling male katydids on visually conspicuous perches in vegetation are often exposed to predators, while females and juveniles tend to be more concealed (Feaver 1977; Spangler 1984; Brockmann 1985). Good examples of this are the meadow katydids, which include the black-sided and short-winged katydids (*Conocephalus* species) that were key subjects in Chapter 7. The riparian competition arenas for these *Conocephalus* males can be under daytime scrutiny by hunting digger wasps, *Tachytes validus* (Sphecidae), if the sandy nesting habitat used by the wasps is within flying distance. Hundreds of paralyzed black-sided and short-winged katydids can be transported by air back to a single colony of 40 to 50 burrowing wasps (Kurczewski and Ginsburg 1971) (Fig. 8.9).[3] As predicted, *T. validus* wasps take more *Conocephalus* males than females (Table 8.2).

Other studies of *Conocephalus* katydids as prey of digger wasps also showed a male bias, most notably by *Sphex* species (Brockmann 1985; H. J. Brockmann, pers. comm. 1982, cited in Gwynne and Dodson 1983; Belavadi and Mohanraj 1996). However, do the prey data from other studies of sphecids as well as other katydid predators support the hypothesis that sexually competitive behavior increases the risk of death? Table 8.2 provides data from studies of katydids that were killed by various vertebrates and insects. The study results are summarized in Table 8.3 as a test of the risk-taking hypothesis. This hypothesis predicts a sex bias in adult katydid prey when the following conditions are fulfilled:

1. Predators are hunting at the same times that sexually competitive individuals are actively seeking mates (right column of Table 8.3). No sex

---

[3] Although it has been suggested that the wasps may home in on song (Vickery and Kevan 1985), it seems unlikely given that sphecids appear to use other sensory modalities to find prey (Evans 1966).

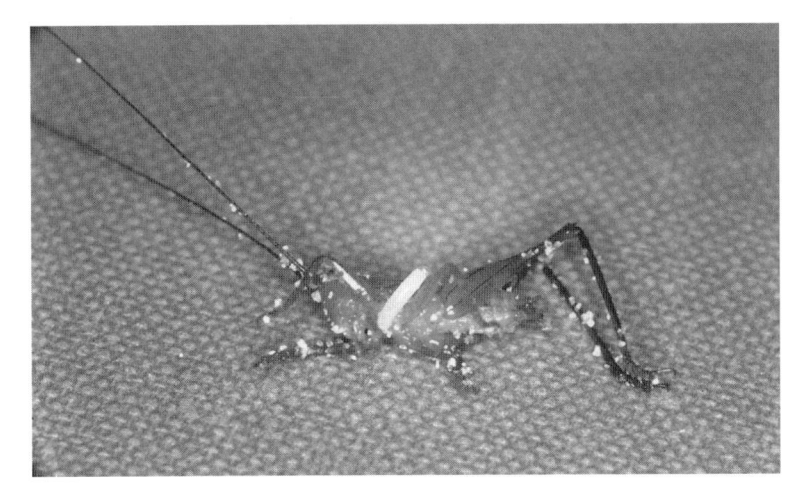

Figure 8.9. Egg of a digger wasp, *Tachytes*, on a paralyzed larval conocephaline excavated from a wasp's burrow. Photo by Howard E. Evans.

bias due to mating activity is expected if hunting occurs at times of prey inactivity, such as when neotropical tamarin monkeys take hidden or camouflaged pseudophyllines (Nickle and Heymann 1996) or day-active sarcophagid flies dispatch the day-inactive *Sciarasaga* (Allen and Pape 1996).

2. Information on the mating system of the katydids being used as prey indicates which of the sexes is sexually competitive. For example, the male-male competitive system in meadow katydids and grass-inhabiting cone-heads (see Chapter 7) predicts a male bias in prey, as we have already noted in the case of wasps preying on *Conocephalus*. Included in Table 8.2 is male-biased predation by ormiine flies on a number of different katydids, including cone-heads *N. triops* (Burk 1982) and *Mygalopsis pauperculus* (Barraclough and Allen 1996) as well as meadow katydids, *Orchelimum* species (which can be sexually active both day and night) (Shapiro 1995). The mating systems of other katydid hosts of ormiines, *Sciarasaga* and *Poecilimon* species, are not well known, but see below for a further discussion of *Poecilimon* mating.

Recall, however, that male katydids are not always competitive; in some populations of Mormon crickets (see Chapter 1), females competed while males were noncompetitive and coy. The coy males called only briefly because it took very little time to attract one or more of the many available females. This reversal in the roles during mating behavior predicts a concomitant reversal in risk taking; more females should be killed than males (Gwynne and Dodson 1983; Elgar and Wedell 1996). Female Mormon crickets did appear to take more risks than males. Males were often hidden

Table 8.2. Biology and sex-bias of katydids killed by predators and parasitoids

Predator	Active (D or N)?	Adult prey	Active (D or N)?	Prey mating system	Duty cycle of prey song	Sample size of prey	% of male prey	Author
Saguinus mystax (a tamarin monkey)	D	Various neotropical species	N	Unknown		132	46	Nickle and Heymann 1996
Saguinus fuscicollis (tamarin)	D	Various neotropical species	N	Unknown		?	34	Nickle and Heymann 1996
Tyto alba (barn owl)	N	*Neoconocephalus triops* (conehead) (Florida)	N	Male coneheads compete (Chapter 7)	Very high (Walker and Greenfield 1983)	5	100	Horner et al. 1974
Otus scops (scops owl)	N	*Tettigonia viridissima* (western Europe)	Mainly D	Males compete (Latimer and Sippel 1987)	Very high (Ragge and Reynolds 1998)	183	54	Heller and Arlettaz 1994
Otus scops (scops owl)	N	*Platycleis albopunctata* (western Europe)	N & D	Unknown	High (Ragge and Reynolds 1998)	154	20	Heller and Arlettaz 1994
Micronycteris hirsuta (bat)	N	*Xestoptera cornea* (Panama)	N	Unknown	Very low (Table 8.1)	776	49	Belwood 1988
Tachytes validus (digger wasp)	D	Two *Conocephalus* species (eastern North America)	D	Males compete (Chapter 7)		Two samples, 91 & 80	60 & 70	Kurczewski and Ginsburg 1971

8.3 Factors That Affect Mating Risk

Species	D/N	Notes	Signal	Sample	%	Reference
Sphex argentatus (digger wasp)	D	Probable male competition (Chapter 7)		?	76	Belavadi and Mohanraj 1996
Sphex ichneumonius (digger wasp)	D & N			Approx. 440	53 (sig)	Brockmann 1985
Many different species (42% meadow katydids) (eastern North America)	D	Male competition in the meadow katydids				
Mormon crickets (*Anabrus simplex*) (western North America)	D	Female competition (Chapter 1)		34	12 (sig)	Gwynne and Dodson 1983
Neoconocephalus robustus (U.S.)	N	Male coneheads compete (Chapter 7)	Very high (Walker 1975)	?	Male bias	Nutting 1953
3 *Orchelimum* spp. (eastern U.S.)	N	Male *Orchelimum* compete (Chapter 7)	Very high (Morris and Walker 1976)	3	100	Shapiro 1995
Neoconocephalus triops (conehead) (eastern U.S.)	N	Male coneheads compete (Chapter 7)	Very high (Walker and Greenfield 1983)	Many large samples	100	Burk 1982; Shapiro 1995
Poecilimon mariannae (Greece)	N		Medium (approx. 1 sec of song every 2.5 sec)	Many large samples	100	Lehmann and Heller 1998

Table 8.2. (continued)

Predator	Active (D or N)?	Adult prey	Active (D or N)?	Prey mating system	Duty cycle of prey song	Sample size of prey	% of male prey	Author
Therobia leonidei (ormiine fly)	N	*Poecilimon veluchianus* (Greece)	N	Similar ratio of mateable males and females indicate equality of sexual competition (Heller and Helversen 1991)	Low (one 80-ms chirp every 2.5 sec)	Many large samples	100	Lakes-Harlan 1992; Lehmann and Heller 1998
Homotrixa alleni (ormiine fly)	N	*Sciarasaga quadrata* (south western Australia)	N	Probable that males compete (G. R. Allen, pers. comm. 1999)	High (Allen 2000)	Many large samples	100	Allen 1995a; Allen 1998
Homotrixa alleni (ormiine fly)	N	*Mygalopsis pauperculus* (south western Australia)	N	Male *Mygalopsis* compete (Dadour and Bailey 1985)	Very high (Bailey 1979)	2 (of 3)	100	Barraclough and Allen 1996

D = day; N = night; sig = significant sex bias.

Table 8.3. A comparative examination of the hypothesis that the competitive sex is at greater risk of predation (when predator and katydid prey share the same activity periods [right column])

	Prey inactive when predator is active	Prey and predator active at same time
Prey bias toward the sex that shows greater mate competition		1. *Ormia* fly on *Neoconocephalus* male[a] 2. *Homotrixa* fly on *Mygalopsis* male 3. *Tachytes* wasp on *Conocephalus* male 4. *Sphex* wasp on *Conocephalus* male[a] 5. Owls on *Conocephalus* male[HDC] 6. *Palmodes* wasp on *Anabrus* female[b]
No prey bias toward the sex that competes	Tamarin on katydids	Owls on *Tettigonia*[HDC]
Prey sex bias when sexes appear to be similar in sexual competitiveness		*Therobia* fly on *Poecilimon* male
No sex bias in prey (competitive sex unknown)		Bat on *Xestopera*[LDC]
Male sex bias (competitive sex unknown)		*Homotrixa* fly on *Sciarasaga*
Female sex bias (sexually competitive sex unknown)	Tamarin on katydids	Owls on *Platycleis*

From data in Table 8.2.
HDC = high duty cycle of male prey's song; LDC = low duty cycle of male prey's song.
[a] Several studies of different host species and several different prey species (see Table 8.2).
[b] The population of *Anabrus* preyed on was role-reversed (female-female competition).

in or near plants and called for very brief periods, while the combative females were seen to hop over open ground toward the caller and often fought with rivals. These movements exposed the females to a similar natural enemy experienced by *Conocephalus* males, digger wasps hunting sexually active individuals. The Mormon cricket hunters, *Palmodes laeviventris*, do not fly with their catch but drag their corpulent prey over the ground to burrows nearby. As expected, there were significantly more females than males in excavated *Palmodes* nests (Gwynne and Dodson 1983) (Fig. 8.10, Table 8.2). There are other hypotheses to explain the preponderance of females being killed. Wasps may select female prey to provide larger meals for their fossorial larvae. This, however, seems to be an unlikely general explanation because other sphecids rarely show a prey bias toward females (Table 8.2) (Gwynne and Dodson 1983).

Taken together, the katydid prey data show a nonsignificant trend (χ^2 statistical test) that the competitive sex is killed more often when predator and prey are active at the same time (first eight cases of right column in Table 8.3): for six of eight cases of katydid prey where there is decent information available on sexual competition in the prey (numbers 1 to 6), there is a prey bias toward the competitive sex. In the remaining two cases, the bias in prey did not agree with available knowledge about the direction of sexual competition.

Figure 8.10. The excavated burrow of a digger wasp, *Palmodes laeviventris*, showing seven paralyzed Mormon crickets in the exposed subterranean chambers. Most of these prey are females. Scale bar = 10 cm.

There are, however, a number of problems with this comparative analysis. First, there may be other explanations for the association between sexual competition and predation that result from taxonomic (phylogenetic) biases in the data. In particular, all five examples of male biased sex ratios where males compete involve conocephaline prey, whereas each of the other categories represents a diversity of katydid groups. Perhaps some unknown aspect of the life history of conocephalines causes males to be more susceptible to predation. There is also a phylogenetic bias in the types of predators; most of the data come from studies of digger wasps and ormiine flies.

Second, the test is conservative in including only one prey genus for each predator genus to test the risk hypothesis. In these examples predation risk can be directly related to some aspect of sexual competition. For instance, parasitoids orienting to male song or digger wasps responding to movements of competitive individuals (the case for all six examples with a prey bias to the more competitive sex). In some of the other examples, the connection between predation and prey behavior is not as obvious, and more importantly, risky behavior by females tends to be overlooked. For example, owls preying on *Platycleis* species (Heller and Arlettaz 1994) may respond to the conspicuous sounds from both sexually active males and ovipositing females (K.-G. Heller, pers. comm. 1999). This raises the question of the riskiness of other sorts of reproductive activities compared to activities associated with sexual competition. Perhaps we can address this question by examining sexual differences in survivorship.

8.4. The Survivorship of the Sexes

Studies of a sex bias in the risk of predation in Table 8.3 examine the effects of only one predator or parasitoid. However, in nature, death usually comes from a plethora of hostile forces. For example, cone-heads, *N. triops*, and oak katydids, *Meconema thalassinum*, have been observed as victims of both insects (Sismondo 1978; Burk 1982) and birds (Smith 1972; Horner et al. 1974), and the well-studied *Poecilimon* species (Heller and Helversen 1990), Mormon crickets, and coulee crickets (see Chapters 1 and 4) are known to fall victim to a spectrum of natural enemies.

The total mortality effects on the sexes can be assessed by examining survivorship data. Two studies of katydids—one on a species with male-male competition (meadow katydids: Feaver 1977) and the other on a species with more similar sexual selection on the sexes (*Poecilimon*: Heller 1992)—provided such data. Feaver's (1977) mark-recapture study of black-legged meadow katydids, *Orchelimum nigripes*, followed cohorts of males and females from when the insects were first marked (with small paint dots) 2 days after their adult molt. The death rates of both sexes was very high (greater than 50%), but in the first 6 days after marking, females had a significantly higher survivorship rate than did males. This result might be predicted for katydids with this male-male competitive mating system because during this 6-day period, the young males engage in hazardous behavior such as moving between territories, interacting with established callers, and at the end of the period, starting to call themselves for the first time (see Chapter 7). However, much of the mortality suffered by young adult black-legged katydids seemed to have little to do with adult risk-taking; up to 60 percent of field-collected young black-legged adults of both sexes died in the lab from horsehair worms (Feaver 1977), parasites that are acquired during the nymphal stages (see Chapter 4). It seems unlikely, although not impossible, that nymphal males are at greater risk of parasitism than females because of some sort of male-specific activity in preparation for the adult mating game. As Feaver (1977) did observe orb-weaving spiders catching black-legged katydids, some of the male-biased mortality may be due to the mobile young adults being killed by other enemies.

What about the mortality patterns for older (calling) *O. nigripes* males and similarly aged females? After all, it is at this stage that males are at risk for attracting ormiine flies (Shapiro 1995). Feaver (1977) found that the older males appeared to survive much better than the females. However, she cautioned that this result may be due to sexual differences in "catchability," which would cause differences in Feaver's recapture rates. Although young females were caught more easily as they moved around the plant tops between potential mates (see Chapter 7), older females (older than 6 days as an adult) became much more secretive, possibly after they had mated (recall that black-legged females mate only once: see Chapter 7).

The rarity of matings by female black-legged katydids appears to be a main cause of the male-male competitive mating system; the excess of sexually active males relative to the number of sexually available females sets up a stiff competition among rival callers (see Chapter 7). In contrast, female *Poecilimon* katydids take on many mates, often copulating as much as once a day (*P. affinis*: Heller and Helversen 1991), as they seek the nutritious spermatophore meals typical of this genus (see Chapter 6). The mating *Poecilimon* male can copulate again after 1 to 4 days (see Chapter 6). This fact and the frequent mating by females result in similar numbers of males and females being available for mating. Indeed, estimates at certain times of the season indicate that there are more sexually available females than males (Heller and Helversen 1991), predicting a low level of male-male competition. Given the possible shortage of males, it may pay for *Poecilimon* females to be less discriminating in mating, much like Mormon crickets (see Chapter 1). Heller at al. (1997) found some evidence of this in the indiscriminate response of *P. ornatus* females to the songs of conspecific males.

If sexual selection on male and female *Poecilimon* is similar, then theory predicts that the risks of death associated with mate acquisition should be similar for both sexes. Heller's (1992) 2-year mark-recapture study examined the survival rates for *P. veluchianus* and *P. affinis* males and females. These two species are similar in habitat use and morphology (both are short-winged and flightless) but differ in their pair-formation system: *P. veluchianus* has the typical katydid system of the male call attracting the female, whereas in *P. affinis*, the female answers the male call and the male moves toward the female sound (see Chapter 5).

The mortality-pattern results were mixed. In both years of the study, the mean survival rate over successive 3-day sampling periods for *P. affinis* showed a tendency for females to survive better than males (although the differences were not statistically significant), whereas the reverse pattern occurred in *P. veluchianus*, with the male survival rate exceeding that of the females (although the difference was statistically significant only in year 2) (Fig. 8.11). Heller's conclusion was that more *P. affinis* males than females died because males of this species suffered the dual risk of calling and moving (recall that the female produces a brief answer call, then stays put to await the male). It is possible, however, that the species-specific pair-forming systems in these insects and the subsequent differences in risk are ultimately the results of species differences in sexual selection pressures on males and females. *P. veluchianus* males may be less competitive than *P. affinis* males and thus take fewer risks because there are few *P. veluchianus* males (and lots of females) available for mating, owing to sexual differences in remating frequencies: the modal time for *P. veluchianus* males to recover from spermatophore production (4 days) is twice that for *P. affinis* males, whereas the intermating interval for females of the two species is reversed (*P. affinis* females remate more quickly) (Heller and Helversen 1991).

8.5. Conclusions and Future Research

The hypothesis that the sex competing for mating opportunities will be subject to greater risk has received some support from studies of individual predator species (Table 8.3). The data are, however, slim. More studies are needed on the effects of individual natural enemies on species for which we have some knowledge of mating behavior that may expose the prey to risk. Studies such as these might answer questions such as why have neotropical katydids under a heavy predation threat from bats apparently evolved much lower calling rates than katydids suffering similar, if not higher, predation rates by eavesdropping ormiine flies? One suggested

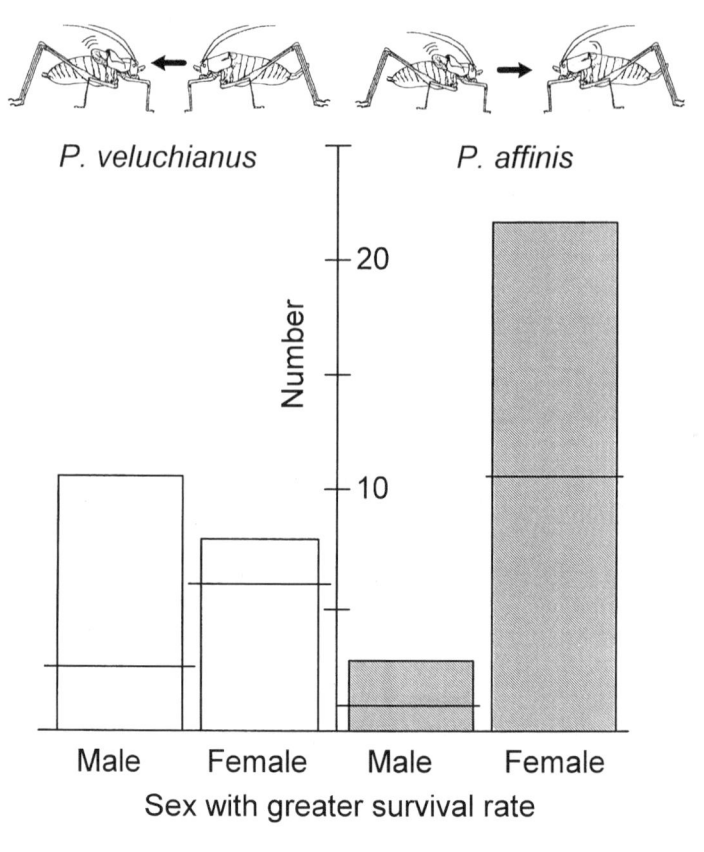

Figure 8.11. Comparison of male and female survival rates for two *Poecilimon* species with contrasting pair-forming systems: *P. veluchianus* females are attracted to the male call, whereas *P. affinis* males move toward the answer song of conspecific females. The lines crossing each bar separate data from two different years. Number = number of 3-day sampling intervals. From Heller and Helversen (1991). *Poecilimon* katydids redrawn from Harz (1957).

answer is that adults of the (usually temperate) species known to suffer fly attack are adapted to be less cautious in singing because their adult lives are so brief (e.g., about 4 weeks for *Poecilimon*: K.-G. Heller, pers. comm. 1999) compared to the tropical species attacked by bats (Heller 1997).

We also need quantitative studies that examine the effects of more than one natural enemy. Feaver (1977) and Heller (1992) inferred death rates from multiple sources using data on the number of males and females disappearing from a cohort of marked adults. However, there are problems with mark-recapture studies of mortality. First is the difficulty of separating sex differences in mortality from sex differences in the likelihood of recapture, particularly when one sex is more cryptic in its behavior.

Second, sexual differences in death due to adult risk-taking may be obscured if a large source of adult mortality is the residual effect of parasitism (e.g., horsehair worms) and disease acquired in the larval stages that are unrelated to sexual selection (although it is possible that male-female differences in, for example, foraging patterns linked to acquiring resources for secondary sexual traits might result in sex differences in susceptibility to parasites and disease). A third problem with mark-recapture studies is the possibility of confounding effects of the marking technique. The problem here is not only that marks might attract visual hunters but also that there might be a differential effect of marking on each sex. An existing sex difference in risks associated with movement might be inflated even when both sexes are marked. For example, Heller and Helverson (1990) and Heller (1992) marked *Poecilimon* species with small reflective "flags" wrapped around a hind femur that may attract predators hunting visually.

Future studies might avoid the problems of mark-recapture methods, by turning to a more traditional observational approach focusing on the sex ratios of prey killed by several predators. Difficulties in actually observing the predation event can be alleviated somewhat by studying enemies that leave some sort of "record" of the attack. Examples that come to mind are parasitoid larvae dissected from prey (Allen 1995a), and prey recovered from webs (Thornhill 1978; Gwynne 1987), roosts (Belwood 1988), or the nests of predators (Gwynne and Dodson 1983).

As we saw in the contents of the nest of one predator, a digger wasp, the low number of male Mormon cricket victims correlated with male prudence in premating activities. It appears that these males have "lost some of that ardour which is usual to their sex, so that they no longer search eagerly for the females" (Darwin 1874). It turns out that the Mormon cricket is not the only katydid species with male coyness. We now move to Chapter 9 for a more detailed look at these katydids and for some general insights from these insects into the factors that control courtship roles in animals.

9

Can Katydids Tell Us Why the Sexes Are Different?

> Which of the two [sexes] takes the initiative here? Have not the parts been reversed?
>
> J. H. Fabre (1917), *The Life of the Grasshopper*

> Why are males masculine, females feminine and occasionally vice-versa?
>
> George C. Williams (1975), *Sex and Evolution*

> Now if we might assume that the males ... have lost some of that ardour which is usual to their sex, so that they no longer search eagerly for the females; or, if we might assume that the females have become much more numerous than the males ... then it is not improbable that the females would have been led to court the males, instead of being courted by them.
>
> Charles Darwin (1874), *The Descent of Man and Selection in Relation to Sex*

9.1. Role Reversals in Courtship

Darwin's (1874) comments on the button quail (*Turnix*) have turned out to be an apt description of the "reversed" courtship behavior of certain species in quite an array of different animals including frogs, fishes, flies, and katydids such as Mormon crickets (see Chapter 1). In these animals (Table 9.1, Fig. 9.1), males are not very eager to search for mates. Instead, females often take this "masculine" competitive role while males assume the "feminine" choosy role (Williams 1975). Research with these exceptional species is central to the study of sexual selection because it allows us to address Williams's question about why typical sexual differences, such as sex-specific flashy traits or weaponry, have evolved. Darwin's quote hints at a possible cause of "role reversal"—females might court their mates if males are in short supply—and theory since Darwin has pointed to sex ratio as one of several potential influences on the supply of mateable males and females. This chapter illustrates how Mormon crickets and other tettigoniids have been front and center in tests of sexual-differences theory because of the flexible nature of the male and female courtship roles (see Chapter 1). It is the experimental work on these katydids that is the main focus of this chapter. Before describing experiments, however, let us examine the natural history of role reversal in the system that has received the most study, the Australian pollen katydid,

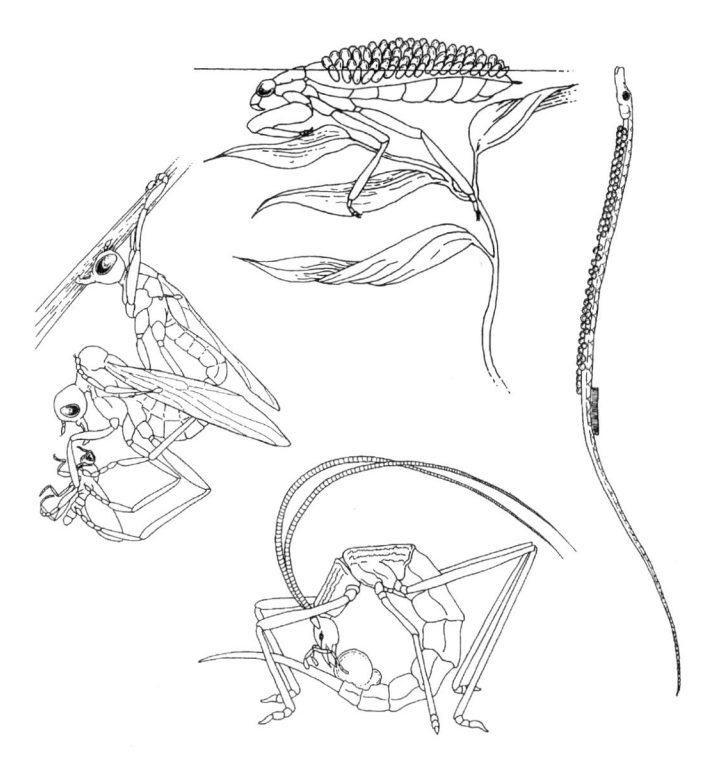

Figure 9.1. Animals with courtship role reversal in which males provide goods and services to their mates. Clockwise from top: male giant water bugs, *Abedus herberti* (Hemiptera: Belostomatidae), and pipefishes, *Nerophis ophidion* (Sygnathidae), brood the eggs on their backs and venters, respectively. Male katydids, *Kawanaphila nartee* (bottom), and dance flies, *Empis borealis* (Diptera: Empididae, left), feed their mates with spermatophores and prey items, respectively. Drawing by Heather Proctor, from Fig. 1, in Gwynne (1991).

Kawanaphila nartee, a species in which, as we have already seen, the male provides a large spermatophore meal (Table 6.3) that enhances his mate's fecundity (Table 6.2). The details of both the natural history and the experimental work with pollen katydids are distilled from a series of studies by biologists working at the University of Western Australia, especially Leigh Simmons, Winston Bailey, and me.

As Darwin's (1874) original motivation in proposing sexual selection theory was to explain sexual dimorphism, I close the chapter (and the book) with a discussion of sex-specific weaponry and other sexually dimorphic devices in katydids. In keeping with the theme of the chapter, this discussion highlights some remarkable devices that may have evolved by sexual selection among competitive females.

Table 9.1. Groups in which courtship-role reversal has been observed

Taxon	Nature of role reversal	Male-male competition?	Number of males and females available for mating	What limits female reproduction?
Birds				
Phalaropus spp. (phalarope) (Reynolds 1985)	Fights among females for prenesting males	No	Female-biased due to polyandry	Egg incubation exclusively by males
Actitis macularia (spotted sandpiper) (Oring and Lank 1982)	Females compete in areas in which males nest; females court males	Some	Female-biased due to polyandry	Egg incubation exclusively by males
Gallinula chloropus (moorhen) (Petrie 1983)	Fights among females for prenesting males	Male fights	Male-biased	Phenotype (quality) of incubating male
Frog				
Dendrobates auratus (poison-arrow frog) (Summers 1989)	Occasional fights between females; females take the active role in courtship	Occasional fights	Possibly female-biased	Male tending of eggs
Fishes				
Nerophis ophidion (pipefish)	Male choice (for larger females) and dominance among females (Berglund et al. 1986a; Rosenqvist 1990)	No	Female-biased due to brood-size limitation in males (Berglund et al. 1986b)	Time taken by male to brood eggs on body
Crustacean				
Pseudosquilla ciliata (stomatopod) (Hatziolos and Caldwell 1983)	Female initiation of courtship and male choice (for larger females)	Some aggression	Unknown	Possibly male-derived nutrients in ejaculate (known in *Squilla*)

Species			Male parental care / nuptial gift
Arachnids *Zygopachylus albomarginatus* (harvestman: Opiliones) (Mora 1990)	Females initiate courtship and engage in aggressive interactions. Males reject certain females	Males take over nests of rivals	No obvious bias; in fact not all males had nests	Males tend eggs in mud nests
Empis borealis and *Rhamphomyia longicauda* (dance flies: Diptera) (Svensson and Petersson 1988; Svensson et al. 1989; Funk and Tallamy 2000)	Females compete in a swarm; males enter the swarm and choose large females	No	Apparently female-biased	Prey-item provided by male (females obtain prey only from males)
Certain populations of *Anabrus simplex*, the Mormon cricket, *Metaballus litus* and *Kawanaphila nartee* (pollen katydids)	In food-limited populations, females fight for access to signaling males; male choice (for larger females) (see text)	Not in these populations	Female bias caused by food limitation	Large spermatophore eaten by female, which increases fecundity
Belostomatids (giant water bugs: Heteroptera)	*Abedus herberti* not completely role-reversed: female initiates mating and male coyness (Smith 1979); *Diplonychus major* females fight to lay eggs (in lab) (Ichikawa 1989)	Male *A. herberti* display to females (Smith 1979)	Female-biased but only at certain times of the season in *D. major* (Ichikawa 1989), *Belostoma flumineum* (Kraus 1989), and *A. indentatus* (Kruse 1990)	Male aeration of eggs laid on his wing covers

Adapted from Gwynne (1991).

9.2. Variable Courtship Behavior in Flower Lovers

Unlike the neotropical bats listening for the cautious calls of false katydids (see Chapter 8), gleaning long-eared bats (*Nyctophilus*) of southwestern Australia's *Eucalyptus*-banksia woodland hear many katydid songs. This is particularly true in early spring when any bat cruising over low vegetation is deluged by a cacophony of ultrasound from dense clusters of the season's first tettigoniid songsters (Gwynne et al. 1988), *Kawanaphila*, literally meaning the "flower lover" but commonly known as pollen katydids (Zaprochilinae, see Chapters 2 and 3). Although the 40- to 60-kHz calls of these insects are undetected by human ears (Gwynne and Bailey 1988; Mason and Bailey 1998), the sounds are pitched right in the frequency ranges used by the local bats to hear their own echolocation clicks (Fullard et al. 1991). Coincidentally, the electronic devices that were originally developed to detect the ultrasonic cries of bats have turned out to be important aids in finding the high-pitched songs of pollen katydids.

Our bat detectors revealed dense populations of one pollen katydid, *K. nartee* (Fig. 9.2A), only in certain locations in the natural bushland of Kings Park, Perth. The park has been our main study site for this small, sticklike flightless species (see Chapter 2; Fig. 4.8), by far the most common katydid in the area (Gwynne et al. 1988). In other places within the park, *K. nartee* appears to be just as common, but far fewer males sing. The striking variation in the number of callers in different areas of the bushland is reminiscent of the differences between the Greystone and Indian Meadows populations of Mormon crickets described in Chapter 1, except that in pollen katydids this variation occurs over a much smaller spatial scale; in *K. nartee*, huge differences in calling activity can occur over just a few dozen meters (Gwynne et al. 1998).

The most vocal populations of pollen katydids occur in September in places where the katydid's food flowers (see Chapter 3) are abundant. Dense choruses of *K. nartee* callers can last for the entire breeding season near plants with season-long flowering such as *Daviesia* and *Jacksonia* bushes (Gwynne et al. 1998). Singing males also are associated with a plant that shows a short spring blooming period, kangaroo paws (*Anigozanthos manglesii*), the "state flower" of Western Australia (Simmons and Bailey 1990). Kangaroo paws appear to be a poorer food source for katydids than flowering bushes, probably because it is pollinated by birds (Hopper 1993) and thus tends to produce less nutritious pollen than flowers visited by pollen-consuming insects. Finally, in areas with few spring flowers of any kind, only the occasional *K. nartee* singer can be heard, even though large numbers of the katydids may be present in these areas. These populations of katydids are doomed to a life of low rations[1] except in the rare spot

---

[1] Females but not males have been observed to move away from sites with less food (Shelly and Bailey 1995) and may use male song (Bailey and Simmons 1991) or perhaps food odors

Figure 9.2. Mating and sexual selection in pollen katydids, *Kawanaphila nartee*. **a.** A male produces a call composed of a series of ticks carried at a very high pure-tone frequency (see Fig. 5.4). The male raises his wings to produce a much abbreviated song, sounding like a single tick to the human ear, when he is interacting with a number of competing females. **b.** A responsive female moving to a calling male holds the tip of her abdomen and her ovipositor in an elevated position, apparently in preparation for a "genital hold" on the male. **c.** A female has mounted and grasped the male in a genital hold. It is at this stage that males at poor food sites appear to weigh and reject lightweight females (mean weight of these females is 66.2 mg: Simmons and Bailey 1990). **d.** A successful, large female with her spermatophylax meal. Successful females averaged 77.4 mg in weight, which is significantly heavier than the mean of rejected females (Simmons and Bailey 1990).

where the pollen famine is dramatically broken by the flowering of a grass-tree, *Xanthorrhoea preissei* (Fig. 9.3), a plant whose floret-crammed stalks can supply abundant food to katydids living close by over the final weeks of their 3-month season. A sudden blooming of grass-trees occurred during our surveys of pollen katydids in 1993, when two flower stalks pushed up from a large skirt of grass-tree leaves (Fig. 9.3) in an area with very few other flowers of any kind. We had detected no callers in the immediate

---

to find local food patches. However, the high risks associated with movement probably mean that hungry pollen katydids do not attempt random searches for food over very long distances.

Figure. 9.3. The open *Eucalyptus-Banksia* forest habitat of pollen katydids with flowering grass-trees in the foreground. Pollen from grass-trees supplies abundant food for the katydids.

area around the stalks, but 2 days after the first florets erupted, and concomitant with an increase in the amount of food in the guts of males (Fig. 9.4), several consistent callers appeared nightly within or close to the grass tree leaves (Gwynne et al. 1998).

A closer look at the behavior of calling male pollen katydids in the areas with few flowers revealed that like Greystone Mormon crickets (see Chapter 1) and Darwin's button quails, responsive females did appear to be "much more numerous than the males" and that roles were reversed: "females . . . court the males, instead of being courted by them" (Darwin 1874). We saw an excess of females in areas with few flowers where the occasional caller only had to sing for about a minute before attracting one or more females. We confirmed this experimentally by noting the response of females to caged singers placed in the site with few flowers (Fig. 9.5). These confined callers usually attracted a female after a minute of singing and, if they were allowed to continue singing for a full 10 minutes, could end up with a dozen or more highly responsive females scrambling around on the surface of the cage (Fig. 9.5). By contrast, females at sites with lots of flowers for the most part were uninterested in singers; caged callers placed in these sites usually failed to attract any females in a 10-minute singing trial. This was true for both the flowering grass-tree and flowering

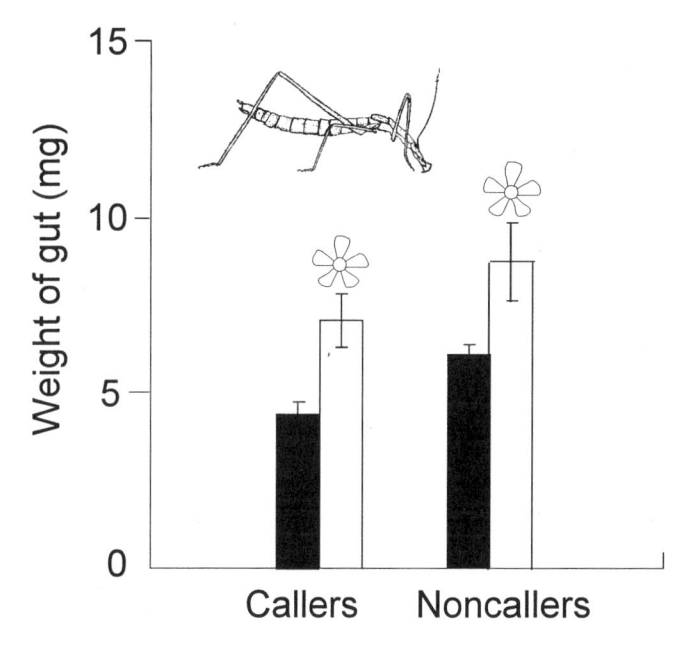

Figure 9.4. After the grass-trees flowered (white bars), there was a significant increase in the gut contents of both calling and noncalling male pollen katydids (callers before flowering were collected not only in the vicinity of the grass-trees but also elsewhere in this site with few flowers). Noncalling males also had fuller guts than did callers, probably because noncallers were unable to mate and so spent more time spent feeding in order to increase the size of their spermatophore glands. Redrawn from Gwynne et al. (1998).

bush sites (Gwynne et al. 1998) (for additional details, see the legend to Fig. 9.5).

The reversal in courtship was evident from the behavior of callers and responsive females at sites with poor food supplies (Simmons and Bailey 1990). First, there was clear evidence of female-female competition when rivals jostled with each other in their race to the caller. Females grappled with front legs and pushed or kicked as each attempted to mount the male (Simmons and Bailey 1990) (Fig. 9.6A). When we allowed females to interact with an experimental calling male, he stopped his continuous stridulation (Fig. 9.2A) as soon as his first suitor appeared. However, if he became separated from the struggling females, as was frequently the case, he would almost always sing again. This time his song was not continuous (Fig. 5.4) but consisted (to our ears, at least) of a single brief "tick." Even this brief burst of sound was sufficient for females to reorient and again hurry toward the male. In nature, separation of the sexes appeared to be caused mainly by the male withdrawing from the skirmishing females. This and his coy clicking delayed mating by up to 42 minutes (mean

Figure 9.5. The top of the cage of an experimental singing pollen katydid (a male fed in the lab for several days before use), showing three of the responsive females attracted when the cage was placed in a site with few flowers. The number of females attracted and the time elapsed before the first female was attracted (see text) were significantly greater and shorter, respectively, when compared to the response of females to caged callers placed at sites where many males called. For example, at the grass trees, three caged callers attracted from one to five females in 10 minutes when flower stalks were immature, but after the stalks flowered, caged callers failed to attract females in six 10-minute trials (Gwynne et al. 1998).

6.5 minutes) and may represent behavior that incites competition among his potential partners (see Chapter 7) (Gwynne and Bailey 1999).

Eventually one of the females mounted the male and engaged genitalia by grasping him in a "genital hold" in which she lowered her ovipositor and pinched the tip of his abdomen in the hinge between the base of her abdomen and ovipositor (Simmons and Bailey 1990) (Fig. 9.2C). But holding the male did not necessarily prevent additional competition. Often a female, occasionally even two females (Fig. 9.6B, C), clambered onto a coupled pair. Movements (up and down, side to side, or back and forth) of the interfering rival usually caused the original pair to break up (Gwynne and Bailey 1999), and in about half of the interactions the interloping female usurped the mounted position and went on to mate with the male (Simmons and Bailey 1990; Gwynne and Bailey 1999). Females also can fight over pollen, and although heavier females won these food fights (Simmons and Bailey 1990), we could find no advantage to body size in fights over males either before or after the mounting stage of mating (Gwynne and Bailey 1999).

When it came to male choice, however, there was an advantage for larger

Figure 9.6. Sexual competition among female pollen katydids. **a.** Three females attracted to a caller interact with him. **b.** A rival female approaches a coupled pair. **c.** A competitive frenzy ensues as three females jostle for position on top of a male.

females. Males favored heavyweights by providing them with a spermatophore (Simmons and Bailey 1990; details in Fig. 9.2 legend). Males rejected smaller females by pulling out from under them in a similar manner to the coy male Mormon crickets at Greystone (Fig. 1.7). Male pollen katydids rejected mounted females in 38 percent of attempted matings at poor pollen sites (kangaroo paws), pulling away an average 50 minutes after mounting began. Even if the spurned females persisted and regained their genital hold on their mates, they were never successful in obtaining a spermatophylax meal (Simmons and Bailey 1990).

Males occasionally spurned females at pollen-rich (grass-tree) sites, but they did not target smaller females for rejection. Furthermore, most courtship at grass-tree sites showed the roles to be typical; most mate rejection was done by females, with 14 percent of females attracted to males turning away before mounting, and of the females that did mount, 25 percent did not grasp their mate in a genital hold. Such reticence by females to link genitalia was never seen at pollen-poor sites (Simmons and Bailey 1990). Choosy females at pollen-rich sites may be going after large males; in lab trials with well-fed individuals, there was a significant bias toward the larger of three experimental males available to the female (Gwynne and Bailey 1988). The lack of a male size advantage in mating pairs of pollen katydids collected from the field (see Chapter 7) may have been because many of the pairs were sampled from kangaroo paws sites where female choice was not expected to occur (Simmons and Bailey 1990).

The other component of typical courtship roles, the direct fights between males, as seen in the Mormon crickets in Indian Meadows (see Chapter 1), has not been seen in pollen katydids. Simmons and Bailey (1990) observed no fights among singing male pollen katydids in and around the basal leaves of flowering grass-trees. However, males did appear to interact acoustically by increasing the rate at which they produced call ticks (Figs. 5.4. and 9.2A) and by moving away when they detected substrate vibrations from a rival's call at increasing intensity (Simmons and Bailey 1992) (see Chapter 7). There is even some evidence that aggression between males can escalate to physical encounters in which males mounted neighboring rivals, thereby preventing calling (Gwynne and Bailey 1988; Bailey and Simmons 1991).

9.3. Courtship Role Reversal in Other Katydids

Evidence of role reversal has now turned up in a number of different katydids. Some information comes from anecdotal accounts, including an observation of a European relative (Tettigoniinae) of Mormon crickets, the female white-faced decticus, which J. H. Fabre (1917) observed to "thrust herself" on a male just before copulation occurred (prompting the quote that opens this chapter). In another long-winged tettigoniine, the Australian *Metaballus litus*, I found female-female competition and male choice at a site in a coastal salt marsh. Like the pollen katydids, *M. litus* (Fig. 9.7) shows remarkable variation in its courtship roles; just a few kilometers inland from the salt marsh where I observed role reversal, I found populations exhibiting more typical roles. Moreover, *M. litus* males that were experimentally transferred to the salt marsh quickly adopted the behavior of their new habitat, including a change in song from a continuous buzz to brief coy "zip" that nevertheless attracted females very quickly (Gwynne 1985) (details in Fig. 9.8 legend).

A third Australian katydid, the listroscelidine *Requena verticalis*, featured in Chapter 6, also has variable courtship roles. Although little is known about the behavior of this species in nature, in the lab Lynam et al. (1992) and Schatral (1993) found that males rejected females, and Simmons (1994a) showed that more males than females rejected their mates when females were parasitized with gregarine protozoans (see Chapter 4). *R. verticalis* males will also reject older females in favor of younger ones,

Figure 9.7. A mated female *Metaballus litus* with her spermatophore (photo by Chris Codd).

although they are unable to differentiate virgins from nonvirgins (Lynam et al. 1992) and appear to use cues of youthfulness to predict the prior mating status of their mates (Simmons et al. 1994). Finally, in bradyporines, Hartley (1993) observed females of two *Uromenus* species competing sexually. Females near a calling male lunged and kicked at each other and engaged in "loud, prolonged stridulation" (recall that in these katydids, females stridulate in "answer" to the calls of males; see Chapter 5). Ephippigerine bradyporines such as *Uromenus* and *Ephippiger* species are record holders for the enormous size of their spermatophylax meals (see Chapter 6), so it is not surprising that they show role reversals. Although there is little field work on these species, lab observations of food-deprived *E. ephippiger* revealed role reversal (Ritchie et al. 1998). As is true for *M. litus* and the pollen katydids, hungry *E. ephippiger* males were coy callers that sang less frequently and at a lower intensity than well-fed, competitive males.

What causes a reversal in the courtship roles in katydids such as *Kawanaphila*, *Metaballus*, and *Anabrus* in some sites but not in others? This question can be answered in two ways. The first deals with proximate cues that individuals use to become adaptively coy or competitive, and the second addresses the ultimate factors that control sexual selection and sexual differences in the katydids.

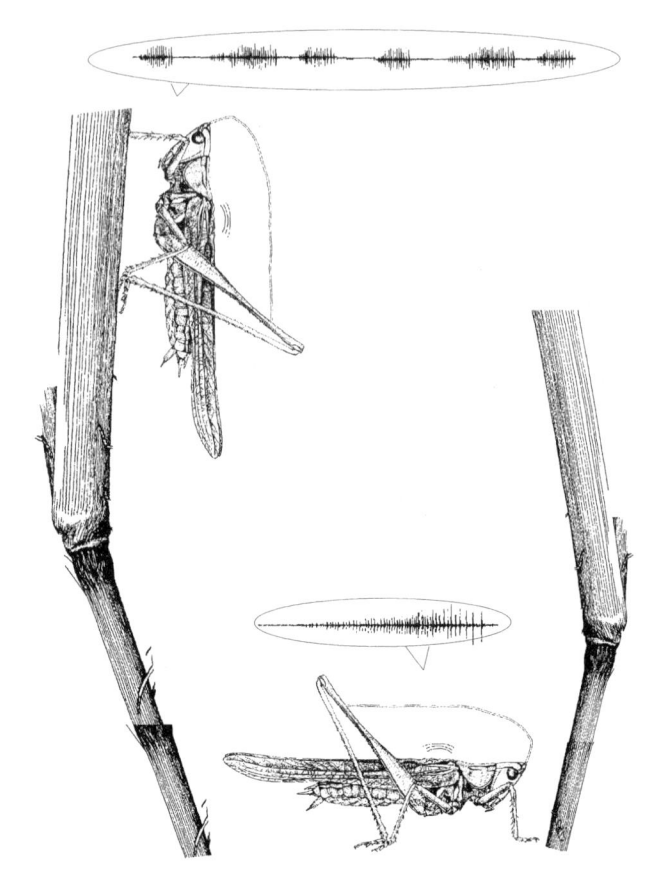

Figure 9.8. Plasticity in the courtship roles of *Metaballus litus*. A male (left) from a site in which the courtship roles are typical competes by singing a continuous song (upper trace) from the tips of long grass stems and occasionally fights with a male attracted to his song. The same male (bottom), however, becomes coy when he is (experimentally) moved to a site where the courtship roles were reversed: he sings from deep in the vegetation using only an occasional single "zip" (lower trace). As is true for Mormon crickets (see Chapter 1), males reject some females that are quickly attracted to this zip song. Rejected females weigh significantly less than acceptable mates (Gwynne 1985). Katydid drawing from Rentz (1985).

9.4. What Cues Do Katydids Use to Alter Their Behavior?

The observation that male pollen katydids coyly withdraw from a rush of responsive females (Gwynne and Bailey 1999) suggests that males become choosy when they frequently encounter responsive females. Alternatively, however, because male choosiness is associated with a lack of decent food, food abundance itself may cue male behavior. As it turns out, both encounter rate and diet appear to be involved.

Experiments by Shelly and Bailey (1992) demonstrated that encounter rate with females was an important cue; male pollen katydids collected from areas with an excess of adult females were more likely to reject small, less-fecund females than were males collected from sites with fewer females. Shelly and Bailey then went on to confirm the encounter-rate hypothesis in a lab experiment showing that males experiencing contact with females on two of the previous four nights were more likely to reject lightweight females in favor of heavyweights than were males that had no contact with females. A later experiment (Kvarnemo and Simmons 1999) showed a similar pattern: mate rejection by male pollen katydids increased with an increase in the number of females previously encountered. However, this occurred only in an experimental treatment in which some of the males had encountered a previous low-quality (low-fecundity) mate and some a high-quality mate.

Both encounter rate and food availability appear to cue the level of male choosiness in *R. verticalis*. Although Shelly (1993) showed that males previously deprived of contact with females were not very choosy, Kvarnemo and Simmons (1998) showed that the male diet also had an effect; males fed a protein-rich diet were significantly less likely to reject females than were males on a poor diet.

Food availability is clearly associated with the roles assumed by mating katydids. Experiments examining this association provided one of the best sets of studies investigating the control of courtship roles of animals. Let us now look at these studies.

9.5. A General Test of Sexual-Differences Theory

9.5.i. Factors predicting courtship roles

As we have seen for several katydids, sexually competitive males and choosy females appear when food is plentiful. When food is in short supply there is a reversal in these courtship roles. Recall for example that in gregarious Mormon crickets, both males and females fight for nutritious food available from the carcasses of dead band-mates (see Chapter 1) (Fig. 9.9). With these observations in mind, I proposed a *food and sexual-differences hypothesis* (Gwynne 1984c), arguing that when nutritious food is scarce for katydids, the supply of mateable males is low both because of the low rate at which males can turn nutrients into spermatophylax meals and because hungry females become more promiscuous as they forage for these meals. Male katydids also might respond to food stress by reducing the size of each spermatophylax meal but only to a point, because with too small an offering, a male runs the risk of compromising his fitness, in terms of his ability either to inseminate or to be a parent of high-quality offspring (see Chapter 6).

Figure 9.9. A pack of hungry Mormon crickets from a band tear apart the body of a road-killed conspecific.

But how does the food and sexual-differences hypothesis fit into a more general theory of factors controlling sexual differences in animals (Fig. 9.10)? After all, courtship role reversals in animals such as Darwin's button quails appear to have little to do with food. However, a factor common to all role-reversed species (Table 9.1) is a sex bias in the numbers of sexually active individuals. As Darwin (1874) may have recognized (see the chapter opening), a relative increase in the number of sexually available females, the *operational sex ratio* of Emlen (1976; Emlen and Oring 1977), should cause females to become more sexually competitive. When the numbers of sexually active individuals from the sexes are similar, both may exhibit choosiness and mate competition. Finally, a large increase in the number of sexually available females should lead to role reversal: when there are more available females than males, males are in a position to choose and females are forced to compete for access to these males. Therefore, sexual selection is more intense on females and less intense on males (Fig. 9.10).

But what are the factors that determine the numbers of males and females available for mating? One obvious factor is the adult sex ratio (Parker and Simmons 1996). Sex-biased predation, for example (see Chapter 8), could change the operational sex ratio. Although there is some evidence that sex ratio biases occur in local populations and can affect courtship roles (for example, see a study of beetles by Lawrence 1986), sex ratios typically are so close to unity that they have little influence on the availability of the sexes for mating.

9.5 General Test of Sexual Differences Theory   237

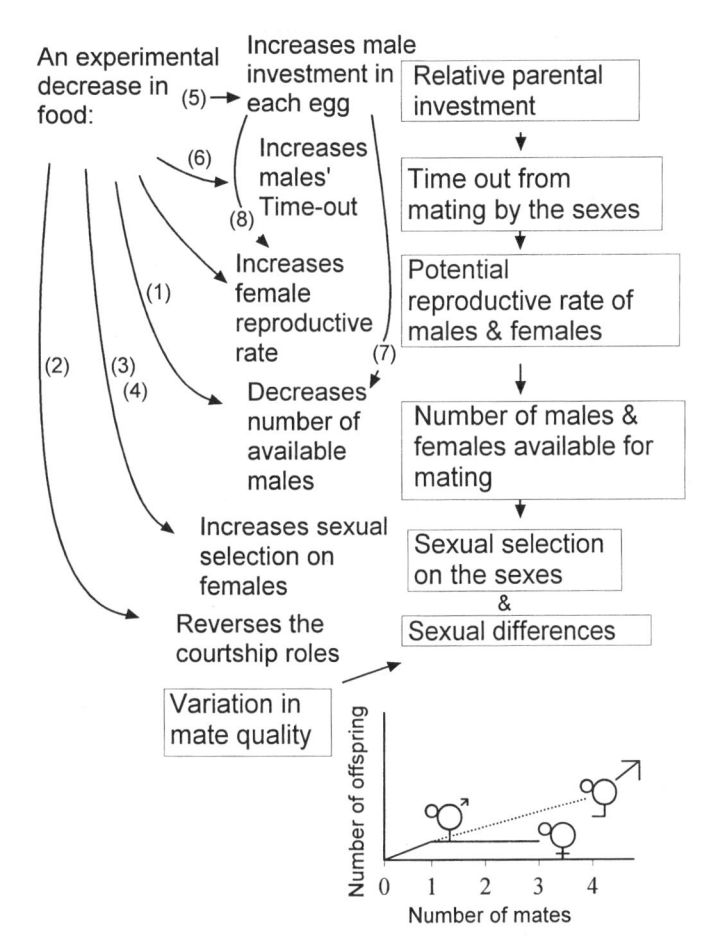

Figure 9.10. The control of sexual differences: theory and predictions. Factors hypothesized to control sexual selection and sexual differences are in the boxes on the right. The left of the figure shows how experimental manipulation of nutritious food available to katydids has been used to test the effects of these factors on sexual selection and sexual differences. Numbers refer to the hypotheses and predictions tested (see text).

The plot in the lower right of the figure represents male (dotted lines) versus female (continuous lines) differences in sexual selection and sexually selected traits. It was Bateman (1948), in a study of *Drosophila melanogaster*, who used the slopes of correlations between number of mates and number of offspring to depict sex differences in sexual selection. Bateman's conclusion is depicted in the two lines. He noted that in "typical" species, such as *Drosophila*, the male (dotted line) has the potential to produce more offspring with each additional female he mates with (number of mates), whereas a female (continuous line) may gain no additional offspring from remating, needing only to mate once to obtain sperm (see Arnold and Duvall 1994). In the plot, I have also depicted Darwin's (1874) process of sexual selection for secondary sexual traits by showing the cartoon male with the larger arrow "tail" obtaining more mates than the male with the smaller tail.

The supply of mateable males and females, however, could be affected by sex-related activities that take one or the other sex away from mating. In most animals, this "time-out" from mating (Clutton-Brock and Parker 1992; Parker and Simmons 1996) is greater for females because they have to collect and provide resources for offspring. Females expend effort in producing large gametes, the eggs (Bateman 1948), and in some species, the additional energies (and associated risks) needed for gestation and maternal care (Williams 1966; Trivers 1972). Thus, the rate at which females can produce offspring is much lower than that of males, so males compete to sire these offspring (Thornhill 1986; Clutton-Brock and Parker 1992).

To summarize the theory (Fig. 9.10), in most species, the typically greater parental investment, and thus time-out from mating, of females relative to males yields a lower relative potential rate at which females can reproduce. This in turn leads to a male bias in the numbers of individuals available for mating and thus to greater sexual selection on males.

This general theory of sexual differences leads to several key predictions, namely, that a courtship role reversal should occur when (A) the male parental contribution is greater than the female's, resulting in (B) a relatively greater time-out from mating activity for males, (C) a relatively lower male reproductive rate, (D) a female bias in the number of individuals available for mating, and (E) a reversal in the strength of sexual selection on the sexes. However, the precise nature of the role reversal can be influenced by additional factors. For example, although sexual competition among females may arise when sexually active males are in short supply, *active* choice of mates by males may be absent if males passively accept winners of competitive bouts among females (Table 7.1). Another factor (ignored in the "operational sex-ratio theory" of role reversal outlined above) influencing the degree of mate choice shown by males is variation in the quality of mates. Variation in the quality of the sex being chosen, in this case females, should influence the degree of choosiness in the opposite sex. The greater the variation in fecundity, or ability to provide parental care, the greater the expected level of choosiness among males (Parker 1983; Owens and Thompson 1994; Johnstone et al. 1996) (Fig. 9.10).

Some of these predictions have been tested empirically, for example, by showing that courtship role reversal (in vertebrate species in which parental care is exclusively by males) is associated with relatively lower male reproductive rates (Clutton-Brock and Vincent 1991). However, work with katydids has provided full experimental tests of the predictions, mainly by manipulation of diet. Let us now turn to how this work addresses specific predictions from the food and sexual-differences hypothesis and thus provides a general test of sexual-differences theory (see predictions 1–8 in Fig. 9.10).

9.5.ii. Food availability and the number of males and females available for mating

The key proposal of the food and sexual-differences hypothesis (prediction 1) is that *a decrease in high-quality food available should result in a greater number of sexually active females than males.*

Our initial observations of role-reversed Mormon crickets and pollen katydids provided some support, as mateable males appeared to be in short supply; recall the extreme scarcity of singing males even though many adult males were present. Furthermore, for pollen katydids, an abundance of ready-to-mate females was revealed when responsive females swarmed onto the cages of our experimental calling males (Gwynne et al. 1998). However, to demonstrate that the apparent female bias in the operational sex ratio is caused by food restriction, it is necessary to manipulate the availability of food.

To accomplish this, we first set up field enclosures (cages) (Fig. 9.11) into which we released individually paint-marked insects. Pollen katydids were the subjects in one experiment (Gwynne and Simmons 1990) and Mormon crickets, the other (Gwynne 1993). In both studies, four of the cages were food restricted, and four were supplied with plenty of food. I will refer to these treatments as "low" and "high" diets, respectively. Importantly, these methods allowed us to control other variables that might influence sexual selection (Sutherland 1987) by keeping sex ratio, density, and ages of individuals similar in all the cages (see legend to Fig. 9.11).

Results with both species were similar in showing that a low diet increased the number of sexually responsive females, as shown in the decreased time it took for a calling male to attract a female. Also, virtually all of these males mated (Fig. 9.12 shows the data for Mormon crickets). We found, as predicted, the opposite effect on the numbers of sexually available males: in both species there were significantly fewer callers in low-diet cages than in high-diet cages (Fig. 9.12).

These sorts of field assessments of available males and females may not provide the precise estimates of the numbers of responsive males and females necessary to test our first prediction, that the presence of more sexually active males than females leads to role reversal. For example, counts of callers will underestimate the number of sexually active males if any males acquire mates without singing (see Chapter 7). Moreover, the number of sexually responsive females would be underestimated if less successful females avoid physical combat by being less responsive to singers.

Maximum mating rates that each sex can achieve, however, can be obtained by giving individual experimental males and females constant access to receptive members of the opposite sex. These experiments were conducted with *R. verticalis* (Gwynne 1990b) and pollen katydids

Figure 9.11. The open *Eucalyptus-Banksia* habitat of pollen katydids, showing three of the field cages used to study the effects of food availability on the operational sex ratio as well as on sexual selection and sexual behavior (Gwynne and Simmons 1990). Similar cage experiments were conducted with Mormon crickets (Gwynne 1993). In both sets of experiments for all cages there was a sex ratio of 1:1, a population density of 50 per 1-m^3 cage for Mormon crickets and 48 per 0.75-m^3 cages for pollen katydids. These densities seem high but are within the normal range observed for both species (up to 75 Mormon crickets have been observed per square meter: Anonymous 1991) Food availability was set at two levels, with the four low-diet replicates (cages) for pollen katydids consisting of cages placed over a patch of kangaroo paws in flower (recall that these flowers supply poor food rations: see text) and the high-diet replicates consisting of cages with artificial grass-tree flowers made from dried grass-tree flower stalks, with patches of pollen glued with honey to the tips of the stalks. Stalks without pollen were placed in low-diet cages. The four low diets for Mormon crickets consisted of (low-protein) rolled oats and the four high diets, a virtual katydid smorgasbord (see Chapter 3) consisting of a mixture of rolled oats, pollen, seeds, and fish food (containing arthropod, meat, and algal products). We observed behavior in the cages for 12 days (pollen katydids) and 17 days (Mormon crickets). Note that to maintain statistical independence, only the first occurrence of an (individually marked) individual was included in a particular behavioral category (e.g., male rejection, female-female competition).

(Simmons 1995c). Both studies showed that the extremes of diet caused large differences in the relative remating rates of the sexes. For *R. verticalis* a high diet increased the mating rate of males to more than twice that of females (operational sex ratio 0.45). In contrast, the mating rate of females on the low diet slightly exceeded that of the males (operational sex ratio 1.05).[2]

---

[2] Mating rates of the sexes are equivalent to the operational sex ratio when the adult sex ratio is 1:1 and maintenance costs of mating activity are very small (Clutton-Brock and Parker 1992).

9.5 General Test of Sexual Differences Theory 241

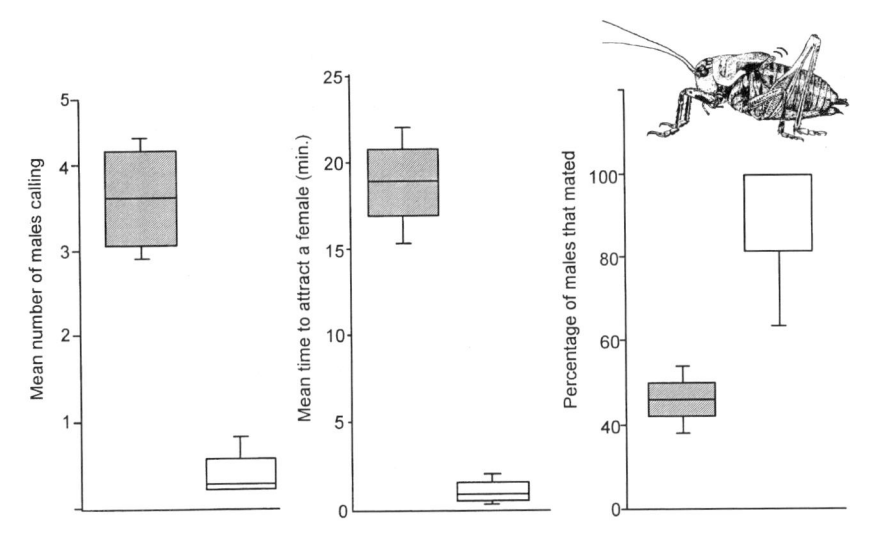

Figure 9.12. Mormon crickets: effect of diet on number of calling males, number of males that mated, and the time it took males to attract their first female (for four high-diet cages (hatched boxes) and low-diet cages (open boxes). Data are represented in "box plots," with each box encompassing interquartile ranges. The cross bars indicate medians and the top and bottom bars, the full range of data. Katydid redrawn from Fig. 2.11.

Simmons (1995c) examined relative mating rates in pollen katydids feeding on various amounts of pollen as part of a study of relationships between several factors hypothesized to control sexual differences (more on this work later in the chapter). As predicted, male mating rates for pollen katydids increased with increased food intake, whereas female mating rates decreased (Fig. 9.13C). Interestingly, as Simmons pointed out, there was a rapid, rather than a gradual, switch in mating rates when food intake fell below 20 to 30 joules of energy per day. The rapid switch in mating rates correlated with the apparently rapid switch in courtship roles seen with decreasing food availability in field populations (Fig. 9.13E) (Simmons and Bailey 1990; Gwynne et al. 1998).

Another interesting result from Simmons's study is that as in the *Requena* experiments, mating rates reversed but only with an extremely low diet (left of the vertical line depicting equal mating rates in Fig. 9.13D). Simmons (1995c) suggested that a more distinct reversal of mating rates might be prevented by male suppression, probably via ejaculatory chemicals, of the female's refractory period, thus preventing her from achieving her preferred (higher) remating rate. There was also a reversal in relative mating rates in data collected from our field cages (Gwynne and Simmons 1990): the male remating rate exceeded the female rate in the high-diet cages (where the insects exhibited typical courtship roles, see next section),

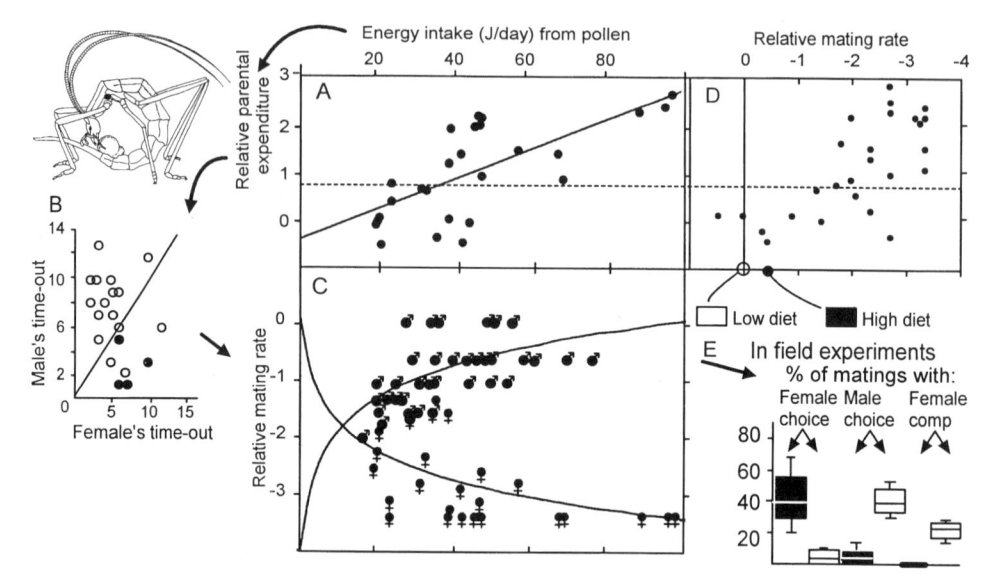

Figure 9.13. The control of sexual differences: experimental tests of sexual differences theory using pollen katydids, *Kawanaphila nartee*. A. Increasing food (pollen) in the diet (and thus energy intake, horizontal axis) increases female parental expenditure in individual eggs relative to that of males (vertical axis, expressed as a natural log of the variable). Note the point at which male investment exceeds female investment, where the regression line (continuous) crosses the dashed line of equality in relative parental investment, occurs between the highest and lowest degrees of food intake of diet. The data were measured for a single reproductive episode of mating and egg production (regression analysis $F_{1,24} = 22.7$, $P < 0.001$, $r^2 = 0.46$) (from Simmons 1995c). B. Diet (and thus relative parental expenditure) in turn controls the relative time-out for each sex from the pool of mating individuals. The line represents equality of male and female time-out and reveals that most females from cages in which they were the main sexual competitors (i.e., low diet, represented by open points) have a relatively shorter time-out. In contrast, females from cages in which males were the main competitors (high diet, represented by filled points) had a relatively longer time-out from mating (from Parker and Simmons 1996: data collected from the field cages used by Gwynne and Simmons 1990). Relative time-out in turn influences the relative mating rates (expressed as the natural log of the variable) of each sex (vertical axis in C and horizontal axis in D). C. The effect of amount of food eaten on relative mating rates of males and females. D. The positive relationship between relative parental expenditure and relative mating rate, with the vertical line representing equality in the mating rates (from Simmons 1995c). The two points on the horizontal axis show the values of relative mating rates for katydids from low-diet (open) and high-diet (closed) field cages used by Gwynne and Simmons (1990) (data presented in Simmons 1995c). E. The consequences of varying food intake on the courtship roles (female and male rejections of mates [Choice] and female-female competition, from Gwynne and Simmons 1990). Filled boxes enclose interquartile ranges of behaviors of katydids from high-diet cages (horizontal line = median, and vertical lines = ranges), and open boxes represent animals from the low-diet cages. All figures were redrawn from their original sources. Katydid redrawn from H. Proctor (Fig. 9.1).

9.5 General Test of Sexual Differences Theory

but rates were reversed in low-diet cages (see points marked on the horizontal axis of Fig. 9.13D). This match of *field* mating rates with courtship roles indicates that estimates of mating rates in nature may be reasonable estimates of maximum mating rates after all.

Another result of these experiments is that the parental spermatophylax meal of pollen katydids, *K. nartee*, is not reduced in its energy content when males are food stressed (Simmons 1994b). Similarly, in *R. verticalis* experimental food stress did not influence spermatophore size until males had mated three times (Fig. 9.14), although well-fed *Requena* males showed decreased spermatophylaxes when their remating interval was decreased (see Chapter 6; Simmons 1995b). Maintaining a full-sized mating meal may be adaptive for males in food-stressed environments, as their spermatophylax nutrients are likely to be particularly important to the fitness of their offspring. Katydid meals that mainly function to increase success in fertilization rather than augment the fitness of offspring (see Chapter 6) might be reduced in size when food is limiting. This may be the case in *Gampsocleis gratiosa* in which the size of spermatophylaxes is reduced when males are food stressed (Zhiyun et al. 1998). The reduction in meal size in

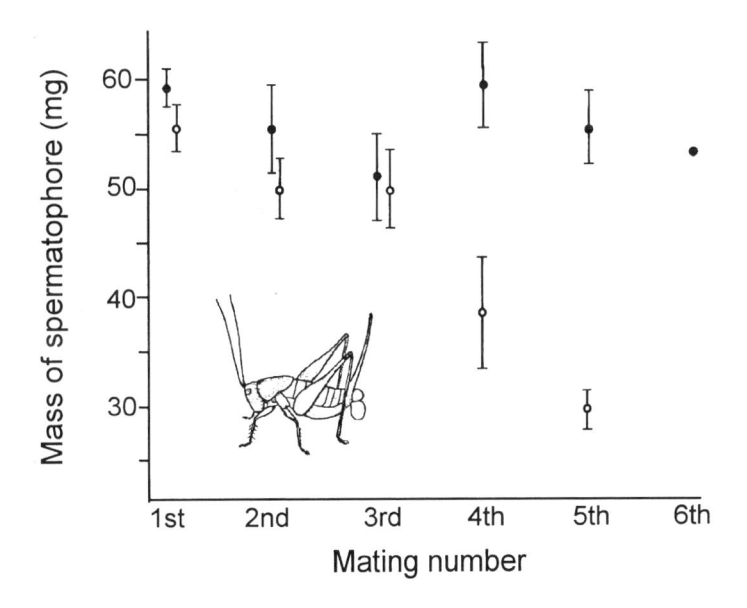

Figure 9.14. The effect of food stress and number of previous matings on spermatophore meal size in *Requena verticalis* (Gwynne 1990b). Shown here are the means and standard error bars for high diet (closed points) and low diet (open points). Analysis of variance showed no significant effect of diet or mating frequency on the size of the spermatophore meal through the first three spermatophores. Katydid redrawn from D. Scott's figure, in Clutton-Brock (1991).

this species, however, still did not allow males to reach the high remating rate achieved by male *K. nartee* in high-diet cages.

9.5.iii. Food availability and the courtship roles

Food stress increases the number of sexually available males relative to females, but does this diet-induced change cause the excess of females to compete for mates and for males to choose? This leads to prediction 2: *food stress causes a role reversal in courtship behavior.*

This prediction is supported by the behavior observed in our caged populations of pollen katydids (Gwynne 1990b) and Mormon crickets (Gwynne 1993). In low-diet cages, we found significantly more reproductive interactions in which males rejected mates while competition occurred between females (Figs. 9.13E and 9.15). Conversely, the incidence of female rejection of mates was significantly higher in high-diet than in low-diet cages. Moreover, obvious cases of male-male aggression (fights) were seen in Mormon crickets in high-diet cages (Fig. 9.15) although not in pollen katydids (Fig. 9.13E). As suggested earlier, intermale competition may be more subtle in pollen katydids.

An experimental reduction in food also caused role reversal in laboratory populations of *E. ephippiger*, from the tribe Ephippigerini, a group that includes species with very large spermatophylax meals (see Chapter 6, Fig. 6.14). Ritchie et al. (1998) found that a low diet caused males to reject females more often than females rejected males, and also caused a complete switch in the direction of sexual competition in that virtually all contests were between females.

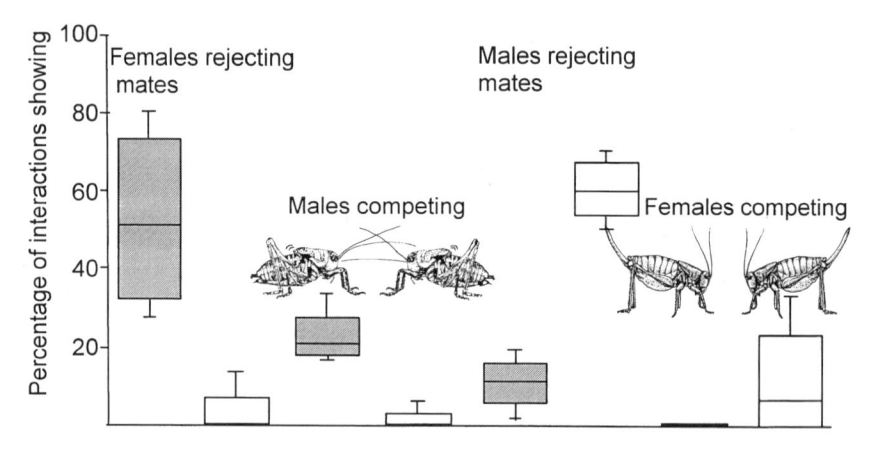

Figure 9.15. Food stress (open boxes) causes a reversal in the courtship roles of Mormon crickets, whereas ad-lib food (filled boxes) results in the typical roles of male-male competition and female mate choice. See legend to Fig. 9.12 for details on box plots. Katydids redrawn from Fig. 2.11 (males) and from artwork by K. Hansen McInnes in Gullan and Cranston (1994) (females).

9.5 General Test of Sexual Differences Theory

Finally, the effects of food stress on male choice and female competition can occur indirectly through parasitism. For example, heavy infections with gregarine protozoans (see Chapter 4) can starve females of the nutrients necessary to achieve high fecundity. In matings with experimentally infected *R. verticalis*, the typical courtship, in which there are more female rejections of males than male rejections of females, reverses; males become the more choosy sex (Simmons 1994a). Possible alternative explanations for increased male choosiness are that males are rejecting sick and potentially infective females, and that the males perceive these females as low-quality mates.

9.5.iv. Food availability, mate quality, and choice by males

When females are forced to compete for limited food, some will be more successful than others. The result is increased variation in female condition, fecundity, and therefore, their quality as mates. The increased variation in mate quality could increase male choosiness (Owens and Thompson 1994). This raises an alternative hypothesis for the male-choice component of role reversal seen in our food-deprivation experiments: the increase in male choosiness in low-diet cages may have been, at least in part, due to increased variation in the quality of females rather than an increase in the relative number of responsive females. After all, the low diet did increase variation in female fecundity and body mass in field-cage experiments (Simmons and Bailey 1990; Gwynne 1993).

Kvarnemo and Simmons (1999) addressed this issue with pollen katydids. First, in an experiment in which each male had encountered a single low-, high-, or medium-quality (fecundity) female, males were, on average, more likely to reject a subsequent female compared to an experimental treatment in which each male had encountered a medium-quality female. However, only in a second experiment could each male assess variance per se by encountering nightly sets of three females varying in quality. This time the results showed no significant difference between the mate rejection rates of these males and males in a low-variance treatment in which males encountered trios of females similar in quality.

More work is needed on the influence of mate quality on male choosiness (see also Kvarnemo and Simmons 1998). This important factor influences male choosiness. After all, there has to be some level of variation in the quality of mates available to males or there would be no benefits to male choosiness in the first place!

9.5.v. Food availability and sexual selection

A courtship role reversal should be associated with a reversal in the intensity of sexual selection; fights by females and choice by males are expected to impose greater sexual selection on females than males. Sexual selection is somewhat of a more difficult process to understand than competition and mate choice. Indeed, whole books have been devoted to defining and measuring sexual selection (e.g., Bradbury and Andersson 1987; Clutton-Brock 1988). There seems to be general agreement that sexual selection occurs on a sex when members of that sex compete for access to mates (in the case of males, fertilization) or to the best mates (see Chapter 1). Therefore, any trait that conveys an advantage in direct sexual competition or that is preferred by the opposite sex is a sexually selected trait. The word *advantage* refers to fitness, which can be measured in success in acquiring mates. Therefore, one way of estimating the "opportunity" for sexual selection in a sex is to measure variation in mating success (Bradbury and Andersson 1987).

Success at mating results in success in producing offspring. Therefore, establishing positive relationships between number of offspring and mating frequency is a convenient way of expressing sexual selection on a sex (Bateman 1948; Arnold and Duvall 1994) (see the plot at bottom of Fig. 9.10). A steeper slope of the relationship for one sex, typically males, than that for the other sex indicates stronger sexual selection on the former sex. This is because a steeper slope can indicate large fitness returns for each mate won in the competition for copulations. Note, however, that a positive correlation between mating frequency and number of offspring may not necessarily be a result of competition for mates. For example, even when female katydids are not competing for mates, each additional mating (spermatophore meal) enhances their fecundity (Table 6.2). Therefore, other sex differences, such as in variation in numbers of mates (Arnold and Wade 1984), may be necessary to demonstrate sexual selection.

If a positive relationship between mating frequency and fecundity indicates sexual selection, a reversal in sexual selection should be the case when the female slope is steeper than the male slope (i.e., the slope of the female line in the plot at the bottom of Fig. 9.10 becomes steeper than that of the male line) (Arnold and Duvall 1994; Lorch 1999). However, no study has yet examined this prediction, particularly in field populations. Although I lack data on male mating frequencies in nature, there are some for female mating frequencies in role-reversed populations that can be compared with data from populations with typical courtship roles. The prediction is that *food stress increases sexual selection on females as revealed in (i) an increased variation in mating success and (ii) an increase in the slope of the fecundity × mating-frequency plot for females* (prediction 3).

The first part of this prediction was supported in cage experiments with pollen katydids: variance in female mating success was greater in low-diet than high-diet cages (Gwynne and Simmons 1990).

The second part of the prediction was not supported in a study of role-reversed and typical-role populations of Mormon crickets observed in nature: although the slopes did not differ, both populations showed positive slopes between mating frequency (determined by counts of spermatodoses, see Chapter 6 and e in "Fig. 15" of Fig. 2.30) and number of mature eggs in the ovary (Gwynne 1984c). One problem with these field data, however, was that fecundity estimates using only ovarian eggs were poor because females may have laid many of their eggs. However, this was not a problem in cage experiments with Mormon crickets because females laid no eggs. Moreover, the fecundity measure I used in the cage studies was total mass of mature ovarian eggs (egg number × egg mass, the latter known to be influenced by spermatophylax feeding: see Chapter 6). Finally, in the second replicate of these experiments (conducted in 1990) I also estimated female mating frequency directly from the observed number of spermatophores received by marked individuals. The results (from the data collected for Gwynne 1993) were equivocal: data from 1989 supported the hypothesis, showing a highly significant relationship in low-diet females (where behavioral role reversal was observed) (Spearman rank correlation $r_s = 0.38$, $P < 0.003$) but no relationship in high-diet females ($r_s = -0.1$, $P > 0.5$); however, in 1990, both low-diet ($r_s = 0.35$, $P < 0.03$) and high-diet females ($r_s = 0.34$, $P < 0.003$) showed a significant positive relationship between mating (spermatodose) number and total fecundity.

Measuring sexual selection using variation in mating frequency and correlations between this variable and fecundity ignores a crucial aspect of the process of sexual selection because we need to know what traits give successful individuals a competitive edge (Darwin 1874). Increased success in sexual competition (moving up the fecundity × mating frequency curve in Fig. 9.10) should be directly related to increased size or elaboration of the sexually selected trait ("secondary sexual character": Hunter 1786) such as a peacock's tail or (in a role-reversed example) bright coloration in female button quails. In Mormon crickets, a less eye-catching trait but one that is nevertheless under sexual selection in females is body size (see Chapter 1); in role-reversed Mormon crickets, as well as *M. litus* (Gwynne 1985) and pollen katydids (Simmons and Bailey 1990), males prefer larger, more fecund females as mates. This brings us to prediction 4: *under food stress larger females are expected to mate more frequently* as *revealed by a body size × mating frequency correlation.*

This prediction was supported in the cage experiments with both pollen katydids and Mormon crickets. In pollen katydids, the pooled data from all replicates within each diet group (see legend to Fig. 9.11) showed that heavyweight females in low-diet cages mated significantly more frequently. On the other hand, there was, as predicted, no mating advantage to body size in high-diet cages (Gwynne and Simmons 1990).

However, a positive correlation between female weight and mating frequency may be due to weight gain from the more frequent spermatophylax meals acquired by large females, rather than large females attracting

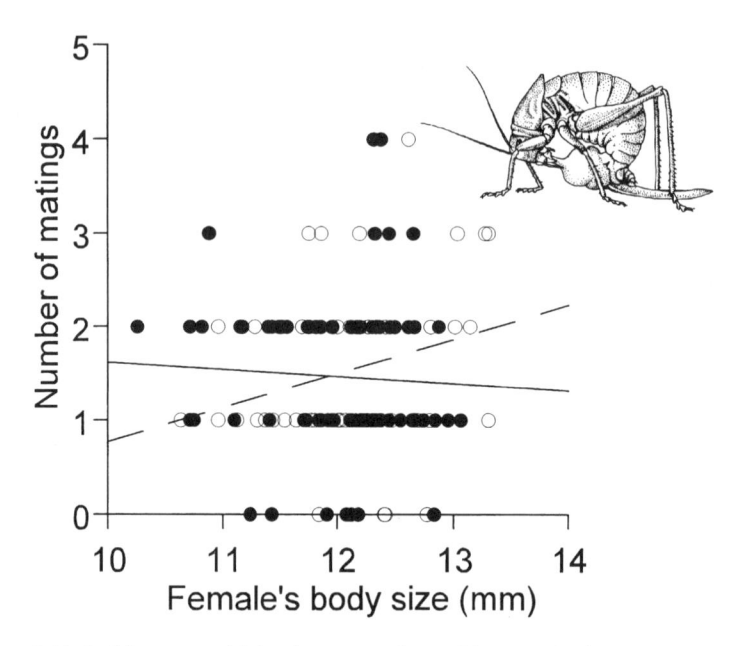

Figure 9.16. In Mormon crickets there was detectable sexual selection for increased female body size in the food-stressed cages where sexual competition among females had been observed. This was revealed in a significant positive regression between female size (pronotum length) and mating frequency (open points and dashed line). No significant regression exists for data from cages with ad-lib food (closed points and continuous line). Because these correlations did not differ between the four replicate cages within each diet group, they were pooled for analysis within each group. Females that had mated were easily spotted by the large spermatophylax "tag." Katydid from artwork by K. Hansen McInnes in Gullan and Cranston (1994).

more matings as posited by the sexual selection hypothesis (Gwynne and Simmons 1990). This cause-and-effect problem was dealt with in work with Mormon crickets because our measure of body size was pronotum length, a parameter that does not change after the adult molt. There was a positive correlation between pronotum length and number of spermatodoses (mating frequency) only in role-reversed populations (Gwynne 1984c), and experimental manipulation of diet (Gwynne 1993) confirmed this. This correlation was found in the low-diet cages where we observed role reversal but not in high-diet cages where the roles were typical (Fig. 9.16). These data support the hypothesis that sexual selection on females favors larger body size in role-reversed populations.

9.5.vi. Food availability and relative parental investment

In theory (Fig. 9.10), a reversal in sexual selection on the sexes is caused by the time-out from mating activity by males exceeding the time-out by females (Parker and Simmons 1996). The greater time-out of males, per se, is argued to be due to male investment in individual offspring (at the cost of investing in other offspring) exceeding female investment (Trivers 1972). Is there any evidence for katydids that *there is a reversal in parental investment by the sexes when food is limited* (prediction 5)?

For katydids there have been no measures of relative parental investment in terms of the cost of producing other offspring. However, because investment by both sexes appears to be similar in currency[3]—both sexes make a prezygotic nutrient contribution to the eggs (Bowen et al. 1984)—it may be possible to estimate relative parental investment, by measuring investment in individual offspring. For pollen katydids in particular, male and female parental investment appears to be similar because nutrition from both pollen and the spermatophylax has similar effects on the number and size of eggs that the female subsequently produces (Simmons and Bailey 1990). Thus, when food is limited, the proportion of materials of male origin (spermatophylax) in individual offspring (eggs) is expected to increase because females would have less material from other sources in their own reserves (Gwynne 1991).

We first examined this prediction by examining investment in a single reproductive event (mating and egg production) using radioisotope-labeled nutrients. ^{14}C- and ^{3}H-labeled proteins identified nutrients of male origin (the spermatophore meal) or female origin (her own reserves) (Simmons and Gwynne 1993). Our subjects were pollen katydids because their diet is so easily manipulated by controlling the amount of pollen they receive. Thus, we could simulate the extremes of diet that would cause typical and reversed courtship roles in nature (Gwynne and Simmons 1990).

Our results showed that the ratio of female to male nutrients did not differ between low and high diets but there was less of each labeled nutrient in eggs of high-diet females due to dilution by the additional (unlabeled) food, and thus greater fecundity, in the high-diet group. These findings support the prediction of a relatively higher parental investment by males in the low-diet group compared to the high-diet group (Simmons and Gwynne 1993). Note that there is no inconsistency between the prediction and our finding of no between-treatment difference in the ratio of labeled male and female nutrients. This is because total relative investment by the sexes included not only our radiolabeled male and female contributions but also unlabeled nutrients, in particular, the large amount received by high-diet females that caused a significant boost in fecundity. Females in both diet groups received equal male input, including unlabeled spermatophore nutrients, as all experimental males were well fed

---

[3] This is in contrast to other animals in which parental investment by the sexes can come in different currencies, e.g., investment in gametes versus in parental care (Knapton 1984).

and investment in the spermatophylax meal by male pollen katydids is not reduced, even when under food stress (Simmons 1994b).

In a less complex method of measuring parental investment in pollen katydids, Simmons (1992, 1995c)[4] estimated male and female energy investment in eggs using calorimetry.[5] Again, investment was examined for a single reproductive event (mating and the batch of eggs laid before the female would have remated in nature). Simmons's results not only supported the prediction that low-diet males invested relatively more in individual offspring (eggs) than did high-diet males (Gwynne 1991) but also showed that when the male's parental expenditure exceeds the female's, courtship role reversal should occur (in this case between the extremes of food intake known to cause reversed and typical courtship roles in field cages: Fig. 9.13A).

Another study of male and female nutrient investment in a female's first reproductive event examined *R. verticalis* using the radiolabeling technique. For this species, Gwynne and Brown (1994) found that in contrast to pollen katydids (Simmons and Gwynne 1993), the proportion of radiolabeled male nutrients per egg actually decreased in low-diet compared to high-diet females, despite the expected "dilution" by unlabeled food in the high-diet, higher-fecundity females. The reduced male investment in eggs from low-diet females (labeled female nutrients per egg were similar between diets) was a result of food-stressed females preferentially retaining male spermatophore nutrients in somatic (e.g., storage) tissues, perhaps because *R. verticalis* females store specialized nutrients provided by males that are not available in the general diet (Gwynne 1988a). For young females in stressful environments, decreased expenditure in current reproduction may be adaptive if important nutrients can be stored for expected future reproduction (Gwynne and Brown 1994). Hence, differing life histories may explain the contrasting results from radiolabeling studies of *K. nartee* and *R. verticalis*. However, food stress should make female remating relatively more likely than male remating, whether they use the additional spermatophylax nutrients to produce current (maturing eggs) or future offspring (storage tissues).

In terms of the control of sexual selection, what is important is that male-derived nutrients used in offspring production are sufficiently costly[6] that they remove the male from the pool of individuals able to mate; that is, they increase male time-out relative to female time-out.

---

[4] With the publication of Simmons's (1992) paper a pollen katydid, *K. nartee*, graced the cover of the journal *Nature* for the second time in 2 years!

[5] Radiolabeling data (Simmons and Gwynne 1993) were used to subtract the portion of the energy in the egg that was derived from spermatophore nutrients.

[6] This is the case whether the male contributions function to increase the fitness of the male's own offspring or to maximize his fertilization rate (see Chapter 6) (Gwynne 1984b; Quinn and Sakaluk 1986).

9.5.vii. Food availability and time-out from mating

A newly mated male takes a time-out from mating by taking a temporary respite from calling, in order to replenish his spermatophore glands. A female time-out occurs when she avoids mating in order to mature eggs. *When the time-out for females exceeds the time-out for males, we expect a reversal in the courtship roles* (prediction 6). Parker and Simmons (1996) supported this prediction using data from experiments in which mating pairs experienced low or high diets in experimental field cages (Gwynne and Simmons 1990) (Figs. 9.11 and 9.13). Parker and Simmons plotted the length of time-outs for the male, the time from mating until he resumes calling, against the length of time-outs for his mate, her refractory period (Fig. 9.13B). They found that the duration of male time-outs decreases and that of female time-outs increases as the amount of food in the diet increases (from right to left in the figure in which the line indicates equality in time-outs for the sexes). Moreover, there appears to be a reversal in the relative time-outs of the sexes that predicts the reversal in the courtship roles: for most of the high-diet females (from cages with katydids showing typical roles: Fig. 9.13E), male time-outs exceed female time-outs, whereas in the four measured low-diet females (from cages with katydids showing role reversal), the reverse is true (Fig. 9.13B).

9.5.viii. Relationships between the factors controlling sexual selection

Our examination of the factors controlling sexual differences so far has examined the effect of variation in diet on individual factors (predictions 1–6 in Fig. 9.10) as well as the consequences for sexual selection and sexual differences in reproductive behavior. But what about the hypothesized relationships between the factors? Simmons (1995c) addressed this question in some elegant experiments with pollen katydids. Parental expenditure should covary with relative mating rates: *as male parental expenditure decreases relative to female expenditure, the mating rates of females relative to males is expected to increase* (prediction 7). Simmons's results (Fig. 9.13D) support this. Furthermore, *as male parental expenditure decreases relative to female expenditure, the reproductive rates of males increase relative to females* (prediction 8). Again the data (Simmons 1995c) support this prediction.

9.6. Reversed Sexual Selection and Darwinian Devices in Females

Experiments with katydids in the laboratory and the field, in general, have been successful in providing the first broad tests of the several interrelated factors that in theory (Fig. 9.10) control sexual differences in sexual selection and the courtship roles of males and females. The feature of the katydid system that makes it so useful in testing these theories is the unique flexibility in the courtship roles. Given the exceptional nature of mating behavior in katydids, can we generalize from our results to the evolution of sexual differences in other animals? Any such generalizations assume that factors observed to vary seasonally or ecologically in katydids, such as relative parental expenditure, have changed in most other animals over evolutionary time to produce species-specific sexual differences not just in behavior but also in structure, as in the brightly colored plumage of female button quails (Darwin 1874).

Darwin's (1874) main focus, on sex differences in structure, is the concluding topic of this chapter. I focus on katydids and address whether there is any evidence that reversals in sexual selection have led to the evolution of reversed sexual dimorphism in any structure. We know that female size (as measured by pronotum length) is under sexual selection in role-reversed populations (see section 9.5.v), and consistent with the hypothesis that this character has undergone sexually selected evolutionary change, it turns out that the pronotum in role-reversed female Mormon crickets in nature is longer than the pronotum in males. In contrast, the size of the sexes is not different at the more eastern site (see Chapter 1) where mating roles are typical (Gwynne 1984c). These between-site differences in size dimorphism may be evolved and inherited because of different long-term selection pressures that appear to have produced different overall life histories in each area (see Chapter 4). Extremely high-density populations of gregarious Mormon crickets (see Chapter 1) have been reported from northwestern Colorado and in many other areas each year for the hundred years or so that records have been kept (Wakeland 1959). Bands of crickets occasionally wander into areas of plentiful food (Gwynne 1984c), making plasticity in the courtship roles adaptive, but these populations usually appear to be starved for food. In contrast, high population densities have never been reported from the areas east of the Rockies in Colorado (Wakeland 1959) where I first observed Mormon crickets with typical courtship roles. However, many more sites need to be sampled in order to test adequately the hypothesis that increased sexual selection on female Mormon crickets is correlated with a female-biased size dimorphism.

For role-reversed katydids such as Mormon crickets and pollen katydids, there is no evidence of the more familiar Darwinian devices such as weaponry and flashy colors, possibly in part because the traits that females assess when choosing mates are expected to be quite different from the traits males assess (more on this topic at the end of the chapter). But Darwin (1874) pointed to another class of sexual dimorphisms: "The sexes, also, often differ in their organs of sense . . . so that the males may quickly discover . . . the females" (see also Thornhill and Alcock 1983; Andersson and Iwasa 1996). The "organs of auditory sense" in pollen katydids do show a striking sexual dimorphism associated with elaboration of the thoracic auditory spiracle, a main input into the auditory system of many katydids. *K. nartee* females have an open spiracle that enters into an enlarged auditory trachea, both structures being typical of both sexes in other katydids (see Chapter 5). In contrast, the auditory spiracle in the male pollen katydid is a closed structure (not much different from an abdominal respiratory spiracle) that leads into unmodified tracheae (Bailey and Simmons 1991) (Fig. 9.17). Furthermore, this dimorphism in structure is associated with a strong sex difference in hearing sensitivity; the male is

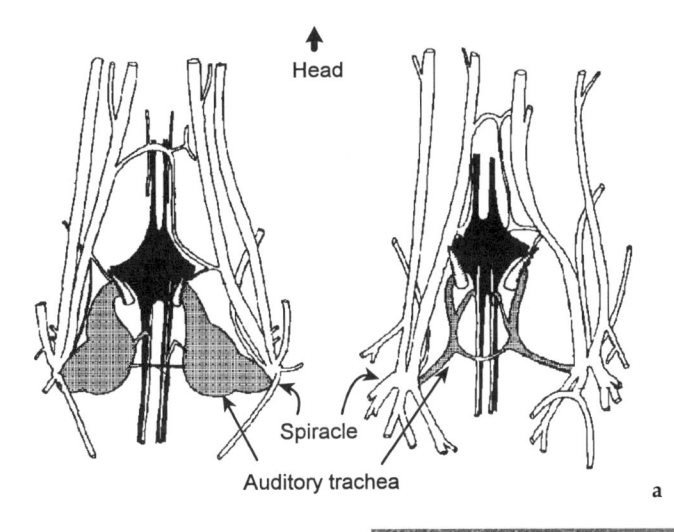

Figure 9.17. Sexual dimorphism in the thoracic auditory structures of pollen katydids, *Kawanaphila nartee*, is seen in both the enlarged auditory tracheal chamber in females (**a**, left) compared to males (**a**, right) and the size of the auditory spiracles (**b**) that enter into these tracheae: females have a large open spiracle whereas males (top) possess a closed structure. From Bailey and Römer (1991).

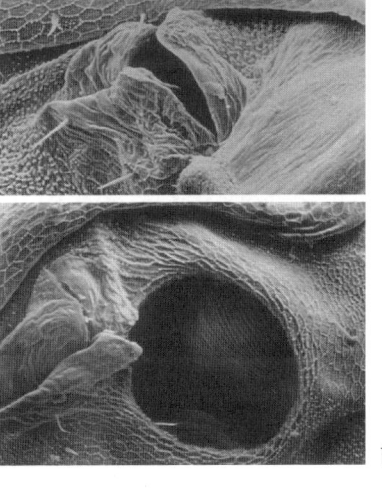

some 10-dB less sensitive to the calling song than is the female (Bailey and Römer 1991).

W. J. Bailey and I tested Darwin's hypothesis as applied to this unique sexual dimorphism in the ears of pollen katydids, a species whose females rather than males "quickly discover" their mates (Gwynne and Bailey 1999). We began by predicting that the size or elaboration of the sensory structure would correlate with mating success in females, on the understanding that females with larger auditory spiracles ought to respond more quickly to calling males. The prediction was based on two aspects of pollen katydid biology. First, as we have seen, in areas with little food there was intense sexual competition among female pollen katydids in reaching a caller, both in the initial sprint of responsive females to a new male caller and in later stages when rival females were often involved in a series of fighting bouts, when the coy male silently withdrew and later "clicked" to reattract females, as noted previously (see section 9.2). Second, female katydids with a larger auditory spiracle are known to be more sensitive to the song of conspecific males (e.g., see Fig. 5.16 for *R. verticalis*) (Bailey 1998a).

Our experimental test with pollen katydids consisted of a series of trials in nature in which two females attracted to a caller were allowed to compete for him. Our winner, designated as the first female to mount and grasp the male in a genital hold (Figs. 9.2 and 9.6), as predicted, had a larger auditory spiracle (Fig. 9.18). We also ran a series of control trials in which females were allowed to struggle for a male later in the mating sequence when the male was silent. In controls, we predicted and found no spiracle size advantage (Gwynne and Bailey 1999) (Fig. 9.18). Surprisingly, there was no overall body size advantage for winning females in either experiment. This finding ruled out the possibility that the large spiracle of females mating with a caller was a result of a correlation with large body size.

This work raised a number of questions that should be addressed in future research. First, if auditory sensitivity is so important to competitive success, why don't all females have equally large spiracles? We suspect that only the most fit females can sustain the large cost of maintaining large open spiracles that are likely to be conduits for water loss and for the entrance of parasitic tracheal mites (see Chapter 4). Second, the species-specific "fixed" nature of the sexual dimorphism in the size of the auditory spiracle in *K. nartee* suggests that most populations of these pollen katydids experience role-reversed sexual selection most of the time. However, it remains possible that the dimorphism was produced by selection not only increasing the size of the female spiracle but also reducing the size of the male's spiracle. Effective hearing may not be as important to the male-male competition in dense singing groups (Simmons and

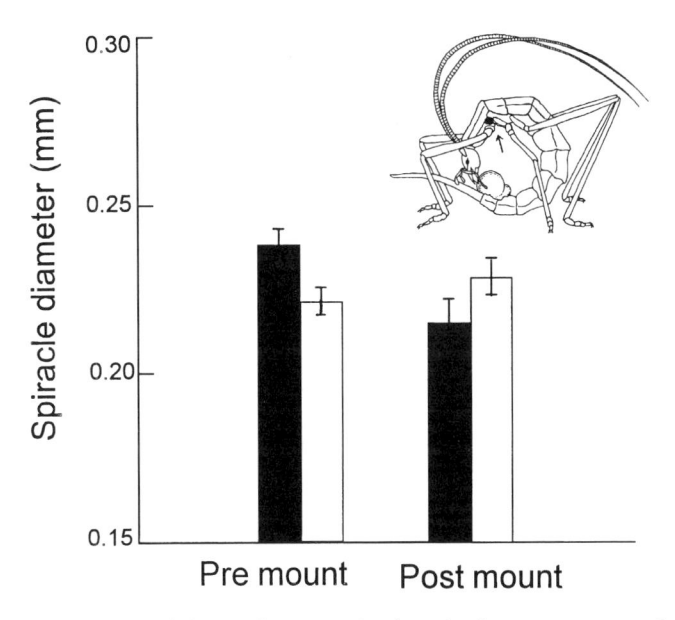

Figure 9.18. Females with larger thoracic spiracles win the race to mount the calling male ("premount" trials) but not in postmount trials in which males did not sing. Bars represent mean diameter of the spiracle for the winners (black) and losers (white). Bars represent standard errors. Redrawn from Gwynne and Bailey (1999). Katydid redrawn from H. Proctor (Fig. 9.1).

Bailey 1992) because males communicate effectively using vibrations (Bailey and Simmons 1991).

In pollen katydids, sexual competition among females no doubt creates sexual selection not only on auditory structures and body size but also on other devices, including what appears to be a "clamp organ" for maintaining a genital hold on males (Figs. 9.2 and 9.19). The amazing diversity of this apparent holding device in other species of *Kawanaphila* (Fig. 9.19), as revealed in a taxonomic revision of pollen katydids (Rentz 1993), reminds us that there can be strong sexual selection on the "terminalia" of insects, usually in males (Eberhard 1985; Arnqvist 1998) but in this case perhaps in females.

Is there evidence in any katydid species for the more familiar Darwinian sexual dimorphisms such as weaponry or other elaborate devices? Darwin's (1874) illustrated examples of sexual structures in animals included the male singing organs of a South American false leaf katydid (Pseudophyllinae: Fig. 7.3). Other sexually selected male devices appear to occur in South American pseudophyllines. These include elongated heads and mandibles in males of genera such as *Gnathoclita* (Fig. 9.19 and Plate 24) (Beier 1960). This sexual dimorphism in head size and shape is

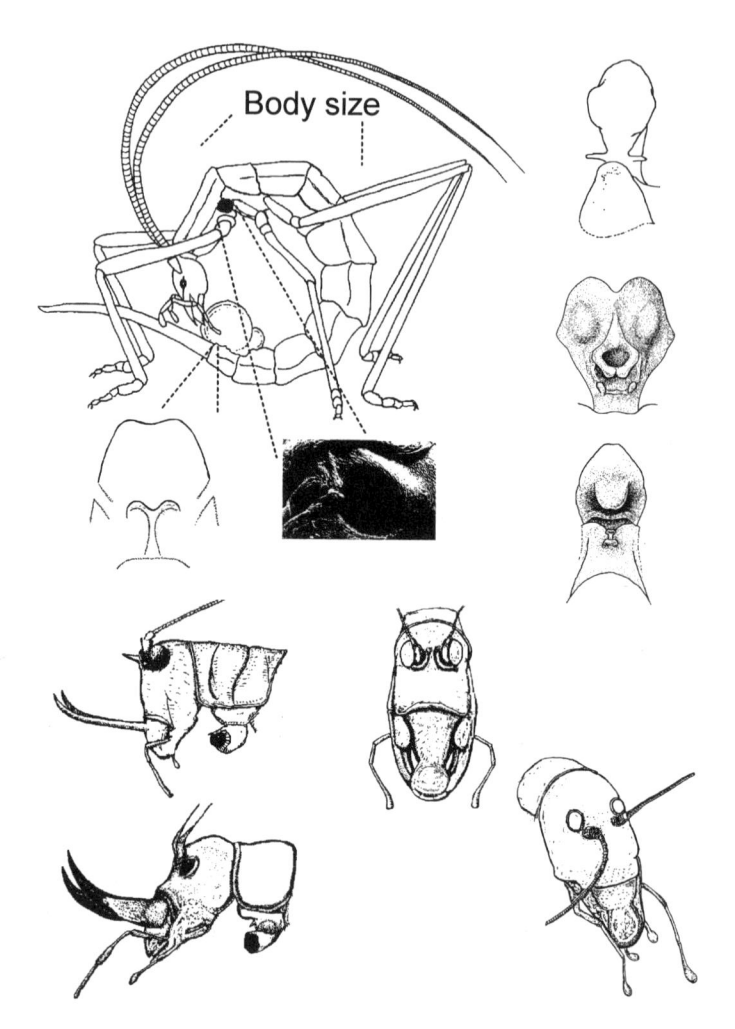

Figure 9.19. Sexual selection and sexual dimorphism in katydids. Top left: *Kawanaphila nartee* female (redrawn from artwork by Heather Proctor [Fig. 9.1]), showing the main features that appear to be under sexual selection: body size, auditory spiracle size, and possibly the elaboration of the organ for "genitally holding" the mounted male. To the right are the elaborate genital holding organs of females in three other species of *Kawanaphila* (redrawn from Rentz 1993). At the bottom of the figure are strikingly similar cephalic stuctures of katydids (top row) and stenopelmatids (weta, bottom row) that have independently evolved in both groups. The left horned katydid is a female *Dicranostomus monocerus* (Pseudophyllinae) (redrawn from Beier 1960). Compare her structures to the male elephant weta, *Manuweta isolata*, below (redrawn from Johns 1997). The extended jaws of the male katydid *Gnathoclita sodalis* (Pseudophyllinae) (redrawn from Beier 1960) are very similar to those of the male tree weta (*Hemideina* species) pictured below the *Gnathoclita* head (redrawn from Meads 1990).

9.6 Reversed Sexual Selection and Darwinian Devices 257

strikingly similar to that observed in related ensiferans, the weta genus *Hemideina* (Stenopelmatidae: see Chapter 2, Fig. 2.7). Large jaws are used in fights between male weta to defend harems of females (Gwynne and Jamieson 1998) (compare the two figures in the lower right of Fig. 9.19).

However, the theme of this chapter is reversed sexual selection. So what about the possibility of sexual weaponry or flashy ornaments in females? From a theoretical viewpoint, ornaments in females are less likely to evolve even in cases of reversed sexual selection because of a fundamental difference in the nature of sexual selection in male-choice versus female-choice systems (Fitzpatrick et al. 1995). Choosy males are much more likely to be seeking the immediate benefits of high fecundity in females, as in Mormon crickets, instead of the genetic benefits that are often raised in discussions about female choice for the "arbitrary" flashy traits of males (Table 7.1). As a result there is expected to be selection against males choosing excessive female ornamentation if the more flashy females have lower fecundity owing to their investments in sexual structures.

Despite the theory, however, there is one striking example of female ornamentation that may be a result of male choice. This occurs in an insect with a similar life history to most katydids in that males feed their mates and appear to have a courtship role reversal. Female dance flies, *Rhamphomyia longicauda* (Diptera: Empididae; Table 9.1), compete in large swarms, and prey-bearing males entering swarms from below choose large-bodied females. Sexual selection on female may have produced bizarre structures that function to exaggerate size when flying in the swarm (Fig. 9.20). These structures include an abdomen that a female inflates with air before joining her flying rivals in a swarm (Funk and Tallamy 2000).

Is there any evidence at all of elaborate sexually selected devices in female katydids that might have evolved through a reversal in sexual selection on the sexes? We know that food-stressed female Mormon crickets and pollen katydids will fight for access to males able to supply spermatophore meals, so could female weaponry or displays evolve in such systems? Again South American pseudophyllines offer some enticing evidence (illustrations can be found in the taxonomic monographs by Beier 1960). For weapons, Beier illustrated several species in which the mandibles of females are extended into tusklike protrusions. The tusks of *Dicranostomus monocerus* females in particular are reminiscent of male tusks in "elephant weta" (Fig. 9.19, lower left), a species with males that joust in apparent contests over females (McIntyre 1991). Could the tusks of *Dicranostomus* females function in battles for access to meal-providing males?

A final structure that may well be a sexually selected device in females concerns sound-producing organs. I have often been asked if there is any evidence for female acoustic displays in role-reversed katydids. I recently

Figure 9.20. a. Two *Rhamphyomya longicauda* females (dance fly: Diptera: Empididae) inflate their abdomens before entering an all-female swarm. **b.** The inflated abdomen and leg hairs of a female in the swarm exaggerate her size to males entering the swarm from below. Photo by David Funk.

came across some enticing evidence—for yet another pseudophylline— when examining katydids collected in Ecuador.[7] *Panoploscleis specularis* females have extraordinary tegminal structures that are even more elaborate than the (typical katydid) tegminal organ of their conspecific males (see Chapter 5). Each female forewing has an extensive membranous area subdivided by five or six tooth-bearing veins that together apparently

[7] Thanks to Nic Tatarnic, Doug Currie, and Chris Darling of the Royal Ontario Museum.

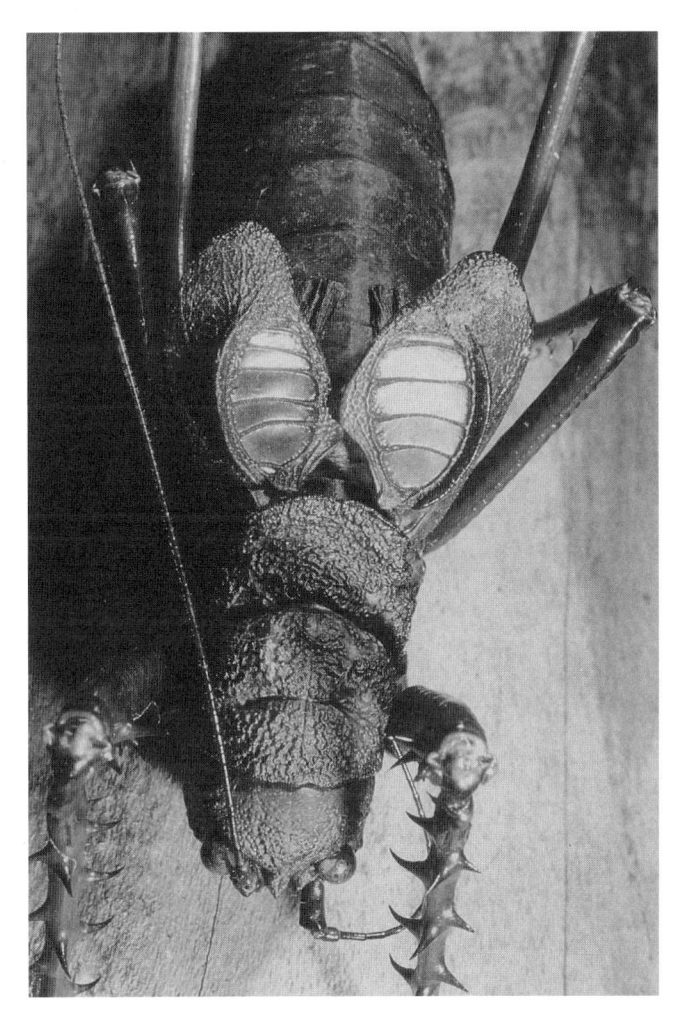

9.6 Reversed Sexual Selection and Darwinian Devices 259

Figure 9.21. A female sexual display in the katydid *Panoploscelis specularis* (Pseudophyllinae)? Here the forewings are spread to show their striking modifications that appear to be song-producing structures. The male of this species has the typical katydid's file-and-scraper tegminal mechanism (see Chapter 5). In contrast, each wing of the female has multiple files! Beier (1960) described each wing's "glassy diaphragm" with transverse veins bordered on its outer edge by an enormous thickened vein. But each of the five or six transverse veins has a single row of teeth on its dorsal surface. There are also numerous teeth on the wing surface posterior to the glassy areas (personal observations of specimens in the Royal Ontario Museum; the same wing overlap and number of veins are seen in Beier's drawings). To produce sounds, a scraper on the right edge of the left wing of this large insect (approximately 7 cm in length) would appear to move over the multiple files on the right wing (the typical overlap for the tegminae of katydids, see Chapter 4). The glassy diaphragms, particularly the large one on the "scraper wing," probably function in radiating sounds.

function in producing and radiating sound (details in Fig. 9.21)! Beier (1960) also noted and illustrated membranous areas and cross-veins. It seems unlikely that this elaborate structure functions as a defense against predators (acoustic devices for defense tend to be simple structures devoid of resonating areas: Ewing 1989). A second hypothesis is that the structure produces a song that answers a calling male. Again, however, the structures that produce such female sounds are simple ones (some European bradyporines: Hartley et al. 1974; see Chapter 5). Could the extravagant forewing devices of *P. specularis* have evolved via sexual selection on females to provide a conspicuous acoustic display, either in answer to the songs of meal-providing males or as an aggressive signal to rival females?

These few observations on the structures of false leaf katydids remind us that much more behavioral research needs to be done, particularly in the neotropics and other areas of the world that have received relatively little attention. As the large amount of ongoing systematic work on the world's katydids reveals more and more diversity, other examples of sexual dimorphism and behavior will undoubtedly come to light, suggesting new and interesting consequences of sexual selection and thus raising even more questions about the sexual patterns of this fascinating and diverse group of insects.

Appendix

List of katydid species mentioned in this book (Chinese common names from Jin 1994)

Genus and species	Common name	Subfamily	Tribe	Location
Acanthodis curvidens (Stål 1875)		Pseudophyllinae	Pleminiini	Costa Rica
Acanthoplus discoidalis (Walker 1869)		Hetrodinae		Africa
A. speiseri Brancsik 1895		Hetrodinae		S. Africa
Acripeza reticulata Guerin 1832	Mountain katydid	Phaneropterinae	Acanthoplini	Eastern Australia
Acrodectes philophagus Rehn & Hebard 1920		Tettigoniinae	Acanthoplini	California
Aganacris insectivora Grant 1958		Phaneropterinae		S. America
Agraecia pulchella Hebard		Conocephalinae	Agraeciini	S. America
Amblycorypha oblongifolia (DeGeer 1773)	Oblong-winged katydid	Phaneropterinae		N. America
A. parvipennis Stål 1876	Western round-winged katydid	Phaneropterinae		N. America
Anabrus simplex (Haldeman 1852)	Mormon cricket	Tettigoniinae		N. America
Ancistrocercus circumdatus (Walker 1870)		Pseudophyllinae	Pleminiini	Neotropics
A. inficitus (Walker 1870)		Pseudophyllinae	Pleminiini	Neotropics
Ancistrura nigrovittata (Brunner V. Wattenwyl 1878)	Green-faced katydid	Phaneropterinae		Macedonia
Anonconotus alpinus Yersin 1858	Small alpine bush-cricket	Tettigoniinae		Europe
Antaxius pedestris (Fabricius 1787)	Pyrenean bush-cricket	Tettigoniinae		Italy
Anthophiloptera dryas Rentz & Clyne 1973	Pollen katydid	Zaprochilinae		Eastern Australia
Apteropedetes anaesegalae (Gurney & Lieberman 1976)		Tettigoniinae		Argentina

261

Appendix (continued)

Genus and species	Common name	Subfamily	Tribe	Location
Arachnoscleis Karny 1912		Listroscelidinae		Colombia
Arethaea grallator (Scudder 1877)		Phaneropterinae		Southern U.S.
Atlanticus testaceous (Scudder 1901)		Tettigoniinae		Eastern U.S.
Austrodectes monticolus Rentz 1985		Tettigoniinae		Australia
Austrosalomona Rentz 1988		Conocephalinae	Agraecini	Australia
Balboa tibialis (Brunner V. Wattenwyl 1895)		Pseudophyllinae		Panama
Barbitistes ocskayi Charpentier 1850		Phaneropterinae		Eastern Europe
B. serricauda (Fabricius 1794)	Saw-tailed bush-cricket	Phaneropterinae		Europe
Belocephalus Scudder 1875	Short-winged cone-head	Conocephalinae	Copiphorini	Southeastern U.S.
Bliastes insularis Bruner 1906		Pseudophyllinae	Cocconotini	S. America
Bradyporus dasypus (Illiger 1800)		Bradyporinae		Eastern Europe
B. multituberculatus		Bradyporinae		Eastern Europe
Bucrates capitatus (De Geer 1773)		Conocephalinae		Panama
Caedicia simplex (Walker 1869)	Kikihipounami	Phaneropterinae		Australia, New Zealand
Callimenellus Walker 1871		Pseudophyllinae	Callimenellini	Eurasia
Chlorodectes montanus Rentz 1985		Tettigoniinae		Australia
Choeroparnops gigliotosi Beier 1960		Pseudophyllinae	Platyphyllini	Ecuador
Clonia multispina Uvarov 1942		Saginae		Africa
Cocconotus wheeleri Hebard 1927		Pseudophyllinae	Cocconotini	Panama
Conocephalus species	Smaller meadow katydids	Conocephalinae	Conocephalini	Worldwide
C. attenuatus (Scudder 1869)	Lance-tailed meadow katydid	Conocephalinae	Conocephalini	Eastern N. America
C. brevipennis (Scudder 1862)	Short-winged meadow katydid	Conocephalinae	Conocephalini	Eastern N. America
C. discolor Thunberg 1815		Conocephalinae	Conocephalini	Western Europe
C. dorsalis Latreille 1804		Conocephalinae	Conocephalini	Europe

C. fasciatus (De Geer 1773)	Slender meadow katydid	Conocephalinae	Conocephalini	N. America
C. longipennis (De Haan 1842)		Conocephalinae	Conocephalini	Malaysia
C. maculatus (Le Guillou 1841)	Cao Zhong, grass katydid	Conocephalinae	Conocephalini	Asia
C. nigropleurum (Bruner 1891)	Black-sided meadow katydid	Conocephalinae	Conocephalini	Eastern N. America
C. semivittatus (Walker 1869)		Conocephalinae	Conocephalini	New Zealand
C. spartinae Fox 1912		Conocephalinae	Conocephalini	Eastern U.S.
C. upoluensis (Karny 1907)		Conocephalinae	Conocephalini	Western Australia
Copiphora brevirostris Stål, 1873		Conocephalinae	Copiphorinae	Colombia
C. rhinoceros Pictet 1888		Conocephalinae	Copiphorinae	Central America
Coptaspis Rettenbacher 1871		Conocephalinae	Agraecini	Australia
Cyrtaspis scutata (Charpentier 1825)		Meconematinae		
Decticita (Caudell 1907)		Tettigoniinae		California
Decticoides brevipennis Ragge 1977	Degasa and wollo cricket	Tettigoniinae		Ethiopia
Decticus albifrons (Fabricius 1775)	White-faced decticus	Tettigoniinae		Europe
D. verrucivorus (Linn. 1758)	Wart biter	Tettigoniinae		Europe
Dicranostomus monocerus Dohrn 1888		Pseudophyllinae		Peru
Docidocercus gigliotosi (Griffini 1896)		Pseudophyllinae		Panama
Ducetia japonica (Thunberg 1815)	Lu Zhong, wing exposed katydid	Phaneropterinae		Asia
Elephantodeta nobilis (Walker 1869)		Phaneropterinae		Australia
Enyaliopsis nyala Glenn 1991		Hetrodinae	Enyaliopsini	Africa
E. nyika Glenn 1991		Hetrodinae	Enyaliopsini	Africa
Ephippiger ephippiger (Fieber 1784)	Saddle-backed bush-cricket	Bradyporinae	Ephippigerini	Western Europe
E. terrestris Yersin 1854		Bradyporinae	Ephippigerini	France
Ephippigerida saussuriana Bolivar 1878		Bradyporinae	Ephippigerini	Spain
E. taeniata (Saussure 1878)	Large striped bush-cricket	Bradyporinae	Ephippigerini	Morocco
Eremopedes Cockerell 1898		Tettigoniinae		Western U.S.
Eubliastes chlorodictyon Montealegre & Morris 1999		Pseudophyllinae	Cocconotini	S. America
Euconocephalus pallidus (Redtenbacher 1891)		Conocephalinae	Conocephalini	Africa

Appendix (continued)

Genus and species	Common name	Subfamily	Tribe	Location
Eugaster spinulosa (Linn. 1764)		Hetrodinae	Eugastrini	Africa
Eupholidoptera megastyla (Ramme 1939)		Tettigoniinae		Greece
Euthypoda acutipennis Karsch 1886		Mecopodinae		W. Africa
Euthyrrachis Brunner 1878		Phaneropterinae		S. America
Gampsocleis burgeri De Haan 1842		Tettigoniinae		Eurasia
G. glabra (Herbst 1786)		Tettigoniinae		Eurasia
G. gratiosa (Brunner V. Wattenwyl 1862)	Jiao Ge-Ge, singing brother	Tettigoniinae		China
Gnathoclita sodalis Brunner V. Wattenwyl 1895		Pseudophyllinae	Eucoconotini	Colombia
G. vorax (Stoll 1813)		Pseudophyllinae	Eucoconotini	Guyana
Gravenreuthia saturata Karsch 1862		Phaneropterinae		Cameroon
Gymnoproctus sculpturatus Schmidt 1990		Hetrodinae		Africa
Haenschiella Beier 1960		Pseudophyllinae	Platyphyllini	S. America
Hemisaga denticulata (White 1841)	Pollen katydid	Austrosaginae		Southwestern Australia
Hexacentrus mundus Walker 1869		Listroscelidinae		Eurasia
H. unicolor (Serville 1831)	Xiao Fang Zhi Niang, small weaving lady	Listroscelidinae		Asia
Holochlora nigrotympana Ingrisch 1990		Phaneropterinae		Thailand
Horatosphaga meruensis (Sjöstedt 1909)		Phaneropterinae		Tanzania
Idiarthron major Hebard 1927		Pseudophyllinae	Cocconotini	Central America
Insara covillae Rehn & Hebard 1914		Phaneropterinae		Central U.S.
Ischnomela pulchripennis Rehn 1906		Pseudophyllinae	Ischnomelini	Central America
Isophya acuminata Brunner V. Wattenwyl 1878		Phaneropterinae		Eurasia

Species	Common name	Subfamily	Tribe	Region
I. kraussi Brunner V. Wattenwyl 1878	Krauss's bush-cricket	Phaneropterinae		Europe
Kawanaphila nartee Rentz 1993		Zaprochilinae		Southwestern Australia
Lacipoda immunda Brunner V. Wattenwyl 1895		Pseudophyllinae	Phyllomimini	Indonesia, Malaysia
Leptophyes albovittata (Kollar 1833)	Striped bush-cricket	Phaneropterinae		Europe
L. boscii Fieber 1853		Phaneropterinae		Europe
L. laticauda (Frivaldsky 1867)	Long-tailed bush-cricket	Phaneropterinae		Europe
L. punctatissima (Bosc. 1792)	Speckled bush-cricket	Phaneropterinae		Europe
Lipotactes minutus Ingrisch 1995		Lipotactinae		Thailand
L. sylvestris Ingrisch 1990		Lipotactinae		Thailand
Macroxiphus sumatranus Ingrisch 1998		Conocephalinae	Agraeciini	Indonesia, Malaysia
Meconema meridionale Costa 1860	Oak bush-cricket	Meconematinae	Meconematini	Western Europe
M. thalassinum (DeGeer 1773)		Meconematinae	Meconematini	Western Europe
Mecopoda elongata (Linn. 1758)	Guo-guoer or Fang Zhi Niang; weaving lady	Mecopodinae		Asia
Metaballus litus Rentz 1985		Tettigoniinae		Western Australia
Metaplastes ornatus (Ramme 1931)		Phaneropterinae		Greece
Metholche nigritarsis Walker 1871		Conocephalinae	Agraeciini	Australia
Metrioptera bicolor (Philippi 1830)	Two-colored bush-cricket	Tettigoniinae		Eastern Europe
M. brachyptera (Linn. 1761)	Bog bush-cricket	Tettigoniinae		Europe
M. roeselii Hagenbach 1822	Roesel's bush-cricket	Tettigoniinae		Europe, N. America
M. saussuriana (Frey-Gessner 1872)	Saussure's bush-cricket	Tettigoniinae		Europe
M. sphagnorum (Walker 1869)	Bog katydid	Tettigoniinae		Canada
Microcentrum rhombifolium Saussure 1859	Angular-winged katydid	Phaneropterinae		Eastern U.S.
Mimetica Pictet 1888		Pseudophyllinae	Pterochrozini	Central, S. America
Mygalopsis marki Bailey 1979		Conocephalinae	Copiphorini	Southwestern Australia
M. pauperculus Walker 1869		Conocephalinae	Copiphorini	Southwestern Australia
M. sandowi Bailey 1979		Conocephalinae	Copiphorini	Southwestern Australia
M. thielei Bailey 1979		Conocephalinae	Copiphorini	Southwestern Australia
Myopophyllom speciosum Beier 1960		Pseudophyllinae		Ecuador
Nastonotus foreli Carl 1921		Pseudophyllinae	Cocconotini	Colombia
Neduba sierranus Rehn & Hebard 1910		Tettigoniinae		Western U.S.

Appendix (continued)

Genus and species	Common name	Subfamily	Tribe	Location
Neoconocephalus affinis Palisot de Beauvois 1805		Conocephalinae	Copiphorini	Central America
N. ensiger (Harris 1841)	Sword-bearer cone-head	Conocephalinae	Copiphorini	Eastern N. America
N. maxillosus (Fabricius 1775)		Conocephalinae	Copiphorinae	Central America
N. punctipes (Redtenbacher 1891)		Conocephalinae	Copiphorinae	Central America
N. robustus (Scudder 1862)		Conocephalinae	Copiphorini	Eastern U.S.
N. spiza Walker & Greenfield 1983		Conocephalinae		Central America
N. triops (Linn. 1758)	Broad-tipped cone-head	Conocephalinae	Copiphorini	Eastern N. America
Orchelimum species	Larger meadow katydids	Conocephalinae	Conocephalini	N. America, Mexico
O. concinnum Scudder 1862	Dusky-faced meadow katydid	Conocephalinae	Conocephalini	Eastern N. America
O. erythrocephalum Davis 1905		Conocephalinae	Conocephalini	Eastern N. America
O. fidicinium Rehn & Hebard 1907		Conocephalinae	Conocephalini	Eastern N. America
O. gladiator Bruner 1891	Gladiator meadow katydid	Conocephalinae	Conocephalini	Eastern N. America
O. nigripes Scudder 1875	Black-footed meadow katydid	Conocephalinae	Conocephalini	Eastern N. America
O. pulchellum Davis 1909	Handsome meadow katydid	Conocephalinae	Conocephalini	Eastern N. America
O. vulgare Harris 1841	Common meadow katydid	Conocephalinae	Conocephalini	Eastern N. America
Orophus conspersus (Brunner v. Wattenwyl 1878)	Esperanza bush katydid	Phaneropterinae		Central, S. America
Oxyaspis Brunner V. Wattenwyl 1895		Pseudophyllinae	Pseudophyllini	Africa
Oxycous lesnei Chopard 1936		Phaneropterinae		Africa
Pachysaga australis (Walker 1869)		Austrosaginae		Southwestern Australia
P. croceopteryx Rentz 1993		Austrosaginae		Southwestern Australia
Panacanthus Walker 1869	Cone-head	Conocephalinae	Copiphorini	Neotropics
Panoploscleis specularis Beier 1950		Pseudophyllinae		S. America
Pantecphylus cerambycinus Karsch 1891		Pseudophyllinae	Pantecphylini	Africa
Peranabrus scabricollis (Thomas 1872)	Coulee cricket	Tettigoniinae		Northwestern U.S.

List of Katydid Species 267

Phaneroptera falcata (Poda 1761)	Sickle-bearing bush-cricket		Phaneropterinae	Eurasia
P. nana Fieber	Sickle-bearing bush-cricket		Phaneropterinae	Europe
Phasmodes ranatriformis Westwood 1845	Stick katydid		Phasmodinae	Southwestern Australia
Phisis pallida (Walker 1869)		Phisidini	Meconematinae	Samoa
Phlugis poecila Hebard 1929			Listroscelidinae	Costa Rica
Pholidoptera griseoaptera (DeGeer 1773)	Dark bush-cricket		Tettigoniinae	Europe
Platycleis affinis Fieber 1853	Tuberous bush-cricket		Tettigoniinae	Southern Europe
P. albopunctata (Goeze 1778)	Grey bush-cricket		Tettigoniinae	Eurasia
P. intermedia (Serville 1839)	Intermediate bush-cricket		Tettigoniinae	Eurasia
P. nigrosinata (Costa 1863)	Black-marked bush-cricket		Tettigoniinae	Italy
P. sepium (Yersin 1854)	Sepia bush-cricket		Tettigoniinae	France
P. tessellata (Charpentier 1825)	Brown-spotted bush-cricket		Tettigoniinae	Europe
Poecilimon affinis (Frivaldsky 1867)			Phaneropterinae	Greece
P. intermedius (Fieber 1853)			Phaneropterinae	Russia
P. jonicus (Fieber 1853)	Ionian bush-cricket		Phaneropterinae	Italy, Greece
P. macedonius Ramme 1926			Phaneropterinae	Macedonia
P. mariannae Willemse 1992			Phaneropterinae	Eurasia
P. ornatus (Schmidt 1849)	Ornate bush-cricket		Phaneropterinae	Yugoslavia
P. schmidtii (Fieber 1853)			Phaneropterinae	Yugoslavia
P. thessalicus Brunner V. Wattenwyl 1891			Phaneropterinae	Greece
P. veluchianus Ramme 1933			Phaneropterinae	Eastern Europe
Polichne parvicauda (Stål 1860)			Phaneropterinae	Australia
Polyancistrus Rehn 1936		Polyancistrini	Pseudophyllinae	Dominican Republic
Polysarcus denticauda (Charpentier 1825)	Saw-tailed bush-cricket		Phaneropterinae	Eurasia
P. scutatus (Brunner V. Wattenwyl 1882)	Shielded bush-cricket		Phaneropterinae	Yugoslavia
Pristonotus tuberosus (Stål 1895)		Pleminiini	Pseudophyllinae	Panama
Promeca perakana Beier 1954			Pseudophyllinae	Indonesia, Malaysia
P. sumatrana Beier 1954			Pseudophyllinae	Sumatra
Pseudosubria Ingrisch 1998			Conocephalinae	Asia
Pseudotettigonia amoena (Henriksen) (fossil)		Agraeciini		Fossil
Pterochroza ocellata (Linn. 1758)		Pterochrozini	Pseudophyllinae	Peru

Appendix (continued)

Genus and species	Common name	Subfamily	Tribe	Location
Pterophylla beltrani Bolívar 1942	Grilleta (false locust)	Pseudophyllinae	Pterophyllini	Mexico
P. camellifolia (Fabricius 1775)	True katydid	Pseudophyllinae	Pterophyllini	Eastern N. America
P. robertsi (Hebard 1941)		Pseudophyllinae	Pterophyllini	Mexico
Pycogaster intermis (Rambur 1839)	Unarmed bush-cricket	Bradyporinae	Bradyporini	Spain
Requena verticalis Walker 1869	Garden katydid	Listroscelidinae		Southwestern Australia
Rhachidorus Herman 1874		Tettigoniinae		Australia
Roxelana crassicornis Stål 1874		Pseudophyllinae	Pterochrozini	S. America
Ruspolia differens Serville 1839	Cone-head	Conocephalinae	Copiphorini	Africa
R. lineosus (Walker 1869)	Cao Zhong, grass katydid	Conocephalinae	Copiphorini	China
R. nitidula (Scopoli 1786)	Cone-head	Conocephalinae	Copiphorini	Africa
Saga campbelli Uvarov 1921		Saginae		Macedonia
S. natoliae Serville 1839		Saginae		Eurasia
S. pedo Pallas 1771		Saginae		Europe, U.S.
Sciarasaga quadrata Rentz 1993		Austrosaginae		Southwestern Australia
Scopiorinus fragilis (Hebard 1927)		Pseudophyllinae	Pterophyllini	Central America
Scudderia curvicauda (De Geer 1773)	Curve-tailed bush-katydid	Phaneropterinae		Eastern N. America
Segestes decoratus Redtenbacher 1872		Mecopodinae	Sexavaini	New Guinea
Segestidea novaeguineae (Brancsik 1897)		Mecopodinae	Sexavaini	New Guinea
S. uniformis Willemse 1940		Mecopodinae	Sexavaini	New Guinea
Sexava (Linn. 1758)		Mecopodinae	Sexavaini	New Guinea
Spelaeala Rehn 1936		Pseudophyllinae	Polyancistrini	Dominican Republic
Sphyrometopa femorata Carl 1908		Conocephalinae	Agraeciini	Costa Rica
Steiroxys Scudder 1894		Tettigoniinae		Western U.S.
Stetharasa exarmata Montealegre & Morris 1999		Pseudophyllinae		S. America

Species	Common name	Subfamily	Tribe	Region
Stictophaula armata Ingrisch 1994		Phaneropterinae		Malaysia
Stilpnochlora marginella (Serville 1839)		Phaneropterinae		S. America
Tabaria Walker 1890		Pseudophyllinae		S. America
Tettigonia cantans (Fuessly 1775)	Wart-biter	Tettigoniinae		Europe
T. caudata (Charpentier 1845)		Tettigoniinae		Hungary
T. viridissima (Linn. 1758)	Great-green grasshopper	Tettigoniinae		Europe
Thliboscelus hypericifolia Stoll 1813		Pseudophyllinae	Pterophyllini	S. America
Tinzeda eburneata (Walker 1869)		Phaneropterinae		Australia
Tonatia sylvicola		Pseudophyllinae		Panama
Torbia Walker 1869		Phaneropterinae		Australia
Tylopsis Karsch 1893		Phaneropterinae		Africa, Eurasia
Tympanophora White 1841		Tympanophorinae		Australia
Tympanophyllum Brunner V. Wattenwyl 1895		Pseudophyllinae	Phyllomimini	Malaysia
Typophyllum Serville 1839		Pseudophyllinae	Pterochrozini	S. America
Uromenus rugosicollis (Serville)	Rough-backed bush-cricket	Bradyporinae	Ephippigerini	Western Europe
U. (Steropleurus) stali (Bolivar 1876)	Stal's bush-cricket	Bradyporinae	Ephippigerini	Western Europe
Usarovites inflatus (Uvarov 1923)	Jie Er, singing sister	Tettigoniinae		China
Veria colorata Walker 1869		Conocephalinae	Agraeciini	Australia
Vestria species (undescribed)		Conocephalinae	Copiphorini	S. America
Vetralla quadrata Walker 1869		Mecopodinae		Malaysia
Xestoptera cornea Brunner V. Wattenwyl 1895		Pseudophyllinae	Pterophyllini	Central America
Xiphidiopsis lita Hebard 1922		Meconematinae		Hawaii
Yersinella raymondi (Yersin 1860)		Tettigoniinae		France
Zabalius apicalis (Bolivar 1886)		Phaneropterinae		Africa
Zeuneria melanopeza Karsch 1888		Phaneropterinae		Cameroon

References Cited

Achmann, R., K.-G. Heller, and J. T. Epplen. 1992. Last-male sperm precedence in the bushcricket *Peocilimon veluchianus* (Orthoptera, Tettigonioidea) demonstrated by DNA fingerprinting. Mol. Ecol. 1: 47–54.

Adamo, S. A., D. Robert, J. Perez, and R. R. Hoy. 1995. The response of an insect parasitoid, *Ormia ochracea* (Tachinidae), to the uncertainty of larval success during infestation. Behav. Ecol. Sociobiol. 36: 111–118.

Ahlén, I. 1981. Ultraljud hos svenska vartbitare. [Ultrasonics in songs of Swedish bush crickets (Orth., Tettigoniidae).] Ent. Tidskr. 102: 27–41.

Alcock, J. A., and D. T. Gwynne. 1991. Evolution of insect mating systems: The impact of individual selectionist thinking. Pages 10–41 in W. Bailey and J. Ridsdill Smith, eds. Reproductive Behaviour in Insects: Individuals and Populations. Chapman and Hall, London.

Alexander, R. D. 1960. Sound communication in Orthoptera and Cicadidae. Pages 38–92 in W. E. Lanyon and W. N. Tavolga, eds. Animal Sounds and Communication. Am. Instit. Biol. Sci. 7., Washington, D.C.

——— 1961. Aggressiveness, territoriality, and sexual behavior in field crickets (Orthoptera: Gryllidae). Behaviour 17: 130–223.

——— 1962. Evolutionary change in cricket acoustical communication. Evolution 16: 443–467.

——— 1975. Natural selection and specialized chorusing behavior in acoustical insects. Pages 35–77 in D. Pimentel, ed. Insects, Science and Society. Academic Press, New York.

Alexander, R. D., and G. Borgia. 1979. On the origin and basis of the male-female phenomenon. Pages 417–440 in M. Blum and N. Blum, eds. Sexual Selection and Reproductive Competition in the Insects. Academic Press, New York.

Alexander, R. D., D. C. Marshall, and J. R. Cooley. 1997. Evolutionary perspectives on insect mating. Pages 4–31 in J. C. Choe and B. J. Crespi, eds. The Evolution of Mating Systems in Insects and Arachnids. Cambridge University Press, Cambridge.

Alexander, R. D., and D. Otte. 1967. The evolution of genitalia and mating behavior in crickets (Gryllidae) and other Orthoptera. Misc. Publ. Mus. Zool. Univ. Mich. 133: 1–62.

Alexander, R. D., A. E. Pace, and D. Otte. 1972. The singing insects of Michigan. Great Lakes Entomol. 5: 33–69.

Allard, H. A. 1928. Our insect instrumentalists and their musical technique. Annu. Report Smithson. Inst. 1928: 563–591.

——— 1929. The last meadow katydid; a study of its musical reactions to light and temperature (Orthoptera: Tettigoniidae). Trans. Am. Entomol. Soc. 55: 155–164.

Allen, G. R. 1995a. The biology of the phonotactic parasitoid, *Homotrixa sp.* (Diptera: Tachinidae), and its impact on the survival of male *Sciarasaga quadrata* (Orthoptera: Tettigoniidae) in the field. Ecol. Entomol. 20: 103–110.

——— 1995b. The calling behavior and spatial-distribution of male bush-crickets (*Sciarasaga quadrata*) and their relationship to parasitism by acoustically orienting tachinid flies. Ecol. Entomol. 20: 303–310.

——— 1998. Diel calling activity and field survival of the bushcricket, *Sciarasaga quadrata* (Orthoptera: Tettigoniidae): A role for sound-locating parasitic flies? Ethology 104: 645–660.

——— 2000. Call structure variability and field survival among bushcrickets exposed to phonotactic parasitoids. Ethology 106: 1–15.

Allen, G. R., D. Kamien, O. Berry, P. Byrne, and J. Hunt. 1999. Larviposition, host cues, and planidial behavior in the sound-locating parasitoid fly *Homotrixa alleni* (Diptera: Tachinidae). J. Ins. Behav. 12: 67–79.

Allen, G. R., and T. Pape. 1996. Description of female and biology of *Blaesoxipha ragg* Pape (Diptera: Sarcophagidae), a parasitoid of *Sciarasaga quadrata* Rentz (Orthoptera: Tettigoniidae) in Western Australia. J. Aust. Entomol. Soc. 35: 147–151.

Ammons, A. R. 1992. Foreward. Page v in V. G. Dethier. Crickets and Katydids, Concerts and Solos. Harvard University Press, Cambridge, Mass.

Ander, K. 1938. Ein abdominales stridulations Organ bei *Cyphoderris* (Prophalangopsidae) und über die systematische Einteilung der Ensiferen (Saltatoria). Opusc. Ent. 3: 32–38.

——— 1939. Comparative anatomical and phylogenetic studies on the Ensifera (Saltatoria). Opusc. Ent. Suppl. II. Lund.

Andersson, M. 1994. Sexual Selection. Princeton University Press, Princeton, N.J.

Andersson, M., and Y. Iwasa. 1996. Sexual selection. Trends Ecol. Evol. 11: 53–58.

Ando, Y. 1991. Photoperiodic control of adult diapause in a subtropical katydid, *Euconocephalus pallidus* Redtenbacher (Orthoptera: Tettigoniidae). Appl. Entomol. Zool. 26: 347–355.

Ando, Y., and J. C. Hartley. 1982. Occurrence and biology of a long-winged form of *Conocephalus discolor*. Entomol. Exp. Appl. 32: 238–241.

Andrade, M. C. B. 1996. Sexual selection for male sacrifice in the Australian redback spider. Science 271: 70–72.

Anonymous. 1991. Mormon cricket *Anabrus simplex* Haldeman. Wyoming Agric. Exp. Sta. Bull. 912.

——— 1990. War on six legs: Nevada's cricket invasion. New York Times, April 8.

Arak, A., and T. Eiriksson. 1992. Choice of singing sites by male bushcrickets (*Tettigonia viridissima*) in relation to signal propagation. Behav. Ecol. Sociobiol. 30: 365–372.

Arak, A., T. Eiriksson, and T. Radesater. 1990. The adaptive significance of acoustic spacing in male bushcrickets *Tettigonia viridissima*: a perturbation experiment. Behav. Ecol. Sociobiol. 26: 1–7.

Arlettaz, R., N. Perrin, and J. Hausser. 1997. Trophic resource partitioning and competition between the two sibling bat species *Myotis myotis* and *Myotis blythii*. J. Anim. Ecol. 66: 897–911.

Arnold, S. J., and D. Duvall. 1994. Animal mating systems: A synthesis based on selection theory. Am. Nat. 143.

Arnold, S. J., and M. Wade. 1984. On the measurement of natural and sexual selection: Applications. Evolution 38: 720–734.

Arnqvist, G. 1998. Comparative evidence for the evolution of genitalia by sexual selection. Nature 393: 784–786.

Austad, S. N., and R. Thornhill. 1986. Female reproductive variation in a nuptial-feeding spider, *Pisaura mirabilis*. Bull. Br. Arach. Soc. 7: 48–52.

Autrum, H. 1940. Über Lautäusserungen und Schallwahrnemung bei Arthropoden. II. Das Richtigungshoren von Locusta und Versuch einer Hörtheorie für Tympanalorgane von Locusidentyp. Z. Vergl. Physiol. 28: 326–352.

Baier, L. J. 1930. Contribution to the physiology of the stridulation and hearing of insects. Zool. Jahrb. 47: 194–199.

Bailey, W. J. 1970. The mechanics of stridulation in bush crickets (Tettigonioidea, Orthoptera). J. Exp. Biol. 52: 495–505.

——1979. A review of Australian Copiphorini (Orthoptera: Tettigoniidae: Conocephalinae). Aust. J. Zool. 27: 1015–1049.

——1985. Acoustic cues for female choice in bushcrickets (Tettigoniidae). Pages 101–110 in K. Kalmring and N. Elsner, eds. Acoustic and Vibrational Communication in Insects. Paul Parey, Hamburg.

——1990. The ear of the bushcricket. Pages 217–247 in W. J. Bailey and D. C. F. Rentz, eds. The Tettigoniidae: Biology, Systematics and Evolution. Crawford House, Bathurst, Australia.

——1991. Acoustic Behaviour of Insects: An Evolutionary Perspective. Chapman and Hall, London.

——1993. The tettigoniid ear (Orthoptera: Tettigoniidae): Multiple functions and structural diversity. Int. J. Ins. Morphol. Embryol. 22: 185–205.

——1998a. Do large bushcrickets have more sensitive ears? Natural variation in hearing thresholds within populations of the bushcricket *Requena verticalis* (Listroscelidinae: Tettigoniidae). Physiol. Entomol. 23: 105–112.

——1998b. The mating biology of *Phasmodes ranatriformis* (Orthoptera: Tettigoniidae), a mute genus of bushcricket from Western Australia. J. R. Soc. West. Aust. 81: 149–155.

Bailey, W. J., R. J. Cunningham, and L. Lebel. 1990. Song power, spectral distribution and female phonotaxis in the bushcricket *Requena verticalis* (Tettigoniidae: Orthoptera): Active female choice of passive attraction? Anim. Behav. 40: 33–42.

Bailey, W. J., and G. Field. 2000. Acoustic satellite behaviour in the Australian bushcricket *Elephantodeta nobilis* (Phaneropterinae; Tettigoniidae; Orthoptera). Anim. Behav. 59: 361–369.

Bailey, W. J., and A. W. R. McCrae. 1978. The general biology and phenology of swarming in the East African tettigoniid *Ruspolia differens* (Serville) (Orthoptera). J. Nat. Hist. 12: 259–288.

Bailey, W. J., and G. K. Morris. 1986. Confusion of phonotaxis by masking sounds in the bushcricket *Conocephalus brevipennis* (Tettigoniidae: Conocephalinae). Ethology 73: 19–28.

Bailey, W. J., and D. Robinson. 1971. Song as a possible isolating mechanism in the genus *Homorocoryphus* (Tettigonioidea, Orthoptera). Anim. Behav. 19: 390–397.
Bailey, W. J., and H. Römer. 1991. Sexual differences in auditory sensitivity: Mismatch of hearing threshold and call frequency in a tettigoniid (Orthoptera, Tettigoniidae: Zaprochilinae). J. Comp. Physiol. [A] 169: 349–353.
Bailey, W. J., and J. D. Sandow. 1983. Mechanisms of defensive stridulation in the bushcricket *Mygalopsis marki* Bailey (Copiphorini, Tettigoniidae). Acta. Zool. 64: 117–122.
Bailey, W. J., and L. W. Simmons. 1991. Male-male behavior and sexual dimorphism of the ear of a zaprochiline tettigoniid (Orthoptera: Tettigoniidae). J. Ins. Behav. 4: 51–65.
Bailey, W. J., and R. O. Stephen. 1978. Directionality and auditory slit function: A theory of hearing in bushcrickets. Science 201: 633–634.
——1984. Auditory acuity in the orientation behaviour of the bushcricket *Pachysagella australis* Walker (Orthoptera, Tettigoniidae, Saginae). Anim. Behav. 32: 816–829.
Bailey, W. J., and D. R. Thiele. 1983. Male spacing behavior in the Tettigonidae: An experimental approach. Pages 163–184 in D. T. Gwynne and G. K. Morris, eds. Orthopteran Mating Systems: Sexual Competition in a Diverse Group of Insects. Westview Press, Boulder, Col.
Bailey, W. J., P. C. Withers, M. Endersby, and K. Gaull. 1993. The energetic costs of calling in the bushcricket *Requena verticalis* (Orthoptera: Tettigoniidae: Listroscelidinae). J. Exp. Biol. 178: 21–37.
Bailey, W. J., and P. B. Yeoh. 1988. Female phonotaxis and frequency discrimination in the bushcricket *Requena verticalis*. Physiol. Entomol. 13: 363–372.
Baker, G. L., and J. L. Capinera. 1997. Nematodes and nematomorphs as control agents of grasshoppers and locusts. Mem. Entomol. Soc. Can. 171: 157–211.
Bancroft, H. H., and A. Bates. 1889. History of Utah. 1540–1886. History Company, San Francisco.
Barendse, W. 1990. Speciation in the genus *Mygalopsis* in southwestern Australia: A new look at distinguishing modes of speciation. Pages 265–279 in W. J. Baily and D. C. Rentz, eds. The Tettigoniidae: Biology, Systematics and Evolution. Crawford House, Bathurst, Australia.
Barendse, W. J. 1986. Speciation in the genus *Mygalopsis* (Orthoptera: Tettigoniidae). Ph.D. thesis, zoology, University of Western Australia, Nedlands.
Barraclough, D. A., and G. R. Allen. 1996. Two new species of *Homotrixa* Villeneuve (Diptera: Tachinidae: Ormiini) from southwestern Australia, with data on biology and ecology. Aust. J. Entomol. 35: 135–145.
Barrientos, L. L. 1998. Mate choice and hybridization experiments between allopatric populations of *Pterophylla beltrani* Bolívar and *P. robertsi* Hebard (Orthoptera: Tettigoniidae: Pseudophyllinae). J. Orthop. Res. 7: 41–49.
Barrientos, L. L., and J. Den Hollander. 1994. Acoustic signals and taxonomy of Mexican *Pterophylla* (Orthoptera: Tettigoniidae: Pseudophyllinae). J. Orthop. Res. 2: 35–40.
Barrientos, L. L., and R. I. H. Jaramillo. 1998. Observaciones sobre la biologia y ecologia de *Pterophylla robertsi* (Hebard) (Orthoptera: Tettigoniidae). J. Orthop. Res. 7: 33–39.
Barrientos, L. L., and M. T. Montes. 1997. Geographic distribution and singing activity of *Pterophylla beltrani* and *P. robertsi* (Orthoptera: Tettigoniidae), under field conditions. J. Orthop. Res. 6: 49–56.
Bartram, J. 1751. Travels in Pensilvania and Canada. J. Whiston and B. White, London.

Bateman, A. J. 1948. Intrasexual selection in *Drosophila*. Heredity 2: 349–368.
Bateman, P. W. 1995. *Psammodromus algirus* (Large Psammodromus) phonotaxis. Herpetol. Rev. 26: 36–37.
——1997. Operational sex ratio, female competition and mate choice in the ephippigerine bushcricket *Steropleurus stali* Bolivar. J. Orthop. Res. 6: 101–104.
——1998a. Does size matter? The function of the large spermatophore of *Steropleurus stali* Bolivar (Orthoptera: Tettigoniidae: Ephippigerinae). J. Orthop. Res. 7: 209–212.
——1998b. Signal reliability and copulation in the bushcricket *Steropleurus stali* Bolivar (Orthoptera: Tettigoniidae: Ephiggerinae [sic]). J. Orthop. Res. 7: 205–107.
Bateman, P. W., and R. Marquez. Manuscript. Female preference for younger males in the bushcricket *Steropleurus stali* (Orthoptera: Tettigioniidae, Ephippigerinae).
Beck, C. W., and L. A. Powell. 2000. Evolution of female mate choice based on male age: Are older males better males? Evol. Ecol. Res. 2: 107–118.
Bei-Benko, G. Y. 1954. Orthoptera II, 2, Tettigoniidae. Pheneropterinae. Fauna USSR. Israel program for scientific translation, Jerusalem 1965.
Beier, M. 1960. Orthoptera: Tettigoniidae (Pseudophyllinae II) 74. W. de Gruyter, Berlin.
Belavadi, V. V., and P. Mohanraj. 1996. Nesting-behavior of the black digger wasp, *Sphex argentatus* Fabricius 1787 (Hymenoptera, Sphecidae) in south India. J. Nat. Hist. 30: 127–134.
Bell, P. D. 1980. Transmission of vibrations along plant stems: Implications for insect communication. J. N. Y. Entomol. Soc. 88: 210–216.
Belovsky, G. E., J. B. Slade, and J. M. Chase. 1996. Mating strategies based on foraging ability: An experiment with grasshoppers. Behav. Ecol. 7: 438–444.
Belwood, J. J. 1988. The influence of bat predation on calling behavior in neotropical forest katydids (Insecta: Orthoptera: Tettigoniidae). Ph.D. thesis, entomology, University of Florida, Gainesville.
——1990. Anti-predator defences and ecology of neotropical forest katydids, especially the Pseudophyllinae. Pages 8–26 in W. J. Bailey and D. C. F. Rentz, eds. The Tettigoniidae: Biology, Systematics and Evolution. Crawford House, Bathurst, Australia.
Belwood, J. J., and G. K. Morris. 1987. Bat predation and its influence on calling behavior in neotropical katydids. Science 238: 64–67.
Bennet-Clark, H. C. 1984. Insect hearing: Acoustics and transduction. Pages 49–82 in T. Lewis, ed. Insect Communication. Academic Press, London.
Berenbaum, M. R. 1995. Bugs in the System. Addison-Wesley, Reading, Mass.
Berg, C. O., and K. Valley. 1985. Nuptial feeding in *Sepedon* spp. (Diptera: Sciomyzidae). Proc. Entomol. Soc. Wash. 87: 622–633.
Berglund, A., G. Rosenqvist, and I. Svensson. 1986a. Mate choice, fecundity and sexual dimorphism in two pipefish species (Syngnathidae). Behav. Ecol. Sociobiol. 19: 301–307.
——1986b. Reversed sex roles and parental energy investment in zygotes of two pipefish (Syngnathidae) species. Mar. Ecol. Prog. Ser. 29: 209–215.
Bertness, M. D., C. Wise, and A. M. Ellison. 1987. Consumer pressure and seed set in a salt marsh perennial plant community. Occologia 71: 190–200.
Birkhead, T. R., and A. P. Møller, eds. 1998. Sperm Competition and Sexual Selection. Academic Press, San Diego.
Bland, R. G., and D. C. F. Rentz. 1991. External morphology and abundance of mouth-

part sensilla in Australian Gryllacrididae, Stenopelmatidae, and Tettigoniidae. J. Morphol. 207: 315–325.

Blatchley, W. S. 1920. Orthoptera of Northeastern America. Nature Publishing, Indianapolis, Ind.

Bleton, A. 1942. Notes sur la biologie au Maroc d'*Eugaster spinulosus* Joh. (Orthoptères, Tettigoniidae). Bull. Soc. Sci. Nat. Maroc. 22: 89–95.

Bockwinkel, G., and K. P. Sauer. 1994. Resource dependence of male mating tactics in the scorpionfly, *Panorpa vulgaris* (Mecoptera, Panorpidae). Anim. Behav. 47: 203–209.

Bohart, R. M., and A. S. Menke. 1976. Sphecid Wasps of the World: A Revision. University of California Press, Berkeley.

Boldyrev, B. T. 1915. Contributions à l'étude de la structure des spermatophores et des particularitès de la copulation chez Locustodea et Gryllodea. Horae Soc. Ent. Rossicae 6: 1–245.

——1927. Copulation and spermatophores of *Gryllomorpha dalmatina* (Oesk.) (Orth. Gryllidae). Eos. 3: 279–288.

——1928. Biological studies on *Bradyporus multituberculatus* F.W. (Orth., Tettig.). Eos. 4: 15–57.

Bowen, B. J., C. G. Codd, and D. T. Gwynne. 1984. The katydid spermatophore (Orthoptera: Tettigoniidae): Male investment and its fate in the mated female. Aust. J. Zool. 32: 23–31.

Bradbury, J. W., and M. B. Andersson. 1987. Sexual Selection: Testing the Alternatives. John Wiley, Chichester, U.K.

Brockmann, H. J. 1985. Provisioning behavior of the great golden digger wasp, *Sphex ichneumoneus* (L.) (Sphecidae). J. Kans. Entomol. Soc. 58: 631–655.

Brooks, D., and D. A. McLennan. 1991. Phylogeny, Ecology and Behavior: A Research Program in Comparative Biology. University of Chicago Press, Chicago.

Brown, R. W. 1956. Composition of Scientific Words. Smithsonian Press, Washington, D.C.

Brown, V. K. 1983. Grasshoppers. Cambridge University Press, Cambridge.

Brown, W. D. 1997a. Courtship feeding in tree crickets increases insemination and female reproductive life-span. Anim. Behav. 54: 1369–1382.

——1997b. Female remating and the intensity of female choice in black-horned tree crickets, *Oecanthus nigricornis*. Behav. Ecol. 8: 66–74.

——1999. Mate choice in tree crickets and their kin. Annu. Rev. Entomol. 44: 371–396.

Brown, W. D., B. J. Crespi, and J. C. Choe. 1997. Sexual conflict and the evolution of mating systems. Pages 352–378 in J. C. Choe and B. J. Crespi, eds. The Evolution of Mating Systems in Insects and Arachnids. Cambridge University Press, Cambridge.

Brown, W. D., and D. T. Gwynne. 1997. Evolution of mating in crickets, katydids, and wetas (Ensifera). Pages 279–312 in S. K. Gangwere, M. C. Mulalirangen, and M. Muralirangen, eds. The Bionomics of Grasshoppers, Katydids and their Kin. CAB International, Oxford.

Brown, W. D., J. Wideman, M. C. B. Andrade, A. C. Mason, and D. T. Gwynne. 1996. Female choice for an indicator of male size in the song of the black-horned tree cricket, *Oecanthus nigricornis* (Orthoptera: Gryllidae: Oecanthinae). Evolution 50: 2400–2411.

Brunellius, G. 1791. De locustarum anatome (in Latin). De Bononiensi Scientarum et Artium Instituto atque Academia Commentarii 7: 198–206.

Brush, J. S., V. G. Gian, and M. D. Greenfield. 1985. Phonotaxis and aggression in the coneheaded katydid *Neoconocephalus affinis*. Physiol. Entomol. 10: 23–32.

Buchler, E. R., and S. B. Childs. 1981. Orientation to distant sounds by foraging big brown bats *Eptesicus fuscus*. Anim. Behav. 29: 428–432.

Burk, T. 1982. Evolutionary significance of predation on sexually-signalling males. Fla. Entomol. 65: 90–104.

—— 1983. Male aggression and female choice in a field cricket (*Teleogryllus oceanicus*): The importance of courtship song. Pages 97–119 in D. T. Gwynne and G. K. Morris, eds. Orthopteran Mating Systems: Sexual Competition in a Diverse Group of Insects. Westview Press, Boulder, Col.

Burley, N. 1988. The differential allocation hypothesis: An experimental test. Am. Nat. 132: 611–628.

Burmeister, W. E. 1836. Manual of Entomology. Edward Churton, London.

Burpee, D. M., and S. K. Sakaluk. 1993. Repeated matings offset costs of reproduction in female crickets. Evol. Ecol. 7: 240–250.

Burr, M. 1897. British Orthoptera. Economic and Educational Museum, Huddersfield, UK.

—— 1936. British Grasshoppers and Their Allies. Janson and Sons, London.

Burr, M., B. P. Campbell, and B. P. Uvarov. 1923. A contribution to our knowledge of the Orthoptera of Macedonia. Trans. Entomol. Soc. (London) 71: 110–169.

Burt, A. 1989. Comparative methods using phylogenetically independent contrasts. Pages 33–53 in P. H. Harvey and L. Partridge, eds. Oxford Surveys in Evolutionary Biology, Vol. 6. Oxford University Press, Oxford.

Buskirk, R. E., C. Frohlich, and K. G. Ross. 1984. The natural selection of sexual cannibalism. Am. Nat. 123: 612–625.

Busnel, R.-G., and B. Dumortier. 1955. Étude du cycle génital du mâle d'*Ephippiger* et son rapport avec le comportement acoustique. Bull. Soc. Zool. Fr. 80: 23–26.

Busnel, R. G., B. Dumortier, and M. C. Busnel. 1956. Recherches sur le comportement acoustique des ephippigéres. Bull. Biol. Fr. Belg. 190: 219–285.

Bussière, L. F. In press. A model of the interaction between "good genes" and direct benefits in courtship feeding animals: When do males of high genetic quality invest less? Phil. Trans. Roy. Soc. Lond.

Butlin, R. K., and G. M. Hewitt. 1986. The response of female grasshoppers to male song. Anim. Behav. 34: 1896–1899.

Butlin, R. K., G. M. Hewitt, and S. F. Webb. 1985. Sexual selection for intermediate optimum in *Chorthippus brunneus* (Orthoptera: Acrididae). Anim. Behav. 33: 1281–1292.

Butlin, R. K., C. W. Woodhatch, and G. M. Hewitt. 1987. Male spermatophore investment increases female fecundity in a grasshopper. Evolution 41: 221–225.

Cabrero, J., L. H. Shapiro, and J. P. M. Camacho. 1999. Male sterility in interspecific meadow katydid hybrids. Hereditas 131: 79–82.

Cade, W. 1975. Acoustically orienting parasitoids: Fly phonotaxis to cricket song. Science 190: 1312–1313.

Cade, W. 1979. The evolution of alternative male reproductive strategies in field crickets. Pages 343–379 in M. Blum and N. Blum, eds. Sexual Selection and Reproductive Competition in Insects. Academic Press, New York.

Cantrall, I. J. 1943. The ecology of the Orthoptera and Dermaptera of the George Reserve, Michigan. Misc. Publ. Mus. Zool. Univ. Mich. 54: 1–182.

—— 1972. *Saga pedo* (Pallas) (Tettigoniidae: Saginae), an Old World katydid, new to Michigan. Great Lakes Entomol. 5: 103–106.

Cappe de Baillon, P. 1922. Contribution anatomique et physiologique a l'étude de la reproduction chez les Locustiens et les Grilloniens. La Cellule 31: 7–41.

Carl, J. 1906. L'Organe stridulateur de Phyllophorae. Arch. Sci. Phys. Nat. (Geneve) 23: 406.

Castner, J. L. 1995. Defensive behavior and display of the leaf-mimicking katydid *Pterochroza ocellata* (L.) (Orthoptera: Tettigoniidae: Pseudophyllinae: Pterochrozini). J. Orthop. Res. 4: 89–92.

Caudell, A. N. 1908a. An old record of observations on the habits of *Anabrus*, and a list of Odonata taken by Dr. H. Skinner, in Carr Canyon, Huachuca Mountains, Arizona. Entomol. News 19: 44–45.

——— 1908b. Orthoptera: Fam. Locustidae: Subfam. Dectininae. Genera Insectorum 72: 1–3.

Cherrill, A. J., and V. K. Brown. 1990. The life cycle and distribution of the wart-biter *Decticus verrucivorus* (L.) (Orthoptera: Tettigoniidae) in a chalk grassland in southern England. Biol. Conserv. 53: 125–143.

Cherrill, A. J., and V. K. Brown. 1992. Ontogenetic changes in the micro-habitat of *Decticus verrucivorus* (Orthoptera: Tettigoniidae) at the edge of its range. Ecography 15: 37–44.

Choe, J. C. 1995. Courtship feeding and repeated mating in *Zorotypus barberi* (Insecta: Zoraptera). Anim. Behav. 49: 1511–1520.

Chopard, L. 1965. Ordre des Orthoptères. I. Sous-Ordre des Ensifera. Pages 617–674 in P.-P. Grassé, ed. Traité de Zoologie, Tome IX. Masson et Cie Éditeurs, Paris.

Chou, I. 1960. A History of Chinese Entomology. Entomotaxonomia, Wugong (Shaanxi), China.

Clutton-Brock, T. H., ed. 1988. Reproductive Success. University of Chicago Press, Chicago.

Clutton-Brock, T. H. 1991. The Evolution of Parental Care. Princeton University Press, Princeton, N.J.

Clutton-Brock, T. H., and G. A. Parker. 1992. Potential reproductive rates and the operation of sexual selection. Q. Rev. Biol. 67: 437–456.

Clutton-Brock, T. H., and A. C. J. Vincent. 1991. Sexual selection and the potential reproductive rates of males and females. Nature 351: 58–60.

Coates, J. 1990. Crickets prepare to darken the land. Chicago Tribune, April 8, page 5.

Cohn, T. J. 1965. The arid-land katydids of the North American genus *Neobarretia* (Orthoptera: Tettigoniidae): Their systematics and a reconstruction of their history. Misc. Publ. Mus. Zool. Univ. Mich. 126.

Colinvaux, P. 1978. Why Big Fierce Animals Are Rare: An Ecologist's Perspective. Princeton University Press, Princeton, N.J.

Comstock, J. 1899. A Manual for the Study of Insects. Comstock Publishing, Ithaca, N.Y.

Côte, I. M., and W. Hunte. 1989. Male and female mate choice in the redlip blenny: Why bigger is better. Anim. Behav. 38: 78–88.

Cowan, F. T. 1929. Life history, habits, and control of the Mormon cricket. Tech. Bull. U.S. Dept. Agric. 161: 1–28.

——— 1932. Mormon Cricket Control in Colorado. Circular 57, U.S. Dept. Agric.

——— 1990. The Mormon Cricket Story. Montana State Univ., Agric. Exp. Sta. Special Rep. 31: 7–42.

Cowan, F. T., and S. C. McCampbell. 1929. The Mormon Cricket and Its Control. Circular 53, U.S. Dept. Agric.

Cox, C. R., and B. J. LeBoeuf. 1977. Female incitation of male competition: A mechanism in sexual selection. Am. Nat. 111: 317–335.

Criddle, N. 1926. The life-history and habits of *Anabrus longipes* Caudell (Orthop). Can. Entomol. 108: 261–265.

Crocker, G., and T. Day. 1987. An advantage to mate choice in the seaweed fly, *Coelopa frigida*. Behav. Ecol. Sociobiol. 20: 295–301.

Cronin, H. 1991. The Ant and the Peacock: Altruism and Sexual Selection from Darwin to Today. Cambridge University Press, Cambridge.

CSIRO, eds. 1991. The Insects of Australia. Melbourne University Press, Melbourne.

Cumming, J. M. 1994. Sexual selection and the evolution of dance fly mating systems (Diptera: Empididae; Empidinae). Can. Entomol. 126: 907–920.

Dadour, I. R. 1990. Dispersal, dispersion, and mating behavior in *Tettigonia cantans* (Orthoptera: Tettigoniidae). J. Ins. Behav. 3: 805–812.

Dadour, I. R., and W. J. Bailey. 1985. Male agonistic behaviour of the bushcricket *Mygalopsis marki* Bailey in response to conspecific song (Orthoptera: Tettigoniidae). Z. Tierpsychol. 70: 320–330.

——1990. The acoustic behaviour of male *Mygalopsis marki* (Copiphorinae). Pages 98–111 in W. J. Bailey and D. C. F. Rentz, eds. The Tettigoniidae: Biology, Systematics and Evolution. Crawford House, Bathurst, Australia.

Dadour, I. R., and M. S. Johnson. 1983. Genetic differentiation, hybridization and reproductive isolation in *Mygalopsis marki* Bailey (Orthoptera: Tettigoniidae). Aust. J. Zool. 31: 353–360.

Dambach, M. 1989. Vibrational responses. Pages 178–197 in F. Huber, T. E. Moore, and W. Loher, eds. Cricket Behavior and Neurobiology. Cornell University Press, Ithaca, N.Y.

Darwin, C. 1859. The Origin of Species by Means of Natural Selection. John Murray, London.

——1874. The Descent of Man and Selection in Relation to Sex. 2nd Edition John Murray, London.

Davey, K. 1960. The evolution of spermatophores in insects. Proc. R. Entomol. Soc. (Lond.) [A] 35: 107–113.

Davies, N. B. 1992. Dunnock Behaviour and Social Evolution. Oxford University Press, Oxford.

Davies, P. M., and I. R. Dadour. 1989. A cost of mating by male *Requena verticalis* (Orthoptera: Tettigoniidae). Ecol. Entomol. 14: 467–469.

Day, T. H., C. S. Crean, A. S. Gilburn, D. M. Shuker, and R. W. Wilcockson. 1996. Sexual selection in seaweed flies: Genetic-variation in male size and its reliability as an indicator in natural populations. Proc. R. Soc. Lond. [B] 263: 1127–1134.

Dean, R. L., and J. C. Hartley. 1977a. Egg diapause in *Ephippiger cruciger* (Orthoptera: Tettigoniidae). I. The incidence, variable duration and elimination of the initial diapause. J. Exp. Biol. 66: 173–183.

——1977b. Egg diapause in *Ephippiger cruciger* (Orthoptera: Tettigoniidae). II. The intensity and elimination of the final egg diapause. J. Exp. Biol. 66: 185–195.

DeFoliart, G. R., M. D. Finke, and M. L. Sunde. 1982. Potential value of the Mormon cricket (Orthoptera: Tettigoniidae) harvested as a high-protein feed for poultry. J. Econ. Entomol. 75: 848–851.

De Geer, C. 1752. Memoires pour servir à l'histoire des insectes. L.L. Grefing, Stockholm.

De Luca, P. A., and G. K. Morris. 1998. Courtship communication in meadow katydids: Female preference for large male vibrations. Behaviour 135: 777–794.

Dethier, V. G. 1992. Crickets and Katydids, Concerts and Solos. Harvard University Press, Cambridge, Mass.

Dewsbury, D. A. 1982. Ejaculate cost and male choice. Am. Nat. 119: 601–610.

Doolan, J. M. 1981. Male spacing and the influence of female courtship behaviour in the bladder cicada, *Cystosoma saundersii* Westwood. Behav. Ecol. Sociobiol. 9: 269–276.

Downhower, J. F., and D. E. Wilson. 1973. Wasps as a defence mechanism of katydids. Am. Midl., Nat. 89: 451–455.

Dugatkin, L. A., and J.-G. Godin. 1998. How females choose their mates. Sci. Am. 278: 56–61.

Duijm, M. 1990. On some song characteristics in *Ephippiger* (Orthoptera, Tettigonioidea) and their geographic variation. Neth. J. Zool. 40: 428–453.

Dumortier, B. 1963. The physical characteristics of sound emissions in Arthropoda. Pages 346–373 in R.-G. Busnel, ed. Acoustic Behaviour in Animals. Elsevier, Amsterdam.

——1967. Essais in vitro pour la rupture de la diapause embryonnaire chez quelques tettigonioides (Insectes-Orthopteres). Ann. Epaphylies 18: 387–400.

Duncan, C. J. 1960. The biology of *Leptophyes punctatissima* (Bosc.) (Orthoptera: Tettigonidae). Entomologist 93: 76–78.

Duncan, J. 1843. Entomology. W.H. Lizars, Edinburgh.

Eberhard, W. G. 1977. Aggressive chemical mimicry by a bolas spider. Science 198: 1173–1175.

——1985. Sexual selection and Animal Genitalia. Harvard University Press, Cambridge, Mass.

——1991. Copulatory courtship and cryptic female choice in insects. Biol. Rev. (Camb.) 66: 1–31.

——1996. Female Control: Sexual Selection by Cryptic Female Choice. Princeton University Press, Princeton, N.J.

Ebner, R. 1951. Uber macropterie bei *Metrioptera* (Orthoptera). Eos. 1950: 267–274.

Eibl, E. 1978. Morphology of the sense organs in the proximal parts of the tibiae of *Gryllus campestris* L. and *Gryllus bimaculatus* de Geer (Insecta, Ensifera). Zoomorphologie 89: 185–205.

Eisner, T. 1988. Insekten als fursorgliche Eltern. Verh. Dtsch. Zool. Ges. 81: 9–17.

Eisner, T., S. R. Smedley, D. K. Young, M. Eisner, B. Roach, and J. Meinwald. 1996a. Chemical basis of courtship in a beetle (*Neopyrochroa flabellata*): Cantharidin as "nuptial gift." Proc. Natl. Acad. Sci. USA 93: 6499–6503.

——1996b. Chemical basis of courtship in a beetle (*Neopyrochroa flabellata*): Cantharidin as precopulatory "enticing agent." Proc. Natl. Acad. Sci. USA 93: 6494–6498.

Elgar, M. A., and N. Wedell. 1996. Role-reversed risky copulation. Trends Ecol. Evol. 11: 189–190.

Eluwa, M. C. 1970. The biology of the West African bush-cricket, *Euthypoda acutipennis* Karsch (Orthoptera: Tettigoniidae). Biol. J. Linn. Soc. 2: 1–39.

——1975a. The egg-laying habits of *Zabalius apicalis* Bolivar (Orthoptera: Tettigoniidae). J. Nat. Hist. 9: 41–49.

——1975b. Studies on the life-history of the African bush-cricket *Zabalius apicalis* Bolivar (Orthoptera: Tettigoniidae). J. Nat. Hist. 9: 33–39.

——1978. Observations on the spermatophore and post-copulatory behaviour of three West African bush-crickets (Orth., Tettigoniidae). Entomol. Mon. Mag. 114: 1–7.

Elzinga, R. J. 1996. A comparative-study of microspines in the alimentary canal of 5 families of Orthoptera (Saltatoria). Int. J. Ins. Morphol. Embryol. 25: 249–260.

Emlen, S. T. 1976. Lek organization and mating strategies in the bullfrog. Behav. Ecol. Sociobiol. 1: 283–313.

Emlen, S. T., and L. W. Oring. 1977. Ecology, sexual selection and the evolution of mating systems. Science 197: 215–222.

Englehardt, V. 1915. On the structure of the alluring gland in *Isophya acuminata* Br-W. Bull. Soc. Entomol. Moscow 1: 58–63.

Evans, H. E. 1963. Wasp Farm. Natural History Press, Garden City, N.Y.

——— 1966. The behavior patterns of solitary wasps. Annu. Rev. Entomol. 11: 123–154.

——— 1985. The Pleasures of Entomology. Smithsonian Press, Washington, D.C.

Evans, H. E., and M. A. Evans. 1970. William Morton Wheeler, Biologist. Harvard University Press, Cambridge, Mass.

Ewing, A. W. 1989. Arthropod Bioacoustics: Neurobiology and Behaviour. Edinburgh University Press, Edinburgh.

Fabre, J. H. 1896. Étude sur les locustiens. Annales des Sciences Naturelles—Zoologie 33: 221–244.

Fabre, J. H. 1917. The Life of the Grasshopper. Hodder and Stoughton, London.

Faure, P. A., and R. R. Hoy. 2000. The sounds of silence: Cessation of singing and song-pausing are ultrasound-induced acoustic startle behaviors in the katydid *Neoconocephalus ensiger* (Orthoptera; Tettigoniidae). J. Comp. Physiol. [A] 186: 129–142.

Feaver, M. 1977. Aspects of the behavioral ecology of three species of *Orchelimum* (Orthoptera: Tettigoniidae). Zoology. Ph.D. thesis, biology, University of Michigan, Ann Arbor.

——— 1983. Pair formation in the katydid *Orchelimum nigripes* (Orthoptera: Tettigoniidae). Pages 205–239 in D. T. Gwynne and G. K. Morris, eds. Orthopteran Mating Systems: Sexual Competition in a Diverse Group of Insects. Westview Press, Boulder, Col.

Feilner, J. 1864. Exploration in upper California in 1860 under the auspices of the Smithsonian Institution. Pages 421–430 in Smithsonian Institution Annual Report for 1864. U.S. Government Printing Office, Washington, D.C.

Field, L. H. 1993. Structure and evolution of stridulatory mechanisms in New Zealand wetas (Orthoptera: Stenopelmatidae). Int. J. Ins. Morphol. Embryol. 22: 163–183.

Field, L. H., and G. R. Sandlant. 1983. Aggression and mating behavior in the Stenopelmatidae (Orthoptera, Ensifera), with reference to New Zealand wetas. Pages 120–146 in D. T. Gwynne and G. K. Morris, eds. Orthopteran Mating Systems: Sexual Competition in a Diverse Group of Insects. Westview Press, Boulder, Col.

Fitzpatrick, S., A. Berglund, and G. Rosenqvist. 1995. Ornaments or offspring—costs to reproductive success restrict sexual selection processes. Biol. J. Linn. Soc. 55: 251–260.

Flook, P. K., S. Klee, and C. H. F. Rowell. 1999. Combined molecular phylogenetic analysis of the Orthoptera (Arthropoda, Insecta) and implications for their higher systematics. Syst. Biol. 48: 233–253.

Flook, P. K., and C. H. F. Rowell. 1997. The phylogeny of the Caelifera (Insecta, Orthoptera) as deduced from mtRNA gene sequences. Mol. Phylog. Evol. 8: 89–103.

——— 1998. Inferences about orthopteroid phylogeny and molecular evolution from small-subunit nuclear ribosomal DNA-sequences. Ins. Mol. Biol. 7: 163–178.

Forster, L. 1992. The stereotyped behaviour of sexual cannibalism in *Latrodectus has-*

selti Thorell (Araneae: Theridiidae), the Australian redback spider. Aust. J. Zool. 40: 1–11.

Fullard, J. H., C. Koehler, A. Surlykke, and N. L. McKenzie. 1991. Echolocation ecology and flight morphology of insectivorous bats (Chiroptera) in South-western Australia. Aust. J. Zool. 39: 427–438.

Fulton, B. B. 1915. The Tree Crickets of New York: Life History and Bionomics. Tech. Bull. N.Y. Exp. Sta. 42.

Funk, D., and D. Tallamy. 2000. Courtship role-reversal and deceptive signals in the long-tailed dance fly *Rhamphomyia longicauda*. Anim. Behav. 59: 411–421.

Galliart, P. L., and K. C. Shaw. 1991. The role of weight and acoustic parameters, including nature of chorusing, in the mating success of males of the katydid, *Amblycorypha parvipennis* (Orthoptera: Tettigoniidae). Fla. Entomol. 74: 453–464.

——1992. The relation of male and female acoustic parameters to female phonotaxis in the katydid, *Amblycorypha parvipennis*. J. Orthop. Res. 1: 110–115.

Gangwere, S. K. 1960. The use of mouthparts of Orthoptera during feeding. Entomol. News 71: 193–206.

——1961. A monograph on food selection in Orthoptera. Trans. Am. Entomol. Soc. 138: 67–203.

——1965. The structural adaptations of mouthparts in Orthoptera and allies. Eos. 41: 67–85.

——1967. The feeding behavior of *Atlanticus testaceus* (Orthoptera: Tettigoniidae). Ann. Entomol. Soc. Am. 60: 74–81.

Gerhardt, H. C. 1987. Evolutionary and neurobiological implications of selective phonotaxis in the green treefrog, *Hyla cinerea*. Anim. Behav. 35: 1479–1489.

——1994a. Reproductive character displacement of female mate choice in the grey treefrog, *Hyla chrysoscelis*. Anim. Behav. 47: 959–969.

——1994b. Selective responsiveness to long-range acoustic-signals in insects and anurans. Am. Zool. 34: 706–714.

Gerhardt, U. 1913. Copulation und spermatophoren von Grylliden und Locustiden I. Zool. Jahrb. (Syst.) 35: 415–532.

——1914. Copulation und Spermatophoren von Grylliden und Locustiden II. Zool. Jahrb. (Syst.) 37: 1–64.

——1921. Neue studien uber copulation und spermatophoren von Grylliden und Locustiden. Acta Zool. 11: 293–327.

Ghiselin, M. 1969. The Triumph of the Darwinian Method. University of California Press, Berkeley.

Gillette, C. P. 1904. Copulation and ovulation in *Anabrus simplex* Hald. Entomol. News 15: 321–325.

Gillott, C., and T. Friedel. 1977. Fecundity-enhancing and receptivity-inhibiting substances produced by male insects: A review. Pages 199–218 in K. G. Adiyodi and R. G. Adiyoti, eds. Advances in Invertebrate Reproduction, Vol. 1. Peralam-kenoth, Karivellur, Kerala.

Glenn, G. S. J. 1991. A systematic revision of *Enyaliopsis* Karsh 1887 (Orthoptera, Tettigoniidae, Hetrodinae). Trans. Am. Entomol. Soc. 117: 67–102.

Goreau, M. 1837. Essai sur la stridulation des insectes. Ann. Soc. Entomol. Fr. 6: 31–75.

Gorochov, A. V. 1988. The classification and phylogeny of grasshoppers (Gryllida—Orthoptera, Tettigonioidea). Pages 145–190 in A. Pomerenko, ed. The Cretaceous Biocoenotic Crisis and the Evolution of Insects. Hayka, Moscow.

——1995. Contribution to the system and evolution of the order Orthoptera. Zool. Zhur. 74: 39–45.

Grafe, T. U. 1997. Costs and benefits of mate choice in the lek-breeding reed frog, *Hyperolius marmoratus*. Anim. Behav. 53: 1103–1117.
Grasse, P. P. 1924. Étude biologique sur *Phaneroptera 4-punctata* Br. et *Ph. falcata* Scop. Bull. Biol. Fr. Belg. 58: 454–472.
Greenfield, M. D. 1983. Unsynchronized chorusing in the coneheaded katydid *Neoconocephalus affinis* (Beauvois). Anim. Behav. 31: 102–112.
——— 1993. Inhibition of male calling by heterospecific signals: Artifact of chorusing or abstinence during suppression of female phonotaxis? Naturwissenschaften 80: 570–573.
——— 1997. Acoustic communication in Orthoptera. Pages 197–230 in S. K. Gangwere and M. C. Muralirangan, eds. Bionomics of Grasshoppers, Katydids, and Their Kin. CAB International, Wallingford, U.K.
Greenfield, M. D., and I. Roizen. 1993. Katydid synchronous chorusing is an evolutionarily stable outcome of female choice. Nature 364: 618–620.
Greenfield, M. D., and K. C. Shaw. 1983. Adaptive significance of chorusing with special reference to the Orthoptera. Pages 1–27 in D. T. Gwynne and G. K. Morris, eds. Orthopteran Mating Systems: Sexual Competition in a Diverse Group of Insects. Westview Press, Boulder, Col.
Greenfield, M. D., M. K. Tourtellot, and W. A. Snedden. 1997. Precedence effects and the evolution of chorusing. Proc. R. Soc. Lond. [B] 264: 1355–1361.
Grove, D. G. 1959. The natural history of the angular-winged katydid. Ph.D. thesis, entomology, Cornell University, Ithaca, N.Y.
Grzeschik, K.-H. 1969. On the systematics, biology and ethology of *Eugaster* Serville (Orthoptera, Tettigoniidae). Form. Funct. 1: 46–144.
Guilding, R. L. 1827. Observations on the crepitaculum and the foramina in the anterior tibiae of some orthopterous insects. Trans. Linn. Soc. 15: 153–155.
Gullan P. J., and P. S. Cranston. 1994. The Insects: An Outline of Entomology. Chapman and Hall, London.
Gwynne, D. T. 1977. Mating behaviour of *Neoconocephalus ensiger* (Orthoptera: Tettigoniidae) with notes on the calling song. Can. Entomol. 109: 237–242.
——— 1981. Sexual difference theory: Mormon crickets show role reversal in mate choice. Science 213: 779–780.
——— 1982. Mate selection by female katydids (Orthoptera: Tettigoniidae, *Conocephalus nigropleurum*). Anim. Behav. 30: 734–738.
——— 1983a. Coy conquistadors of the the sagebrush. Nat. Hist. 92: 70–75.
——— 1983b. Male nutritional investment and the evolution of sexual differences in Tettigonidae and other Orthoptera. Pages 337–366 in D. T. Gwynne and G. K. Morris, eds. Orthopteran Mating Systems: Sexual Competition in a Diverse Group of Insects. Westview Press, Boulder, Col.
——— 1984a. Courtship feeding increases female reproductive success in bushcrickets. Nature 307: 361–363.
——— 1984b. Male mating effort, confidence of paternity and insect sperm competition. Pages 117–149 in R. L. Smith, ed. Sperm Competition and the Evolution of Animal Mating Systems. Academic Press, New York.
——— 1984c. Sexual selection and sexual differences in Mormon crickets (Orthoptera: Tettigoniidae, *Anabrus simplex*). Evolution 38: 1011–1022.
——— 1985. Role-reversal in katydids: Habitat influences reproductive behaviour (Orthoptera: Tettigoniidae: *Metaballus* sp.). Behav. Ecol. Sociobiol. 16: 355–361.
——— 1986. Courtship feeding in katydids (Orthoptera: Tettigoniidae): Investment in offspring or in obtaining fertilizations? Am. Nat. 128: 342–352.

―――1987. Sex-biased predation and the risky mate-locating behaviour of male ticktock cicadas (Homoptera: Cicadidae). Anim. Behav. 35: 571–576.
―――1988a. Courtship feeding and the fitness of female katydids (Orthoptera: Tettigoniidae). Evolution 42: 545–555.
―――1988b. Courtship feeding in katydids benefits the mating male's offspring. Behav. Ecol. Sociobiol. 23: 373–377.
―――1989. Does copulation increase the risk of predation? Trends Ecol. Evol. 4: 54–56.
―――1990a. The katydid spermatophore: Evolution of a parental investment. Pages 27–40 in W. J. Bailey and D. C. F. Rentz, eds. The Tettigoniidae: Biology, Systematics and Evolution. Crawford House, Bathurst, Australia.
―――1990b. Testing parental investment and the control of sexual selection in katydids: The operational sex ratio. Am. Nat. 136: 474–484.
―――1991. Sexual competition among females: What causes courtship-role reversal? Trends Ecol. Evol. 6: 118–121.
―――1993. Food quality controls sexual selection in Mormon crickets by altering male mating investment. Ecology 74: 1406–1413.
―――1994. Variation in bushcricket nuptial gifts may be due to common ancestry rather than ecology as taxonomy and diet are almost perfectly confounded. Behav. Ecol. 6: 458.
―――1995. Phylogeny of the Ensifera (Orthoptera): A hypothesis supporting multiple origins of acoustical signalling, complex spermatophores and maternal care in crickets, katydids, and weta. J. Orthop. Res. 4: 203–218.
―――1997a. The evolution of edible "sperm sacs" and other forms of courtship feeding in crickets, katydids and their kin (Orthoptera: Ensifera). Pages 110–129 in J. C. Choe and B. J. Crespi, eds. The Evolution of Mating Systems in Insects and Arachnids. Cambridge University Press, Cambridge.
―――1997b. Glandular gifts. Sci. Am. 277: 46–51.
―――1998. Genitally does it. Nature 393: 734–735.
Gwynne, D. T., and W. J. Bailey. 1988. Mating system, mate choice and ultrasonic calling in a zaprochiline katydid (Orthoptera: Tettigoniidae). Behaviour 105: 202–223.
―――1999. Female-female competition in katydids: Sexual selection for increased sensitivity to a male signal? Evolution 53: 546–551.
Gwynne, D. T., W. J. Bailey, and A. Annells. 1998. The sex in short supply for matings varies over small spatial scales in a katydid (*Kawanaphila nartee*, Orthoptera: Tettigoniidae). Behav. Ecol. Sociobiol. 42: 157–162.
Gwynne, D. T., B. J. Bowen, and C. G. Codd. 1984. The function of the katydid spermatophore and its role in fecundity and insemination. Aust. J. Zool. 32: 15–22.
Gwynne, D. T., and W. D. Brown. 1994. Mate feeding, offspring investment, and sexual differences in katydids (Orthoptera: Tettigoniidae). Behav. Ecol. 5: 267–272.
Gwynne, D. T., and G. N. Dodson. 1983. Nonrandom provisioning by the digger wasp, *Palmodes laeviventris* (Hymenoptera: Sphecidae). Ann. Entomol. Soc. Am. 76: 434–436.
Gwynne, D. T., and E. D. Edwards. 1986. Ultrasound production by genital stridulation in *Syntonarcha iriastis* (Lepidoptera: Pyralidae): Long-distance signalling by male moths? Zool. J. Linn. Soc. 88: 363–376.
Gwynne, D. T., and I. Jamieson. 1998. Sexual selection and sexual dimorphism in a harem-polygynous insect, the alpine weta (*Hemideina maori*, Orthoptera Stenopelmatidae). Ethol. Ecol. Evol. 10: 393–402.

Gwynne, D. T., and G. K. Morris. 1986. Heterospecific recognition and behavioral isolation in acoustic Orthoptera (Insecta). Evol. Theor. 8: 33–38.

Gwynne, D. T., and L. W. Simmons. 1990. Experimental reversal of courtship roles in an insect. Nature 346: 172–174.

Gwynne, D. T., and A. W. Snedden. 1995. Paternity and female remating in *Requena verticalis* (Orthoptera: Tettigoniidae). Ecol. Entomol. 20: 191–194.

Gwynne, D. T., P. Yeoh, and A. Schatral. 1988. The singing insects of King's Park and Perth gardens. West. Aust. Nat. 17: 25–81.

Haes, E. C. M., A. J. Cherrill, and V. K. Brown. 1990. Meteorological correlates of wartbiter (Orthoptera: Tettigoniidae) abundance. Entomologist 109: 93–99.

Hamilton, W. D., and M. Zuk. 1982. Heritable true fitness and bright birds: A role for parasites? Science 218: 384–387.

Hancock, J. L. 1904. The oviposition and carnivorous habits of the green meadow grasshopper (*Orchelimum glaberrimum* Burmeister). Psyche 11: 69–71.

——— 1916. Pink katy-dids and the inheritance of pink coloration. Entomol. News 27: 70–82.

Hansen, R. M., and D. N. Ueckert. 1970. Dietary similarity of some primary consumers. Ecology 51: 640–648.

Harris, A. 1993. Weta driven to water. Otago Daily News, Otago, New Zealand, November 8.

Harris, T. W. 1842. A Treatise on Some Insects of New England Which Are Injurious to Vegetation. White and Potter, Boston.

Hartley, J. C. 1964. The structure of eggs of the British Tettigoniidae (Orthoptera). Proc. R. Entomol. Soc. (Lond.) [A] 39: 111–117.

——— 1967. Laboratory culture of a tettigoniid, *Homorocoryphus nitidulus vicinus* (Wlk.) (Orthoptera). Bull. Entomol. Res. 57: 203–205.

——— 1986. Melanisation in the bush cricket *Polysarcus (Orphania) denticaudus* (Charp.) (Orth., Tettigoniidae). Entomol. Mon. Mag. 122: 1–4.

——— 1990. Egg biology of the Tettigoniidae. Pages 41–70 in W. J. Bailey and D. C. F. Rentz, eds. The Tettigoniidae: Biology, Systematics and Evolution. Crawford House, Bathurst, Australia.

——— 1993. Acoustic behaviour and phonotaxis in the duetting ephippigerines, *Steopleurus nobrei* and *Steropleurus stali* (Tettigoniidae). Zool. J. Linn. Soc. 107: 155–167.

Hartley, J. C., and Y. Ando. 1988. Egg development patterns in diapausing and nondiapausing species of *Ruspolia*. Entomol. Exp. Appl. 49: 203–212.

Hartley, J. C., and M. M. Bugren. 1986. Colour polymorphism in *Ephippiger ephippiger* (Orthoptera, Tettigoniidae). Biol. J. Linn. Soc. 27: 191–199.

Hartley, J. C., and R. L. Dean. 1974. *Ephippiger cruciger* as a laboratory insect (Orthoptera, Tettigoniidae). J. Nat. Hist. 8: 349–354.

Hartley, J. C., D. J. Robinson, and A. C. Warne. 1974. Female response song in the Ephippigerines *Steropleurus stali* and *Platystolus obvius* (Orthoptera, Tettigoniidae). Anim. Behav. 22: 382–389.

Harvey, P. H., and M. D. Pagel. 1991. The Comparative Method in Evolutionary Biology. Oxford University Press, Oxford.

Harz, K. 1957. Die Geradflugler Mitteleuropas. VEB Gustav Fischer Verlag, Jena, Germany.

——— 1969. The Orthoptera of Europe, Vol. 1. W. Junk N. V., The Hague.

Hatziolos, M. E., and R. L. Caldwell. 1983. Role reversal in courtship in the stomatopod *Pseudosquilla ciliata* (Crustacea). Anim. Behav. 31: 1077–1087.

Hayashi, F. 1992. Large spermatophore production and consumption in dobsonflies *Protohermes* (Megaloptera, Corydalidae). Jpn. J. Entomol. 60: 59–66.
——1993. Male mating costs in two insect species (*Protohermes*, Megaloptera) that produce large spermatophores. Anim. Behav. 45: 343–349.
——1998. Multiple mating and lifetime reproductive output in female dobsonflies that receive nuptial gifts. Ecol. Res. 13: 283–289.
Hayes, M. P., and D. C. F. Rentz. 1986. Observations on the biology of the neotropical katydid *Haemodiasma tesselata* (Orthoptera: Tettigoniidae). Entomol. News 97: 222–224.
Heath, J. E., and R. K. Josephson. 1970. Body temperature and singing in the katydid *Neoconocephalus robustus* (Orthoptera, Tettigoniidae). Biol. Bull. 138: 272–285.
Hedrick, A. V. 1988. Female choice and the heritability of attractive male traits: An empirical study. Am. Nat. 132: 267–276.
Hedrick, A. V., and L. M. Dill. 1993. Mate choice by female crickets is influenced by predation risk. Anim. Behav. 46: 193–196.
Heinrich, R., M. Jatho, and K. Kalmring. 1993. Acoustic transmission characteristics of the tympanal tracheae of bushcrickets (Tettigoniidae). II: Comparative studies of the tracheae of seven species. J. Acoust. Soc. Am. 93: 3481–3489.
Helfert, B. V., and K. Sänger. 1995. Ant-mimicking in larvae of *Macroxiphus siamsensis* (Orthoptera: Tettigoniidae) [German with English abstract]. Z. Arbeit Österr. Ent. 47: 41–48.
Heller, K.-G. 1986. Warm-up and stridulation in the bushcricket, *Hexacentrus unicolor* Serville (Orthoptera, Conocephalidae, Listroscelidinae). J. Exp. Biol. 126: 97–109.
——1988. Bioakustik der Europäischen Laubheuschrecken. Verlag Josef Margraf, Weikersheim, Germany.
——1992. Risk shift between males and females in the pair-forming behavior of bushcrickets. Naturwissenschaften 79: 89–91.
——1995. Acoustic signaling in palaeotropical bush-crickets (Orthoptera, Tettigonioidea, Pseudophyllidae): Does predation pressure by eavesdropping enemies differ in the palaeotropics and neotropics? J. Zool. 237: 469–485.
——1996. Unusual abdomino-alary, defensive stridulatory mechanism in the bushcricket *Pantecphylus cerambycinus* (Orthoptera, Tettigonioidea, Pseudophyllidae). J. Morphol. 227: 81–86.
——1997. Geld oder Leben—die unterschiedlichen Kosten des Gastangs bei Laubheuschrecken. Jahrb. Akad. Wissensch. Gott. 1997: 132–152.
Heller, K.-G., and R. Arlettaz. 1994. Is there a sex ratio bias in the bushcricket prey of the scops owl due to predation on calling males? J. Orthop. Res. 2: 41–42.
Heller, K.-G., S. Faltin, P. Fleischmann, and O. von Helversen. 1998. The chemical composition of the spermatophore in some species of phaneropterid bushcrickets (Orthoptera: Tettigonioidea). J. Ins. Physiol. 44: 1001–1008.
Heller, K.-G., and D. von Helversen. 1986. Acoustic communication in phaneropterid bushcrickets: Species-specific delay of female stridulatory response and matching male sensory time window. Behav. Ecol. Sociobiol. 18: 189–198.
——1991. Operational sex ratio and individual mating frequencies in two bushcricket species (Orthoptera, Tettigonioidea, *Poecilimon*). Ethology 89: 211–228.
——1993. Calling behavior in bushcrickets of the genus *Poecilimon* with differing communication systems (Orthoptera: Tettigonioidea, Phaneropteridae). J. Ins. Behav. 6: 361–377.
Heller, K.-G., and O. von Helversen. 1990. Survival of a Phaneropterid bush-cricket

studied by a new marking technique (Orthoptera: Phaneropteridae). Entomol. Gener. 15: 203–208.

Heller, K.-G., O. von Helversen, and M. Sergejeva. 1997. Indiscriminate response behaviour in a female bushcricket: Sex role reversal in selectivity of acoustic mate recognition? Naturwissenschaften 84: 252–255.

Heller, K.-G., and K. Reinhold. 1994. Mating effort function of the spermatophore in the bushcricket *Poecilimon veluchianus* (Orthoptera, Phaneropteridae): Support from a comparison of the mating behaviour of two subspecies. Biol. J. Linn. Soc. 53: 153–163.

Helversen, D. von, and O. von Helversen. 1991. Pre-mating sperm removal in the bushcricket *Metaplastes ornatus* Ramme 1931 (Orthoptera, Tettigonioidea, Phaneropteridae). Behav. Ecol. Sociobiol. 28: 391–396.

Hill, K. G., and B. P. Oldfield. 1981. Auditory function in Tettigoniidae (Orthoptera: Ensifera). J. Comp. Physiol. [A] 142: 169–180.

Hjermann, D. O., and R. A. Ims. 1996. Landscape ecology of the wart-biter *Decticus verrucivorus* in a patchy landscape. J. Anim. Ecol. 65: 768–780.

Hockham, L. R., and K. Vahed. 1997. The function of mate guarding in a field cricket (Orthoptera: Gryllidae; *Teleogryllus natalis* Otte and Cade). J. Ins. Behav. 10: 247–256.

Hopper, S. D. 1993. Kangaroo Paws and Cats Paws: A Natural History and Field Guide. West Australia Dept. Conservation and Land Management, Perth, Australia.

Horapollo, and A. T. Cory. 1840. The Hieroglyphics of Horapollo Nilous. W. Pickering, London.

Horner, J., R. Wallace, and D. W. Johnston. 1974. Food of the barn owl at Gainesville. Florida. Fla. Field Nat. 2: 28–29.

Howard, D. J., G. L. Waring, C. A. Tibbets, and P. G. Gregory. 1993. Survival of hybrids in a mosaic hybrid zone. Evolution 47: 789–800.

Howard, R. D., H. H. Whiteman, and T. I. Schueller. 1994. Sexual selection in American toads: A test of a good-genes hypothesis. Evolution 48: 1286–1300.

Hsu, Y. C. 1928–29. Crickets in China. Peking Soc. Nat. Hist. Bull. 3: 5–45.

Hubbell, S. 1993. Broadsides from the Other Orders: A Book of Bugs. Random House, New York.

Hudson, L. 1972. New Zealand Tettigoniidae (Orthoptera). J. R. Soc. N. Z. 2: 249–255.

Hunt, J., and G. R. Allen. 1998. Fluctuating asymmetry, call structure and the risk of attack from phonotactic parasitoids in the bushcricket *Sciarasaga quadrata* (Orthoptera: Tettigoniidae). Oecologia 116: 356–364.

Hunter, J. 1786. Observations on certain parts of the animal oeconomy. London.

Ichikawa, N. 1989. Breeding strategy of the male brooding water bug, *Diplonychus major* Esaki (Heteroptera: Belostomatidae): Is male back space limiting? J. Ethol. 7: 133–140.

Ingrisch, S. 1978. Labor- und Freilanduntersuchungen zur Dauer der postembryonalen Entwicklung einiger mitteleuropaeischer Laubheuschrecken (Orthoptera: Tettigoniidae) und ihre Beeinflussung durch Temperatur und Feuchte. Zool. Anz. 200: 309–320.

―― 1981. Zur vikariierenden Verbreitung von *Tettigonia viridissima* und *T. cantans* in Hessen (Orthoptera: Tettigoniidae). Mitt Dtsch. Ges. Allg. Angew Entomol.. 3: 155–159.

―― 1984. The influence of environmental factors on dormancy and duration of egg development in *Metrioptera roeseli* (Orthoptera: Tettigoniidae). Oecologia 61: 254–258.

———1986a. The pleurennial life cycles of the European Tettigoniidae (Insecta: Orthoptera). 1. The effect of temperature on embryological development and hatching. Oecologia 70: 606–616.

———1986b. The pleurennial life cycles of the European Tettigoniidae (Insecta: Orthoptera). 3. The effect of drought and the variable duration of the initial diapause. Oecologia 70: 624–630.

———1986c. The pleurennial life cycles of the European Tettigoniidae (Insecta: Orthopterta). 2. The effect of photoperiod on the induction of an initial diapause. Oecologia 70: 617–623.

———1988. Water uptake and drought resistance of the eggs of European bush crickets (Orthoptera: Tettigoniidae). Zool. Jb. Physiol. 92: 117–170.

———1990. Significance of seasonal adaptations in Tettigoniidae. Bol. San. Veg. Plagas (Fuera de serie) 20: 209–218.

———1995. Revision of the Lipotactinae, a new subfamily of Tettigonioidea (Ensifera). Entomol. Scand. 26: 273–320.

———1996. Evidence of an embryonic diapause in a tropical Phaneropterinae (Insecta: Ensifera, Tettigonioidea). Trop. Zool. 9: 431–439.

———1998a. Monograph of the Oriental Agraeciini (Insecta, Ensifera, Tettigoniidae): Taxonomic revision, phylogeny, biogeography, stridulation, and development. Courier Forsch. Senckenberg 206: 1–391.

———1998b. Überliegende Eier bei einer tropischen Laubheuschrecke (Ensifera: Tettigoniidae). Ent. Zeit. 108: 45–48.

Ingrisch, S., and G. Köhler. 1998. Die Heuschrecken Mitteleuropas. Westarp Wissenschaften, Die Neue Brehm-Bücherei 629, Magdeburg, Germany.

Irish, J. 1992. The Hetrodinae (Orthoptera: Bradyporidae) of southern Africa: Systematics and phylogeny. Navors. Nas. Mus. Bloemfontein 9: 393–434.

Isely, F. B. 1941. Researches concerning Texas Tettigoniidae. Ecol. Monogr. 11: 457–475.

Ivy, T. M., J. C. Johnson, and S. K. Sakaluk. 1999. Hydration benefits to courtship feeding in crickets. Proc. R. Soc. Lond. [B] 266: 1523–1527.

Jaeger, E. C. 1955. A Source Book of Scientific Names and Terms. Charles Thomas, Springfield, Ill.

Jago, N. D. 1997. Crop-centred integrated pest management in grasshoppers and other Orthoptera. Pages 443–480 in S. K. Gangwere, M. C. Mulalirangen, and M. Muralirangen, eds. The Bionomics of Grasshoppers, Katydids and their Kin. CAB International, Oxford.

Jeram, S., W. Rossler, A. Cokl, and K. Kalmring. 1995. Structure of atympanate tibial organs in legs of the cave-living Ensifera, *Troglophilus neglectus* (Gryllacridoidea, Raphidophoridae). J. Morphol. 223: 109–118.

Jia, F.-Y., and M. D. Greenfield. 1997. When are good genes good? Variable outcomes of female choice in wax moths. Proc. R. Soc. Lond. [B] 264: 1057–1063.

Jin, X.-B. 1993. Katydids and crickets do have commercial value! An insight into the insect market in China. Metalepea 14: 4.

———1994. Chinese cricket culture. URL in Sept. 2000: www.insects.org/ced3/chinese_crcul.html.

Johns, P. M. 1997. The Gondwanaland weta: Family Anostostomatidae (formerly in Stenopelmatidae, Henicinae or Minermidae): Nomenclatural problems, world checklist, new genera and species. J. Orthop. Res. 6: 125–138.

Johnstone, R. A., J. D. Reynolds, and J. C. Deutsch. 1996. Mutual mate choice and sex-differences in choosiness. Evolution 50: 1382–1391.

Jones, M. D. R. 1966. The acoustic behaviour of the bush cricket *Pholidoptera griseoaptera* 1. Alternation, synchronism and rivalry between males. J. Exp. Biol. 45: 15–30.
Judd, W. W. 1947. A comparative study of the proventriculus of orthopteroid insects with reference to its use in taxonomy. Can. J. Res. 26: 93–159.
Kaitala, A., and C. Wiklund. 1994. Polyandrous female butterflies forage for matings. Behav. Ecol. Sociobiol. 35: 385–388.
Kalmring, K., E. Hoffmann, M. Jatho, T. Sickmann, and M. Grossbach. 1996. The auditory-vibratory sensory system of the bush-cricket *Polysarcus denticauda* (Phaneropterinae, Tettigoniidae). 2. Physiology of receptor-cells. J. Exp. Zool. 276: 315–329.
Kalmring, K., A. Keuper, and W. Kaiser. 1990. Aspects of acoustic and vibratory communication in seven European bushcrickets. Pages 191–216 in W. Bailey and D. C. F. Rentz, eds. The Tettigoniidae: Biology, Systematics and Evolution. Crawford House, Bathurst, Australia.
Kalmring, K., W. Rossler, and C. Unrast. 1994. Complex tibial organs in the forelegs, midlegs, and hindlegs of the bushcricket *Gampsocleis gratiosa* (Tettigoniidae): Comparison of the physiology of the organs. J. Exp. Zool. 270: 155–161.
Kaltenbach, A. P. 1990. The predatory Saginae. Pages 280–302 in W. J. Bailey and D. C. F. Rentz, eds. The Tettigoniidae: Biology, Systematics and Evolution. Crawford House, Bathurst, Australia.
Kamp, J. W. 1973. Numerical classification of the Orthopteroids, with special reference to the Grylloblattodea. Can. Entomol. 105: 1235–1249.
Kasuya, E., and N. Sato. 1998. Effects of the consumption of male spermatophylax on the oviposition schedule of females in the decorated cricket, *Gryllodes sigillatus*. Zool. Sci. 15: 127–130.
Kathirithamby, J., and W. D. Hamilton. 1995. Exotic pests and parasites. Nature 374: 769–770.
Kathirithamby, J., S. Simpson, T. Solulu, and R. Caudwell. 1998. Strepsiptera parasites—novel biocontrol tools for oil palm integrated pest management in Papua New Guinea. Int. J. Pest Manage. 44: 127–133.
Kessel, E. L. 1955. Mating activities of balloon flies. Syst. Zool. 4: 97–104.
Kevan, D. K. M. 1977. The higher classification of the Orthopteroid insects: A general view. Mem. Lyman Entomol. Mus. Res. Lab. 4: 1–31.
——1979. The place of grasshoppers and crickets in amerindian cultures. Proc. Pan. Acrid. Soc. 1979: 8–74.
——1982. Orthoptera. Pages 352–383 in S. P. Parker, ed. Synopsis and Classification of Living Organisms. McGraw Hill, New York.
——1986. A rational for the classification of orthopteroid insects—the saltatorial orthopteroids or grigs—one order or two? Proc. Pan. Acrid. Soc. 1985: 49–67.
——1989. Grigs that dig and grasshoppers that grovel. Rev. Ecol. Biol. Sol. 26: 267–289.
Kevan, D. K. M., E. J. LeRoux., and C. d'Ornellas. 1962. Further observations on *Metrioptera (Roeseliana) roeseli* (Hegenbach, 1822) in Quebec, with notes on the genus *Metrioptera* Wesmael, 1938 (Orthoptera: Tettigoniidae: Decticinae). Ann. Entomol. Soc. Que. 7: 70–86.
Key, K. H. L. 1957. Kentromorphic phases in three species of Phasmatodea. Aust. J. Zool. 5: 247–284.
Khalifa, A. 1949. Spermatophore production in Trichoptera and some other insects. Trans. R. Entomol. Soc. (Lond.) 100: 449–471.

Kindvall, O. 1996. Habitat heterogeneity and survival in a bush-cricket metapopulation. Ecology 77: 207–214.
Kindvall, O., and I. Ahlen. 1992. Geometrical factors and metapopulation dynamics of the bush cricket, *Metrioiptera bicolor* Philippi (Othoptera: Tettigoniidae). Conserv. Biol. 6: 520–529.
Kirby, W., and W. Spence. 1818. An Introduction to Entomology, Vol. 2. Longman, Hurst, Rees, Orme, and Brown, London.
——1826. An Introduction to Entomology, Vol. 4. Longman, Hurst, Rees, Orme, and Brown, London.
Kirkpatrick, M. 1987. Sexual selection by female choice in polygynous animals. Annu. Rev. Ecol. Syst. 18: 43–70.
——1996. Good genes and direct selection in the evolution of mating preferences. Evolution 50: 2125–2140.
Kirkpatrick, M., and M. Ryan. 1991. The evolution of mating preferences and the paradox of the lek. Nature 350: 33–38.
Knapton, R. W. 1984. Parental investment: The problem of currency. Can. J. Zool. 62: 2673–2674.
Kraus, W. F. 1989. Is male back space limiting? An investigation into the reproductive demography of the giant water bug, *Abedus indentatus* (Heteroptera: Belostomatidae). J. Ins. Behav. 2: 623–648.
Kruse, K. P. 1990. Male backspace availability in the giant waterbug (*Belostoma flumineum* Say). Behav. Ecol. Sociobiol. 26: 281–289.
Kurczewski, F., and S. E. Ginsburg. 1971. Nesting behavior of *Tachytes* (*Tachyplena*) *validus*. J. Kans. Entomol. Soc. 44: 113–131.
Kvarnemo, C., and L. W. Simmons. 1998. Male potential reproductive rate influences mate choice in a bushcricket. Anim. Behav. 55: 1499–1506.
——1999. Variance in female quality, operational sex ratio and male mate choice in a bushcricket. Behav. Ecol. Sociobiol. 45: 245–252.
Lakes, R., and T. Schikorski. 1990. Neuroanatomy of tettigoniids. Pages 166–190 in W. J. Bailey and D. C. F. Rentz, eds. The Tettigoniidae: Biology, Systematics and Evolution. Crawford House, Bathurst, Australia.
Lakes-Harlan, R. 1992. Utrasound-sensitive ears in a parasitoid fly. Naturwissenschaften 79: 224–226.
Lakes-Harlan, R., W. J. Bailey, and T. Schikorski. 1991. The auditory system of an atympanate bushcricket *Phasmodes ranatriformes* (sic) (Westwood) (Tettigoniidae: Orthoptera). J. Exp. Biol. 158: 307–324.
Lakes-Harlan, R., H. Stolting, and A. Stumpner. 1999. Convergent evolution of insect hearing organs from a preadaptive structure. Proc. R. Soc. Lond. [B] 266: 1–7.
Lakes-Harlan, R., A. Stumpner, and G. Allen. 1995. Functional adaptations of the auditory system of two parasitoid fly species *Therobia leonidei* and *Homotrixa* spec. In M. Burrows, T. Matheson, P. Newland, and H. Schuppe, eds. Nervous Systems and Behaviour. Thieme Verlag, Stuttgart.
Lande, R. 1981. Models of speciation by sexual selection on polygenic traits. Proc. Natl. Acad. Sci. USA 78: 3721–3725.
Lang, A. 1996. Silk investment in gifts by males of the nuptial feeding spider *Pisaura mirabilis* (Araneae, Pisauridae). Behaviour 133: 697–716.
Lange, C. E., C. M. MacVean, J. E. Henry, and D. A. Sreett. 1995. *Heterovesicula cowani* N. G., N. Sp. (Heterovesiculidae N. Fam), a Microsporidian parasite of Mormon crickets, *Anabrus simplex*, 1852 (Orthoptera: Tettigoniidae). J. Euk. Microbiol. 42: 552–558.

La Rivers, I. 1945. The wasp *Chlorion laeviventris* as a natural control of the Mormon cricket. Am. Midl. Nat. 33: 743–763.
Latimer, W. 1980. Song and spacing in *Gampsocleis glabra* (Orthoptera, Tettigoniidae). J. Nat. Hist. 14: 201–213.
Latimer, W., and W. B. Broughton. 1984. Acoustic interference in bush crickets; a factor in the evolution of singing insects? J. Nat. Hist. 18: 599–616.
Latimer, W., and A. Schatral. 1983. The acoustic behaviour of the bushcricket *Tettigonia cantans*. I. Behavioural responses to sound and vibration. Behav. Proc. 8: 113–124.
Latimer, W., and M. Sippel. 1987. Acoustic cues for female choice and male competition in *Tettigonia cantans*. Anim. Behav. 35: 887–900.
Lawrence, W. S. 1986. Male choice and competition in *Tetraopes tetraopthalmus*: Effects of local sex ratio variation. Behav. Ecol. Sociobiol. 18: 289–296.
Lehmann, G. U. C., and G.-C. Heller. 1998. Bushcricket song structure and predation by the acoustically orienting parasitoid fly *Therobia leonidei* (Diptera: Tachinidae: Ormiini). Behav. Ecol. Sociobiol. 43: 239–245.
——Manuscript. Sexy sons are favoured as parasitoid hosts.
Leopold, R. A. 1976. The role of male accessory glands in insect reproduction. Annu. Rev. Entomol. 21: 199–221.
Leroy, Y. 1969. Quelques aspects de la reproduction des Tettigonioidea de Trinidad [Orth.]. Ann. Soc. Ent. Fr. (N.S.) 5: 775–798.
Leung, B., and M. R. Forbes. 1996. Fluctuating asymmetry in relation to stress and fitness: Effects of trait type as revealed by meta-analysis. Ecoscience 3: 400–413.
Lewis, D. B. 1974. The physiology of the tettigoniid ear. J. Exp. Biol. 60: 821–869.
Li, S.-Z. 1578. Ben Chao Gang Mo. [Encyclopedia of Chinese Medicine]. (As cited in Jin 1994).
Libersat, F., and R. R. Hoy. 1991. Ultrasonic startle behavior in bushcrickets (Orthoptera; Tettigoniidae). J. Comp. Physiol. [A] 169: 507–514.
Littlejohn, M. J. 1981. Reproductive isolation: A critical review. Pages 298–334 in W. R. Atchley and D. S. Woodruff, eds. Evolution and Speciation: Essays in Honor of M. J. D. White. Cambridge University Press, Cambridge.
Lloyd, J. E. 1973a. Model for the mating protocol of synchronously flashing fireflies. Nature 245: 268–270.
——1973b. Firefly parasites and predators. Coleop. Bull. 27: 91–106.
——1976. Coxo-sternal stridulation in *Phyllophora* (Orthoptera: Tettigonioidea: Phyllophoridae). Entomol. News 87: 285–286.
——1979a. Mating behavior and natural selection. Fla. Entomol. 62: 17–34.
——1979b. Sexual selection in luminescent beetles. Pages 293–342 in M. S. Blum and N. A. Blum, eds. Sexual Selection and Reproductive Competition in Insects. Academic Press, New York.
Lloyd, J. E., and A. B. Gurney. 1975. Labral stridulation in a katydid (a coconut-infesting "treehopper") (Orthoptera: Tettigoniidae: Mecopodinae). Entomol. News 86: 47–50.
Lockwood, J. A., and D. C. F. Rentz. 1996. Nest construction and recognition in a gryllacridid: The discovery of pheromonally mediated autorecognition in an insect. Aust. J. Zool. 44: 129–141.
Löhe, G., and H.-U. Kleindienst. 1994. The role of the medial septum in the acoustic trachea of the cricket *Gryllus bimaculatus*, II. Influence on directionality of the auditory system. J. Comp. Physiol. [A] 174: 601–606.

Loher, W. 1984. Behavioral and physiological changes in cricket females after mating. Adv. Invert. Reprod. 3: 189–201.
Loher, W., and K. Edson. 1973. The effect on mating on egg production and release in the cricket *Teleogryllus commodus*. Entomol. Exp. Appl. 16: 483–490.
Loher, W., I. Ganjian, I. Kubo, D. Stanley-Samuelson, and S. S. Tobe. 1981. Prostaglandins: Their role in egg-laying of the cricket *Teleogryllus commodus*. Proc. Natl. Acad. Sci. USA 78: 7835–7838.
Lorch, P. 1999. Life history and sexual selection. Ph.D. thesis, zoology, University of Toronto, Toronto.
Lorch, P., and D. T. Gwynne. 2000. Radio telemetric evidence of migration in the gregarious but not the solitary morph of the Mormon cricket (*Anabrus simplex*: Orthoptera: Tettigoniidae). Naturwissenschaften 87: 370–372.
Lymbery, A., and W. Bailey. 1980. Regurgitation as a possible anti-predator defensive mechanism in the grasshopper *Goniaea* sp. (Acrididae, Orthoptera). J. Aust. Entomol. Soc. 19: 129–130.
Lymbery, A. J. 1984. The plasticity of coloration in the bush cricket *Mygalopsis marki* (Orthoptera: Tettigoniidae). Ph.D. thesis, zoology, University of Western Australia, Nedlands.
——1987. Seasonal populations in the life cycle of *Mygalopsis marki* Bailey (Orthoptera: Tettigonidae). J. Aust. Entomol. Soc. 26: 323–330.
——1992. The environmental control of colouration in a bushcricket, *Mygalopsis marki* Bailey (Orthoptera: Tettigoniidae). Biol. J. Linn. Soc. 45: 71–89.
Lynam, A. J., S. Morris, and D. T. Gwynne. 1992. Differential mating success of virgin female katydids *Requena verticalis* (Orthoptera: Tettigoniidae). J. Ins. Behav. 5: 51–59.
Lysaght, A. M. 1925. The eggs of the katydid, *Caedicia simplex* (Walker). N. Z. J. Sci. Technol. 7: 372.
——1931. Biological notes on a long-horned grasshopper, *Caedicia simplex* (Walk.). N. Z. J. Sci. Technol. 12: 296–303.
MacVean, C. 1989. Microbial control, diet composition, and damage potential of the Mormon cricket. Ph.D. thesis, entomology, Colorado State University, Fort Collins.
——1990. Mormon crickets: A brighter side. Rangelands 12: 234–235.
MacVean, C. M. 1987. Ecology and management of the Mormon cricket, *Anabrus simplex* Haldeman. Pages 116–136 in J. L. Capinera, ed. Integrated Pest Management on Rangeland: A Shortgrass Prairie Perspective. Westview Press, Boulder, Col.
MacVean, C. M., and J. L. Capinera. 1991. Pathogenicity and transmission of *Nosema locustae* and *Variomorpha sp.* (Protozoa: Microsporidia) in Mormon crickets (*Anabrus simplex*, Orthoptera: Tettigoniidae): A laboratory evaluation. Invert. Pathol. 57: 23–36.
——1992. Field evaluation of two microsporidian pathogens, an entomopathogenic nematode, and carbaryl for suppression of the Mormon cricket, *Anabrus simplex* Hald. (Orthoptera: Tettigoniidae). Biol. Contr. 2: 59–65.
Madsen, D. B. 1989. A grasshopper in every pot. Nat. Hist.: 22–25.
Maekawa, K., O. Kitade, and T. Matsumoto. 1999. Molecular phylogeny of orthopteroid insects based on the mitochondrial cytochrome oxidase II gene. Zool. Sci. 16: 175–184.
Manley, G. V. 1985. The predatory status of *Conocephalus longipennis* (Orthoptera: Tettigoniidae) in rice fields of West Malaysia. Entomol. News 96: 167–170.
Mann, T. 1984. Spermatophores. Springer Verlag, Berlin.

Markl, H. 1983. Vibrational communication. Pages 332–353 in F. Huber and H. Markl, eds. Neuroethology and Behavioral Physiology. Springer-Verlag, Berlin.

Marshall, J. A., and E. C. M. Haes. 1988. Grasshoppers and Allied Insects of Great Britain and Ireland. Harley Books, Colchester, U.K.

Mason, A. C. 1996. Territoriality and the function of song in the primitive acoustic insect *Cyphoderris monstrosa* (Orthoptera, Haglidae). Anim. Behav. 51: 221–224.

Mason, A. C., and W. J. Bailey. 1998. Ultrasound hearing and male-male communication in Australian katydids (Tettigoniidae: Zaprochilinae) with sexually dimorphic ears. Physiol. Entomol. 23: 139–149.

Mason, A. C., G. K. Morris, and P. Wall. 1991. High ultrasonic hearing and tympanal slit function in rainforest katydids. Naturwissenschaften 78: 365–367.

Mathis, A. 1991. Large male advantage for access to females: Evidence of male-male competiton and female discrimination in a territorial salamander. Behav. Ecol. Sociobiol. 29: 133–138.

Mbata, K. J. 1992a. The biology and host plant specificity of *Acanthoplus speiseri* Brancsik (Orthoptera: Tettigoniidae: Hetrodinae), a pest of grain crops. J. Entomol. Soc. S. Afr. 55: 99–106.

——— 1992b. Some observations on the reproductive behaviour of *Acanthoplus speiseri* Brancsik (Orthoptera: Tettigoniidae: Hetrodinae). Ins. Sci. Applic. 13: 19–26.

——— 1992c. Structural deviations in the prothoracic tracheal apparatuses of the armoured ground crickets, *Acanthoplus speiseri* Brancsik., and *Enyaliopsis* cf. *matebelensis* Per. (Orthoptera: Tettigoniidae, Hetrodinae) from the "typical" Hetrodinae condition of Zeuner (1936a) and their importance to tettigoniid classification. Zool. J. Linn. Soc. 104: 57–65.

McCafferty, A. R., and W. W. Page. 1978. Factors influencing the production of long winged *Zonocerus variegatus*. J. Ins. Physiol. 24: 465–472.

McIntyre, M. E. 1991. The "elephant weta." Aust. Nat. Hist. 23: 602–603.

Meads, M. 1990. The Weta Book: A Guide to the Identification of Wetas. DSIR Land Resources, Lower Hutt, New Zealand.

Meixner, A. J., and K. C. Shaw. 1979. Spacing and movement of singing *Neoconocephalus nebrascensis* males (Tettigoniidae: Copiphorinae). Ann. Entomol. Soc. Am. 72: 602–606.

——— 1986. Acoustic and associated behavior of the coneheaded katydid *Neoconocephalus nebrascensis* (Orthoptera: Tettigoniidae). Ann. Entomol. Soc. Am. 79: 554–565.

Melander, A. L., and M. A. Yothers. 1917. The coulee cricket. Washington State College Agric. Exp. Sta. Bull. 137.

Meng, Z. L. 1993. Zhong Guo Ming Chong Yu Hu Lu. [Chinese Singing Insects and Gourds.] Tianjin. (As cited in Jin 1994).

Michelsen, A. 1992. Hearing and sound communication in small animals: Evolutionary adaptations to the laws of physics. Pages 61–67 in D. B. Fay and A. N. Popper, eds. The Evolutionary Biology of Hearing. Springer-Verlag, New York.

Michelsen, A., K.-G. Heller, A. Stumpner, and K. Roherseitz. 1994. A new biophysical method to determine the gain of the acoustic trachea in bushcrickets. J. Comp. Physiol. [A] 175: 145–151.

Michelsen, A., and G. Löhe. 1995. Tuned directionality in cricket ears. Nature 375: 639.

Mills, H. B., and O. B. Hitchcock. 1938. Control of Mormon crickets in Montana. Montana Exp. Serv. Bull. 160.

Moffett, T., E. Wotton, K. Gesner, T. Penny, and T. T. D. Mayerne. 1634. Insectorum

sive minimorum animalium theatrum: Olim ab Edoardo Wottono, Conrado Gesnero, Thomaque Pennio inchoatum. Ex officinâ typographicâ Thom Cotes.; et venales extant apud Guliel. Hope . . . , Londini.

Møller, A. P., and R. Thornhill. 1997. A meta-analysis of the heritability of developmental stability. J. Evol. Biol. 10: 1–16.

———1998. Male parental care, differential parental investment by females and sexual selection. Anim. Behav. 55: 1507–1515.

Montealegre Z, F. 1997. Estudio de la fauna de Tettigoniidae (Orthoptera: Ensifera) de Valle de Cauda. Biòl. Ent. thesis, Universidad del Valle, Cali, Colombia.

Montealegre Z, F., and G. K. Morris. 1999. Songs and systematics of some Tettigoniidae from Colombia and Ecuador. 1. Pseudophyllinae. J. Orthop. Res. 8: 163–236.

Mora, G. 1990. Paternal care in a neotropical harvestman, *Zygopachylus albomarginis* (Arachnida, Opiliones: Gonyleptidae). Anim. Behav. 39: 582–593.

Morris, G. K. 1967. Song and aggression in Tettigoniidae. Ph.D. thesis, entomology, Cornell Unversity, Ithaca, N.Y.

———1970. Sound analyses of *Metrioptera sphagnorum* (Orthoptera: Tettigoniidae). Can. Entomol. 102: 363–368.

———1971. Aggression in male conocephaline grasshoppers (Tettigoniidae). Anim. Behav. 19: 132–137.

———1972. Phonotaxis of male meadow grasshoppers (Orthoptera: Tettigoniidae). J. N. Y. Entomol. Soc. 80: 5–6.

———1980. Calling display and mating behaviour of *Copiphora rhinoceros* Pictet (Orthoptera: Tettigoniidae). Anim. Behav. 28: 42–51.

———1999. Song in arthropods. Pages 508–517 in E. Knobil and J. D. Neill, eds. Encyclopedia of Reproduction. Academic Press, San Diego.

Morris, G. K., R. B. Aiken, and G. E. Kerr. 1975a. Calling songs of *Neduba macneilli* and *N. sierranus* (Orthoptera: Tettigoniidae: Decticinae). J. N. Y. Entomol. Soc. 83: 229–234.

Morris, G. K., and M. Beier. 1982. Song structure and description of some Costa Rican katydids (Orthoptera: Tettigoniidae). Trans. Am. Entomol. Soc. 108: 287–314.

Morris, G. K., and J. H. Fullard. 1983. Random noise and congeneric discrimination in *Conocephalus* (Orthoptera: Tettigoniidae). Pages 73–96 in D. T. Gwynne and G. K. Morris, eds. Orthopteran Mating Systems: Sexual Competition in a Diverse Group of Insects. Westview Press, Boulder, Col.

Morris, G. K., and D. T. Gwynne. 1978. Geographical distribution and biological observations of *Cyphoderris* (Orthoptera: Haglidae) with a description of a new species. Psyche 85: 147–167.

Morris, G. K., G. E. Kerr, and J. H. Fullard. 1978. Phonotactic preferences of female meadow katydids (Orthoptera: Tettigoniidae: *Conocephalus nigropleurum*). Can. J. Zool. 56: 1479–1487.

Morris, G. K., G. E. Kerr, and D. T. Gwynne. 1975b. Calling song function in the bog katydid, *Metrioptera sphagnorum* (F. Walker) (Orthoptera, Tettigoniidae): Female phonotaxis to normal and altered song. Z. Tierpsychol. 37: 502–514.

———1975c. Ontogeny of phonotaxis in *Orchelimum gladiator* (Orthoptera: Tettigoniidae: Concephalinae). Can. J. Zool. 53: 1127–1130.

Morris, G. K., D. E. Klimas, and D. A. Nickle. 1989. Acoustic signals and systematics of false-leaf katydids from Ecuador (Orthoptera, Tettigoniidae, Pseudophyllinae). Trans. Am. Entomol. Soc. 114: 215–264.

Morris, G. K., and A. C. Mason. 1995. Covert stridulation: Novel sound generation by a South American katydid. Naturwissenschaften 82: 96–98.

Morris, G. K., A. C. Mason, P. Wall, and J. Belwood. 1994. High ultrasonic and tremulation signals in neotropical katydids (Orthoptera: Tettigoniidae). J. Zool. (Lond.) 233: 129–163.

Morris, G. K., and R. E. Pipher. 1967. Tegminal amplifiers and spectrum consistencies in *Conocephalus nigropleurum* (Bruner), Tettigoniidae. J. Ins. Physiol. 13: 1075–1085.

—— 1972. The relation of song structure to tegminal movement in *Metrioptera sphagnorum* (Orthoptera: Tettigoniidae). Can. Entomol. 104: 977–985.

Morris, G. K., and T. J. Walker. 1976. Calling songs of *Orchelimum* meadow katydids (Tettigoniidae). I. Mechanism, terminology, and geographic distribution. Can. Entomol. 108: 785–800.

Myers, J. G. 1929. Insect Singers: A Natural History of the Cicadas. George Routledge, London.

Naskrecki, P. 1994. The Mecopodinae of southern Africa (Orthoptera: Tettigonioidea: Tettigoniidae). J. Afr. Zool. 108: 279–320.

—— 1996. Systematics of the southern African Meconematinae (Orthoptera: Tettigoniidae). J. Afr. Zool. 110: 159–193.

Naskrecki, P., and D. Otte. 1999. An Illustrated Catalog of Orthoptera, Vol. 1. Tettigonioidea (CD ROM and www site—http://viceroy.eeb.uconn.edu). Orthopterist's Society at the Academy of Natural Sciences of Philadelphia, Publications on Orthopteran Diversity, Philadelphia.

Nickle, D. A., and T. C. Carlysle. 1975. Morphology and function of female sound-producing structures in ensiferan Orthoptera with special emphasis on the Phaneropterinae. Int. J. Ins. Morphol. Embryol. 4: 159–168.

Nickle, D. A., and J. L. Castner. 1995. Strategies utilized by katydids (Orthoptera: Tettigoniidae) against diurnal predators in rainforests of northeastern Peru. J. Orthop. Res. 4: 75–88.

Nickle, D. A., J. L. Castner, S. R. Smedley, A. B. Attygalle, J. Meinwald, and T. Eisner. 1996. Glandular pyrazine emission by a tropical katydid: An example of chemical aposematism? (Orthoptera: Tettigoniidae: Copiphorinae: *Vestria* Stål). J. Orthop. Res. 5: 221–223.

Nickle, D. A., and E. W. Heymann. 1996. Predation on Orthoptera and other orders of insects by tamarin monkeys, *Saguinus mystax mystax* and *Saguinus fuscicollis nigrifrons* (Primates, Callitrichidae), in north-eastern Peru. J. Zool. (Lond.) 239: 799–819.

Nickle, D. A., and P. A. Naskrecki. 1997. Recent developments in the sytematics of Tettigoniidae and Gryllidae. Pages 41–58 in S. K. Gangwere, M. C. Muralirangan, and M. Muralirangan, eds. The Bionomics of Grasshoppers, Katydids and Their Kin. CAB International, Wallingford, U.K.

Nielsen, E. T. 1971. Stridulatory activity of *Eugaster* (Orthoptera, Ensifera). Entomol. Exp. Appl. 14: 234–244.

—— 1972. Precrepuscular stridulation in *Tettigonia viridissima* (Orthoptera ensifera). Entomol. Scand. 3: 156–158.

Nielsen, E. T., and H. Dreisig. 1970. The behavior of stridulation in Orthoptera Ensifera. Behaviour 37: 205–252.

Nocke, H. 1974. The tympanal trachea as an integral part of the ear in *Acripeza reticulata* Guerin (Orthoptera, Tettigonioidea). Z. Naturforsch. 29: 652–654.

—— 1975. Physical and physiological properties of the tettigoniid ("grasshopper") ear. J. Comp. Physiol. [A] 100: 25–57.

Nutting, W. L. 1953. The biology of *Euphasiopteryx brevicornis* (Townsend) (Diptera:

Tachinidae), parasitic in the cone-headed grasshoppers (Orthoptera: Copiphorinae). Psyche 60: 69–81.

Oda, K., and M. Ishii. 1998. Factors affecting color polymorphism in the meadow grasshopper, *Conocephalus maculatus* (Orthoptera: Tettigoniidae). Appl. Entomol. Zool. 33: 455–460.

O'Donnell, S. 1993. Interactions of predaceous katydids (Orthoptera: Tettigoniidae) with neotropical social wasps (Hymenoptera: Vespidae): Are wasps a defense mechanism or prey? Entomol. News 104: 39–42.

Oldfield, B. P. 1982. Tonotopic organization of auditory receptors in Tettigoniidae (Orthoptera: Ensifera). J. Comp. Physiol. [A] 147: 461–469.

Oring, L. W., and D. B. Lank. 1982. Sexual selection, arrival times, philopatry and site fidelity in the polyandrous spotted sandpiper. Behav. Ecol. Sociobiol. 10: 185–191.

Otte, D. 1977. Communication in Orthoptera. Pages 334–361 in T. Sebeok, ed. How Animals Communicate. Indiana University Press, Bloomington, Ind.

——— 1992. Evolution of cricket songs. J. Orthop. Res. 1: 25–47.

——— 1997. Orthoptera Species File. 7. Tettigonioidea. Academy of Natural Sciences, Philadelphia.

Owen, D. F. 1965. Swarming and polymorphism in the African edible grasshopper, *Homorocoryphus nitidulus* (Tettigonioidea, Conocephalidae). Acta Trop. 22: 55–61.

——— 1969. Green-brown polymorphism in African Tettigoniidae. Rev. Zool. Bot. Afr. 80: 346–351.

——— 1988. A plague of Orthoptera on El Hierro, Canary Islands. Entomol. Gaz. 39: 269–270.

Owens, I. P. F., and D. B. A. Thompson. 1994. Sex-differences, sex-ratios and sex-roles. Proc. R. Soc. Lond. [B] 258: 93–99.

Pardo, M. C., J. P. M. Camacho, and G. M. Hewitt. 1994. Dynamics of ejaculate transfer in *Locusta migratoria*. Heredity 73: 190–197.

Parker, G. A. 1970. Sperm competition and its evolutionary consequences in the insects. Biol. Rev. (Camb.) 45: 525–567.

——— 1983. Mate quality and mating decisions. Pages 141–166 in P. Bateson, ed. Mate Choice. Cambridge University Press, Cambridge.

Parker, G. A., and L. W. Simmons. 1989. Nuptial feeding in insects: Theoretical models of male and female interests. Ethology 82: 3–26.

——— 1996. Parental investment and the control of sexual selection: Predicting the direction of sexual competition. Proc. R. Soc. Lond. [B] 263: 315–321.

Parsons, K. A., and A. A. de la Cruz. 1980. Energy flow and grazing behavior of conocephaline grasshoppers in a *Juncus roemerianus* marsh. Ecology 61: 1045–1050.

Paterson, H. E. H. 1982. Perspective on speciation by reinforcement. S. Afr. J. Sci. 78: 53–57.

Peckham, G. W., and E. G. Peckham. 1898. On the Instincts and Habits of the Solitary Wasps. Bull. Wisc. Geol. Nat. Hist. Surv. 2: 1–245.

——— 1895. The sense of sight in spiders with some observations on the color sense. Trans. Wisc. Acad. Sci. Arts Lett. 10: 231–261.

Pemberton, R. W. 1990. The selling of *Gampsocleis gratiosa* Brunner (Orthoptera: Tettigoniidae) as singing pets in China. Pan-Pacific Entomol. 66: 93–95.

Perez-Gelabert, D. E., and W. L. Grogan. 1999. *Forcipomyia* (*microhelea*) *tettigonaris* (Diptera: Ceratopogonidae) parasitizing katydids (Orthoptera: Tettigoniidae) in the Dominican Republic. Entomol. News 110: 311–314.

Petrie, M. 1983. Female moorhens compete for small fat males. Science 220: 413–415.

Poinar, G. O. 1991a. Nematoda and Nematomorpha. Pages 249–283 in J. H. Thorpe and A. P. Covich, eds. Ecology and Classification of North American Invertebrates. Academic Press, New York.

Poinar, G. O. J. 1991b. Hairworm (Nematomorpha: Gordioidea) parasites of New Zealand wetas (Orthoptera: Stenopelmatidae). Can. J. Zool. 69: 1592–1599.

Poulton, E. B. 1898. Natural selection and the cause of mimetic resemblance and common warning colours. J. Linn. Soc. (Lond.) 26: 558–612.

Power, J. H. 1958. On the biology of *Acanthoplus bechuanus* Per. (Orthoptera: Tettigoniidae). J Entomol. Soc. S. Afr. 2: 376–381.

Price, D. K., and T. F. Hansen. 1998. How does offspring quality change with age in male *Drosophila melanogaster*? Behav. Genet. 28: 395–402.

Pritchard, G. 1967. Laboratory observations on the mating behaviour of the island fruit fly *Rioxa pornia* (Diptera: Tephritidae). J. Aust. Entomol. Soc. 6: 127–132.

Proctor, H. C. 1991. Courtship in the water mite *Neumania papillator*: Males capitalize on female adaptations for predation. Anim. Behav. 42: 589–598.

Proctor, H. C., R. L. Baker, and D. T. Gwynne. 1995. Mating behaviour and spermatophore morphology: A comparative test of the female-choice hypothesis. Can. J. Zool. 73: 2010–2020.

Quinn, J. S., and S. K. Sakaluk. 1986. Prezygotic male reproductive effort in insects: Why do males provide more than sperm? Fla. Entomol. 69: 84–94.

Qvarnstrom, A., and E. Forsgren. 1998. Should females prefer dominant males? Trends Ecol. Evol. 13: 498–501.

Raffelson, J. J. 1989. Predators of the range. Rangelands 11: 26–27.

Ragge, D. R. 1955. The Wing-Venation of the Orthoptera Saltatoria. British Museum of Natural History, London.

——— 1965. Grasshoppers, Cockroaches and Crickets of the British Isles. Warne, London.

Ragge, D. R., and W. J. Reynolds. 1998. The Songs of the Grasshoppers and Crickets of Western Europe. Harley Books, Colchester, U.K.

Ramsay, G. W. 1964. Moult number in Orthoptera (Insecta). N. Z. J. Sci. 7: 644–666.

Ramsey, G. W. 1955. The exoskeleton and musculature of the head, and the life cycle of *Deinacrida rugosa* Buller, 1870. M.Sc. thesis, Victoria University, Wellington, New Zealand.

Rasa, O. A. E., S. Bisch, and T. Teichner. 1998. Female mate choice in a subsocial beetle: Male phenotype correlates with helping potential and offspring survival. Anim. Behav. 56: 1212–1220.

Redak, R. A., J. L. Capinera, and C. D. Bonham. 1992. Effects of sagebrush removal and herbivory by Mormon crickets (Orthoptera: Tettigoniidae) on understory plant biomass and cover. Ann. Entomol. Soc. Am. 21: 94–102.

Rehn, J. A. G., and M. Hebard. 1915. Studies in American Tettigoniidae (Orthoptera): A synopsis of the species of the genus *Conocephalus* (*Xiphidium* of authors) found in North America north of Mexico. Trans. Am. Entomol. Soc. 41: 155–224.

Reinhold, K. 1994. Inheritance of body and testis size in the bushcricket *Poecilimon veluchianus* Ramme (Orthoptera: Tettigoniidae) examined by means of subspecies hybrids. Biol. J. Linn. Soc. 52: 305–316.

——— 1998. Light effects on larval hatching in the bushcricket species *Poecilimon veluchianus* (Orthoptera: Phaneropteridae). Entomol. Gener. 22: 205–209.

——— 1999. Paternal investment in *Poecilimon veluchianus* bushcrickets: Beneficial effects of nuptial feeding on offspring viability. Behav. Ecol. Sociobiol. 45: 293–299.

Reinhold, K., and K.-G. Heller. 1993. The ultimate function of nuptial feeding in the

bushcricket *Poecilimon veluchianus* (Orthoptera: Tettigoniidae: Phaneropterinae). Behav. Ecol. Sociobiol. 32: 55–60.
Reinhold, K., and D. von Helversen. 1997. Sperm number, spermatophore weight and remating in the bush-cricket *Poecilimon veluchianus*. Ethology 103: 12–18.
Reinhold, R. 1983. Houston is at five plagues and counting. New York Times, November 13.
Rentz, D. C. 1963. Biological observations on the genus *Decticita* (Orthoptera: Tettigoniidae). Wass. J. Biol. 21: 91–94.
—— 1972a. The lock and key as an isolating mechanism in katydids. Am. Sci. 60: 750–755.
—— 1972b. Taxonomic and faunistic comments on decticine katydids with the description of several new species (Orthoptera: Tettigoniidae: Decticinae). Proc. Acad. Natl. Sci. Phila. 124: 41–77.
—— 1975. Two new katydids of the genus *Melanonotus* from Costa Rica with comments on their life history strategies (Tettigoniidae: Pseudophyllinae). Entomol. News 86: 129–140.
—— 1979. Comments on the classification of the orthopteran family Tettigoniidae, with a key to subfamilies and description of two new subfamilies. Aust. J. Zool. 27: 991–1013.
—— 1983. *Orophus conspersa* (Esperanza bush Katydid). Pages 749–750 in D. H. Janzen, ed. Costa Rican Natural History. University of Chicago Press, Chicago.
—— 1985. A Monograph of the Tettigoniidae of Australia, Vol. 1: The Tettigoniinae. CSIRO, Melbourne.
—— 1991. Orthoptera. Pages 369–393 in CSIRO, ed. Insects of Australia. Melbourne University Press, Melbourne.
—— 1993. A Monograph of the Tettigoniidae of Australia, Vol. 2: The Austrosaginae, Phasmodinae and Zaprochilinae. CSIRO, Melbourne.
—— 1995. Do the spines on the legs of katydids have a role in predation? (Orthoptera: Tettigoniidae: Listroscelidinae). J. Orthop. Res. 4: 199–200.
—— 1996. Grasshopper Country: The Abundant Orthopteroid Insects of Australia. University of New South Wales Press, Sydney.
Rentz, D. C., and D. Clyne. 1983. A new genus and species of pollen- and nectar-feeding katydids from eastern Australia. J. Aust. Entomol. Soc. 22: 155–160.
Rentz, D. C., and D. Colless. 1990. A classification of shield-backed katydids (Tettigoniidae). Pages 352–377 in W. J. Bailey and D. C. F. Rentz, eds. The Tettigoniidae: Biology, Systematics and Evolution. Crawford House, Bathurst, Australia.
Rentz, D. C., and A. B. Gurney. 1985. The shield-backed katydids of South America (Orthoptera: Tettigoiidae, Tettigoniinae) and a new tribe of Conocephalinae with genera in Chile and Australia. Entomol. Scand. 16: 69–119.
Reynolds, J. D. 1985. Philandering phalaropes. Nat. Hist. 94: 58–65.
Rheinlaender, J., and H. Römer. 1990. Acoustic cues for sound localisation and spacing in orthopteran insects. Pages 248–264 in W. J. Bailey and D. C. F. Rentz, eds. The Tettigoniidae: Biology, Systematics and Evolution. Crawford House, Bathurst, Australia.
Ribi, W. A., and L. Ribi. 1979. Natural history of the Australian digger wasp *Sphex cognatus* Smith (Hymenoptera, Sphecidae). J. Nat. Hist. 13: 693–701.
Rice, W. R. 1996. Sexually antagonistic male adaptation triggered by experimental arrest of female evolution. Nature 381: 232–234.
Richards, A. M. 1973. A comparative study of the biology of the giant wetas *Deinacrida*

heteracantha and *D. fallai* (Orthoptera: Henicidae) from New Zealand. J. Zool. (Lond.) 169: 195–236.

Ridley, M. 1988. Mating frequency and fecundity in insects. Biol. Rev. (Camb.) 63: 509–549.

——1989. The timing and frequency of mating in insects. Anim. Behav. 37: 535–545.

Riek, E. F. 1976. The Australian genus *Tympanophora* White (Orthoptera: Tettigoniidae: Tympanophorinae). J. Aust. Entomol. Soc. 15: 161–171.

Riley, C. V. 1874. Katydids. Pages 150–169. Report on the noxious, beneficial, and other insects of the State of Missouri. State Entomologist Office, Mo.

Riley, C. V., A. S. Packard, and C. Thomas. 1880. Second Report of the United States Entomological Commission for the Years 1878 and 1879 relating to the Rocky Mountain Locust and the Western Cricket. U.S. Government Printing Office, Washington, D.C.

Ristich, S. S. 1953. A study of the prey, enemies, and habits of the great-golden digger wasp *Chlorion ichneumoneum* (L.). Can. Entomol. 85: 374–386.

Ritchie, M. G., I. D. Couzin, and W. A. Snedden. 1995. What's in a song? Female bushcrickets discriminate against the song of older males. Proc. R. Soc. Lond. [B] 262: 21–27.

Ritchie, M. G., D. Sunter, and L. R. Hockham. 1998. Behavioral components of sex role reversal in the tettigoniid bushcricket *Ephippiger ephippiger*. J. Ins. Behav. 11: 481–491.

Robert, D., J. Amoroso, and R. R. Hoy. 1992. The evolutionary convergence of hearing in a parasitoid fly and its cricket host. Science 258: 1135–1137.

Robert, D., R. S. Edgecomb, M. P. Read, and R. R. Hoy. 1996. Tympanal hearing in tachinid flies (Diptera, Tachinidae, Ormiini): the comparative morphology of an innovation. Cell Tissue Res. 284: 435–448.

Robert, D., and R. R. Hoy. 1994. Overhearing cricket love songs. Nat. Hist. 103: 49–50.

Robinson, D. J. 1980. Acoustic communication between the sexes of the bush cricket, *Leptophyes punctatissima*. Physiol. Entomol. 5: 183–189.

——1990. Acoustic communication between the sexes in bushcrickets. Pages 112–129 in W. Bailey and D. C. F. Rentz, eds. The Tettigoniidae: Biology, Systematics and Evolution. Crawford House, Bathurst, Australia.

Robinson, D. J., and J. C. Hartley. 1978. Laboratory studies of a tettigoniid (Insecta: Orthoptera) *Ruspolia differens* (Serville): Color polymorphism. J. Nat. Hist. 12: 81–86.

Robinson, D. J., J. Rheinlaender, and J. C. Hartley. 1986. Temporal parameters of male-female sound communication in *Leptophyes punctatissima*. Physiol. Entomol. 11: 317–323.

Robinson, M. H. 1969. The defensive behaviour of some orthopteroid insects from Panama. Trans. R. Entomol. Soc. Lond. 121: 281–303.

——1973. The evolution of cryptic postures in insects, with special reference to some New Guinea Tettigoniids (Orthoptera). Psyche 80: 159–165.

Robinson, M. H., and T. Pratt. 1975. The phenology of *Hexacentrus mundus* at Wau, Papua New Guinea (Orthoptera: Tettigoniidae). Psyche 82: 315–322.

Roeder, K. D. 1966. Acoustic sensitivity of the noctuoid tympanal organ and its range for the cries of bats. J. Comp. Physiol. 140: 101–111.

Römer, H. 1993. Environmental and biological constraints for the evolution of long-range signalling and hearing in acoustic insects. Philos. Trans. R. Soc. [B] 340: 179–185.

Römer, H., and W. Bailey. 1998. Strategies for hearing in noise: Peripheral control over auditory-sensitivity in the bush-cricket *Sciarasaga quadrata* (Austrosaginae, Tettigoniidae). J. Exp. Biol. 201: 1023–1033.

Römer, H., W. Bailey, and I. Dadour. 1989. Insect hearing in the field. III. Masking by noise. J. Comp. Physiol. [A] 164: 609–620.

Römer, H., and W. J. Bailey. 1986. Insect hearing in the field. II. Male spacing behaviour and correlated acoustic cues in the bushcricket *Mygalopsis marki*. J. Comp. Physiol. [A] 159: 627–638.

Römer, H., M. Spickermann, and W. Bailey. 1998. Sensory basis for sound intensity discrimination in the bush-cricket *Requena verticalis* (Tettigoniidae, Orthoptera). J. Comp. Physiol. [A] 182: 595–607.

Room, P. M., C. H. Perry, and P. T. Bailey. 1984. A population study of the coconut pest *Segestidea uniformis* (Willemse) (Orthoptera: Tettigoniidae) on an equatorial island. Bull. Entomol. Res. 34: 439–451.

Rosenqvist, G. 1990. Male mate choice and female-female competition for mates in the pipefish *Nerophis ophidion*. Anim. Behav. 39: 1110–1115.

Russell, L. J., et al. (committee). 1831. Insect Miscellanies. Charles Knight, London.

Rust, J., A. Stumpner, and J. Gottwald. 1999. Singing and hearing in a tertiary bushcricket. Nature 399: 650–650.

Rutowski, R. L., G. W. Gilchrist, and B. Terkanian. 1987. Female butterflies mated with recently mated males show reduced reproductive output. Behav. Ecol. Sociobiol. 20: 319–322.

Ryan, M. J. 1990. Sexual selection, sensory systems and sensory exploitation. Oxford Surv. Evol. Biol. 7: 157–195.

——— 1998. Sexual selection, receiver biases, and the evolution of sex differences. Science 281: 1999–2003.

Ryan, M. J., and A. S. Rand. 1993. Species recognition and sexual selection as a unitary problem in animal communication. Evolution 47: 647–657.

Ryan, M. J., M. D. Tuttle, and L. K. Taft. 1981. The costs and benefits of frog chorusing behavior. Behav. Ecol. Sociobiol. 8: 273–278.

Sakaluk, S. K. 1984. Male crickets feed females to ensure complete sperm transfer. Science 223: 609–610.

——— 1986. Sperm competition and the evolution of nuptial feeding behavior in the cricket, *Gryllodes supplicans* (Walker). Evolution 40: 584–593.

——— 1990. Sexual selection and predation: Balancing reproductive and survival needs. Pages 63–90 in D. L. Evans and J. O. Schmidt, eds. Insect Defenses. State University of New York Press, Albany.

——— 1991. Post-copulatory mate guarding in decorated crickets. Anim. Behav. 41: 207–216.

Sales, G., and D. Pye. 1974. Ultrasonic Communication in Animals. Chapman and Hall, London.

Samways, M. J. 1997. Conservation biology of Orthoptera. Pages 481–496 in S. K. Gangwere, M. C. Mulalirangen, and M. Muralirangen, eds. The Bionomics of Grasshoppers, Katydids and Their Kin. CAB International, Oxford.

Sandow, J. D. 1980. The morphology, distribution and acoustic repertoire of the genus *Mygalopsis* (Orthoptera: Tettigoniidae). M.Sc. thesis, zoology, University of Western Australia, Nedlands.

Sandow, J. D., and W. J. Bailey. 1978. An experimental study of defensive stridulation in *Mygalopsis ferruginea* Redtenbacher (Orthoptera: Tettigoniidae). Anim. Behav. 26: 1004–1011.

Sänger, K. 1980. Zur Phaenologie einiger Saltatoria (Insecta: Orthoptera) im pannonischen Raum Oesterreichs. Zool. Anz. 204: 165–176.

——1984. Population density as a reason for macropterous ecomorphoses of *Tessellana vittata* (Charp.) (Orthoptera, Tettigoniidae). Zool. Anz. 213: 68–76.

Savalli, U. M. B. F. C. W. 1998. Sexual selection and the fitness consequences of male body-size in the seed beetle *Stator limbatus*. Anim. Behav. 55: 473–483.

Schatral, A. 1990. Body size, song frequency and mating success of male bushcrickets *Requena verticalis* (Orthoptera, Tettigoniidae, Listroscelidinae) in the field. Anim. Behav. 40: 982–984.

——1993. Diet influences male-female interactions in the bushcricket *Requena verticalis* (Orthoptera: Tettigoniidae). J. Ins. Behav. 6: 379–388.

Schatral, A., and W. J. Bailey. 1991. Song variability and the response to conspecific song and to song models of different frequency contents in males of the bushcricket *Requena verticalis* (Orthoptera: Tettigoniidae). Behaviour 116: 163–179.

Schatral, A., W. Latimer, and B. Broughton. 1984. Spatial dispersion and agonistic contacts of male bush crickets in the biotope. Z. Tierpsychol. 65: 201–214.

Schmidt, G. H. 1990. A new species of *Gymnoproctus* Karsch 1887 (Grylloptera Tettigoniidae Hetrodinae) from Kenya with some notes on its biology. Trop. Zool. 3: 121–137.

Schul, J. 1994. A case of interspecific hybridization in the genus *Tettigonia* (Saltatoria, Ensifera). Entomol. Gener. 19: 185–190.

——1998. Song recognition by temporal cues in a group of closely related bushcricket species (genus *Tettigonia*). J. Comp. Physiol. [A] 183: 401–410.

Schul, J., D. von Helversen, and T. Weber. 1998. Selective phonotaxis in *Tettigonia cantans* and *T. viridissima* in song recognition and discrimination. J. Comp. Physiol. [A] 182: 687–694.

Schweis, G. C., L. M. Burge, and G. M. Shogren. 1939. Mormon cricket control in Nevada, 1935–1938. Bull. State Dept. Agric. 1 & 2.

Searcy, W. A., and M. Andersson. 1986. Sexual selection and the evolution of song. Annu. Rev. Ecol. Syst. 17: 507–533.

Shapiro, L. H. 1995. Parasitism of *Orchelimum* katydids (Orthoptera: Tettigoniidae) by *Ormia lineafrons* (Diptera: Tachinidae). Fla. Entomol. 78: 615–616.

——1996. Hybridization and geographic variation in two meadow katydid contact zones. Ph.D. thesis, biology, State University of New York at Stony Brook, Stony Brook.

——1998. Hybridization and geographic variation in two meadow katydid contact zones. Evolution 52: 784–796.

——1999. Reproductive costs to heterospecific mating between two hybridizing katydids (Orthoptera: Tettigoniidae). Ann. Entomol. Soc. Am. 93: 440–446.

Sharov, A. G. 1968. Phylogeny of the Orthopteroidea. Trans. Paleontol. Inst. Acad. Sci. 118: 1–216.

Shaw, K. C. 1968. An analysis of the phonoresponse of males of the true katydid *Pterophylla camellifolia* (Fabricius) (Orthoptera: Tettigoniidae). Behaviour 31: 203–260.

Shaw, K. C., R. J. Bitzer, and R. C. North. 1982. Spacing and movement of *Neoconocephalus ensiger* males (Conocephalinae: Tettigoniidae). J. Kans. Entomol. Soc. 55: 581–592.

Shaw, K., and P. Galliart. 1987. Acoustic and mating behavior of a Mexican katydid, *Pterophylla beltrani* (Orthoptera: Tettigoniidae). Fla. Entomol. 70: 354–368.

Shaw, K. C., R. C. North, and A. J. Meixner. 1981. Movement and spacing of singing *Amblycorypha parvipennis* males. Ann. Entomol. Soc. Am. 74: 436–444.

Shelly, T. E. 1993. Effects of female deprivation on mating propensity and mate selectivity by male *Requena verticalis* (Orthoptera: Tettigoniidae). J. Ins. Behav. 6: 689–698.
Shelly, T. E., and W. J. Bailey. 1992. Experimental manipulation of mate choice by male katydids: The effect of female encounter rate. Behav. Ecol. Sociobiol. 30: 277–282.
——1995. Movement in a zaprochiline katydid (Orthoptera: Tettigoniidae): Sex-specific response to food plant distribution. Fla. Entomol. 78: 251–258.
Shelly, T. E., and M. D. Greenfield. 1985. Alternative mating strategies in a desert grasshopper: A transitional analysis. Anim. Behav. 33: 1211–1222.
Sickmann, T., K. Kalmring, and A. Muller. 1997. The auditory-vibratory system of the bush-cricket *Polysarcus denticauda* (Phaneropterinae, Tettigoniidae). 1. Morphology of the complex tibial organs. Hear. Res. 104: 155–166.
Sillén-Tullberg, B., and O. Leimar. 1988. The evolution of gregariousness in distasteful insects as a defense against predators. Am. Nat. 132: 723–734.
Simmons, L. W. 1986. Female choice in the field cricket *Gryllus bimaculatus* (De Geer). Anim. Behav. 34: 1463–1470.
——1988. The contribution of multiple mating and spermatophore consumption to the lifetime reproductive success of female field crickets (*Gryllus bimaculatus*). Ecol. Entomol. 13: 57–69.
——1990. Nuptial feeding in tettigoniids: Male costs and the rates of fecundity increase. Behav. Ecol. Sociobiol. 27: 43–47.
——1992. Quantification of role reversal in relative parental investment in a bush cricket. Nature 358: 61–63.
——1993. Some constraints on reproduction for male bushcrickets, *Requena verticalis* (Orthoptera: Tettigoniidae): Diet, size and parasite load. Behav. Ecol. Sociobiol. 32: 135–140.
——1994a. Courtship role reversal in bush crickets: Another role for parasites? Behav. Ecol. 5: 259–266.
——1994b. Reproductive energetics of the role reversing bushcricket, *Kawanaphila nartee* (Orthoptera: Tettigoniidae: Zaprochilinae). J. Evol. Biol. 7: 189–200.
——1995a. Courtship feeding in katydids (Orthoptera: Tettigoniidae): Investment in offspring and in obtaining fertilizations. Am. Nat. 146: 307–315.
——1995b. Male crickets tailor their spermatophores in relation to their remating intervals. Funct. Ecol. 9: 881–886.
——1995c. Relative parental expenditure, potential reproductive rates and the control of sexual selection in katydids. Am. Nat. 145: 797–808.
Simmons, L. W., and W. J. Bailey. 1990. Resource influenced sex roles of zaprochiline tettigoniids (Orthoptera: Tettigoniidae). Evolution 44: 1853–1868.
——1992. Agonistic communication between males of a zaprochiline katydid (Orthoptera: Tettigoniidae). Behav. Ecol. 4: 364–368.
Simmons, L. W., L. Beesley, P. Lindhjem, D. Newbound, J. Norris, and A. Wayne. 1999. Nuptial feeding by male bushcrickets: An indicator of male quality? Behav. Ecol. 10: 263–269.
Simmons, L. W., M. Craig, T. Llorens, M. Schinzig, and D. Hosken. 1993. Bushcricket spermatophores vary in accord with sperm competition and parental investment theory. Proc. R. Soc. Lond. [B] 251: 183–186.
Simmons, L. W., and D. T. Gwynne. 1991. The refractory period of female katydids (Orthoptera: Tettigoniidae): Sexual conflict over the remating interval? Behav. Ecol. 2: 276–282.
——1993. Reproductive investment in bushcrickets: The allocation of male and female nutrients to offspring. Proc. R. Soc. Lond. [B] 252: 1–5.

Simmons, L. W., T. Llorens, M. Schinzig, D. Hosken, and M. Craig. 1994. Sperm competition selects for male mate choice and protandry in the bushcricket, *Requena verticalis* (Orthoptera: Tettigoniidae). Anim. Behav. 47: 117–122.

Simmons, L. W., and M. T. Siva-Jothy. 1998. Sperm competition in insects: Mechanisms and the potential for selection. Pages 341–434 in T. R. Birkhead and A. P. Møller, eds. Sperm Competition and Sexual Selection. Academic Press, San Diego.

Simmons, L. W., R. J. Teale, M. Maier, R. J. Standish, W. J. Bailey, and P. C. Withers. 1992. Some costs of reproduction for male bushcrickets, *Requena verticalis* (Orthoptera: Tettigoniidae): Allocating resources to mate attraction and nuptial feeding. Behav. Ecol. Sociobiol. 31: 57–62.

Sismondo, E. 1978. *Meconema thalassinum* (Orthoptera: Tettigoniidae), prey of *Sphex ichneumoneus* (Hymenoptera: Sphecidae) in Westchester County, New York. Entomol. News 89: 244.

Skaife, S. H., A. Bannister, and J. Ledger. 1979. African Insect Life. C. Struik, Capetown.

Smalley, A. E. 1960. Energy flow of a salt marsh grasshopper population. Ecology 41: 672–677.

Smith, R. L. 1979. Paternity assurance and altered roles in the mating behaviour of a giant water bug, *Abedus herberti* (Heteroptera: Belostomatidae). Anim. Behav. 27: 716–725.

Smith, R. L., ed. 1984. Sperm Competition and the Evolution of Animal Mating Systems. Academic Press, New York.

Smith, S. C. H. 1972. Goldcrests feeding young with oak bush crickets. Br. Birds 65: 33.

Smyth, J. F. D. 1784. A Tour of the United States of America, Vol. 2. G. Robinson, J. Robinson and J. Sewell, London.

Snead, J. S., and J. Alcock. 1985. Aggregation formation and assortative mating in two meloid beetles. Evolution 39: 1123–1131.

Snedden, W. A., and M. D. Greenfield. 1998. Females prefer leading males: Relative call timing and sexual selection in katydid choruses. Anim. Behav. 56: 1091–1098.

Snedden, W. A., and S. Irazusta. 1994. Attraction of female sagebrush crickets to male song: The importance of field bioassays. J. Ins. Behav. 7: 233–236.

Snodgrass, R. E. 1905. The coulee cricket of central Washington (*Peranabus scabricollis* Thomas). J. N. Y. Entomol. Soc. 13: 74–85.

Solulu, T. M., S. J. Simpson, and J. Kathirithamby. 1998. The effect of strepsipteran parasitism on a tettigoniid pest of oil palm in Papua New Guinea. Physiol. Entomol. 23: 388–398.

Soper, R. S., G. E. Shewell, and D. Tyrrell. 1976. *Colcondamyia auditrix* Nov. sp. (Diptera: Sarcophagidae), a parasite which is attracted by the mating song of its host, *Okanagana rimosa* (Homoptera: Cicadidae). Can. Entomol. 108: 61–68.

Spangler, H. G. 1984. Silence as a defense against predatory bats in two species of calling insects. Southwest. Nat. 29: 481–488.

Spencer, A. M. 1995. Sexual maturity in the male tree weta *Hemideina crassidens* (Orthoptera: Stenopelmatidae). M.Sc. thesis, biology, Victoria University of Wellington, Wellington, New Zealand.

Spencer, H. G., D. M. Lambert, and B. H. McArdle. 1987. Reinforcement, species, and speciation: A reply to Butlin. Am. Nat. 130: 958–962.

Spooner, J. D. 1964. The Texas bush katydid—its sounds and their significance. Anim. Behav. 12: 235–244.

———1968a. Collection of male phaneropterine katydids by imitating sounds of the female. J. Georgia Entomol. Soc. 3: 45–46.
———1968b. Pair-forming acoustic systems of phaneropterine katydids (Orthoptera, Tettigoniidae). Anim. Behav. 16: 197–212.
———1973. Sound production in *Cyphoderris monstrosa* (Orthoptera: Prophalangopsidae). Ann. Entomol. Soc. Am. 66: 4–5.
———1995. Pair-forming phonotaxic strategies of phaneropterine katydids (Tettigoniidae, Phaneropterinae). J. Orthop. Res. 4: 127–130.
Stawell, R. 1921. Fabre's Book of Insects. Dodd, Mead and Co., New York.
Stephen, R. O., and W. J. Bailey. 1982. Bioacoustics of the ear of the bushcricket *Hemisaga* (Saginae). J. Acoust. Soc. Am. 72: 13–25.
Stevens, E. D., and R. K. Josephson. 1977. Metabolic rate and body temperature in singing katydids. Physiol. Zool. 50: 31–42.
Stevens, S. S., and F. Warshofsky. 1965. Sound and Hearing. Time, Morristown, N.J.
Stiedl, O., and K. Kalmring. 1989. The importance of song and vibratory signals in the behaviour of the bushcricket *Ephippiger ephippiger* Fieber (Orthoptera, Tettigoniidae): Taxis by females. Oecologia 80: 142–144.
Stiling, P., B. V. Brodbeck, and D. R. Strong. 1991. Population increases of planthoppers on fertilized salt-marsh cord grass may be prevented by grasshopper feeding. Fla. Entomol. 74: 88–97.
Stiling, P. D., and D. R. Strong. 1982. Egg parasitism in the grasshopper *Orchelimum fidicinium*. Fla. Entomol. 65: 285–286.
Suga, N. 1966. Ultrasonic production and its reception in some neotropical Tettigoniidae. J. Ins. Physiol. 12: 1039–1050.
Summers, K. 1989. Sexual selection and intra-female competition in the green poison-dart frog, *Dendrobates auratus*. Anim. Behav. 37: 797–805.
Sutherland, W. J. 1987. Random and deterministic components of variance in mating success. Pages 209–219 in J. W. Bradbury and M. B. Andersson, eds. Sexual Selection: Testing the Alternatives. John Wiley, Chichester, U.K.
Svard, L., and C. Wiklund. 1988. Fecundity, egg weight and longevity in relation to multiple matings in females of the monarch butterfly. Behav. Ecol. Sociobiol. 23: 39–43.
Svensson, B. G., and E. Petersson. 1988. Non-random mating in the dance fly *Empis borealis*: The importance of male choice. Ethology 79: 307–316.
Svensson, B. G., E. Petersson, and E. Forsgren. 1989. Why do males of the dance fly *Empis borealis* refuse to mate? The importance of female age and size. J. Ins. Behav. 2: 387–395.
Svensson, B. G., E. Petersson, and M. Frisk. 1990. Nuptial gift size prolongs copulation duration in the dance fly *Empis borealis*. Ecol. Entomol. 15: 225–229.
Swan, L. A., and C. S. Papp. 1972. The Common Insects of North America. Harper & Row, New York.
Sword, G. A. 1999. Density-dependent warning coloration. Nature 397: 217.
Sword, G. A., S. J. Simpson, O. T. M. El Hadi, and H. Wilps. 2000. Density-dependent aposematism in the desert locust. Proc. R. Soc. Lond. [B] 267: 63–68.
Tejedo, M. 1993. Do male natterjack toads join larger breeding choruses to increase mating success? Copeia 1: 75–80.
Thiele, D., and W. J. Bailey. 1980. The function of sound in male spacing behaviour in bush-crickets (Tettigoniidae, Orthoptera). Aust. J. Ecol. 5: 275–286.
Thorne, G. 1940. The hairworm, *Gordius robustus* Leidy, as a parasite of the Mormon cricket, *Anabrus simplex* Haldeman. J. Wash. Acad. Sci. 30: 219–231.

Thornhill, R. 1976a. Sexual selection and nuptial feeding behavior in *Bittacus apicalis* (Insecta: Mecoptera). Am. Nat. 110: 529–548.

——1976b. Sexual selection and paternal investment in insects. Am. Nat. 110: 153–163.

——1978. Some arthropod predators and parasites of adult scorpionflies (Mecoptera). Environ. Entomol. 7: 714–716.

——1981. *Panorpa* (Mecoptera: Panorpidae) scorpionflies: Systems for understanding resource-defense polygyny and alternative male reproductive effort. Annu. Rev. Ecol. Syst. 12: 355–386.

——1983. Cryptic female choice and its implications in the scorpionfly *Harpobittacus nigriceps*. Am. Nat. 122: 765–788.

——1986. Relative parental contribution of the sexes to their offspring and the operation of sexual selection. Pages 113–135 in M. H. Nitecki and J. A. Kitchell, eds. Evolution of Animal Behavior: Paleontological and Field Approaches. Oxford University Press, New York.

Thornhill, R., and J. Alcock. 1983. The Evolution of Insect Mating Systems. Harvard University Press, Cambridge, Mass.

Thornhill, R., and D. T. Gwynne. 1986. The evolution of sexual differences in insects. Am. Sci. 74: 382–389.

Tinkham, E. R. 1944. Biological, taxonomic and faunistic studies on the shield-back katydids of the North American deserts. Am. Midl. Nat. 31: 257–328.

——1948. Faunistic and ecological studies on the Orthoptera of the Big Bend Region of Trans-Pecos Texas. Am. Midl. Nat. 40: 521–663.

Togashi, I. 1980. Food habits of Umaoi and shrub katydids. Kontyu, Tokyo 48: 29.

Trivers, R. L. 1972. Parental investment and sexual selection. Pages 136–179 in B. Campbell, ed. Sexual Selection and the Descent of Man, 1871–1971. Aldine, Chicago.

Tuckerman, J. F., D. T. Gwynne, and G. K. Morris. 1992. Reliable acoustic cues for female mate preference in a katydid (*Scudderia curvicauda*, Orthoptera: Tettigoniidae). Behav. Ecol. 4: 106–113.

Turnbow, J. H., and K. M. O'Neill. Manuscript. Behavioral thermoregulation in the Mormon cricket, *Anabrus simplex* Haldeman (Orthoptera: Tettigoniidae). J. Therm. Biol.

Turnbow, J. H., K. M. O'Neill, and E. A. Oma. Manuscript. Temperature-dependent mycosis of Mormon crickets (Orthoptera: Tettigoniidae) infected with *Beauveria bassana*. Biol. Contr.

Tuttle, M. D., M. J. Ryan, and J. J. Belwood. 1985. Acoustical resource partitioning by two species of phyllostomid bats (*Trachops cirrhosus* and *Tonatia sylvicola*). Anim. Behav. 33: 1369–1371.

Tyus, H. M., and W. L. Minckley. 1988. Migrating Mormon crickets, *Anabrus simplex* (Orthoptera: Tettigoniidae), as food for stream fishes. Great Basin Nat. 48: 25–30.

Ueckert, D. N., and R. M. Hansen. 1970. Seasonal dry-weight composition in diets of Mormon crickets. J. Econ. Entomol. 63: 96–98.

Urquhart, F. A., and J. R. Beaudry. 1953. A recently introduced species of European grasshopper. Can. Entomol. 85: 78–79.

Uvarov, B. I. 1977. Grasshoppers and Locusts, Vol. 2. Centre for Overseas Pest Research, London.

Vahed, K. 1996. Prolonged copulation in oak bushcrickets (Tettigoniidae: Meconematinae; *Meconema thalassinum* and *M. meridionale*). J. Orthop. Res. 5: 199–204.

——1997. Copulation and spermatophores in the Ephippigerinae (Orthoptera:

Tettigoniinae): Prolonged copulation is associated with a smaller nuptial gift in *Uromenus rugosicollis* Serv. J. Orthop. Res. 6: 83–89.

——1998a. The function of nuptial feeding in insects: Review of empirical-studies. Biol. Rev. (Camb.) 73: 43–78.

——1998b. Sperm precedence and the potential of the nuptial gift to function as paternal investment in the tettigoniid *Steropleurus stali* Bolivar (Orthoptera: Tettigoniidae: Ephippigerinae). J. Orthop. Res. 7: 223–226.

Vahed, K., and F. S. Gilbert. 1996. Differences across taxa in nuptial gift size correlate with differences in sperm number and ejaculate volume in bush-crickets (Orthoptera, Tettigoniidae). Proc. R. Soc. Lond. [B] 263: 1257–1265.

——1997. No effect of nuptial gift consumption on female reproductive output in the bush-cricket *Leptophyes laticauda* Friv. Ecol. Entomol. 22: 479–482.

Verdier, M. 1958. Modifications pigmentaires liecs a la densitie chez les Tettigonides. Bull. Soc. Zool. Fr. 90: 252–253.

Vickery, V. R. 1965. Factors governing the distribution and dispersal of the recently introduced grasshoppers *Metrioptera roeseli* (Hgb.) (Orthoptera: Ensifera). Ann. Soc. Entomol. Que. 10: 165–171.

Vickery, V. R., and D. K. M. Kevan. 1983. A Monograph of the Orthopteroid Insects of Canada and Adjacent Regions. Lyman Mus. Res. Lab. Memoir 13.

——1985. The Grasshoppers, Crickets and Related Insects of Canada and Adjacent Regions. Canadian Government Services, Ottawa.

Viscuso, R., L. Narcisi, L. Sottile, and N. Barone. 1998. Structure of spermatodesms of Orthoptera Tettigonioidea. Tissue Cell 30: 453–463.

Waage, J. K. 1979. Dual function of the damselfly penis: Sperm removal and transfer. Science 203: 916–918.

Wagner, W. E. J. 1996. Convergent song preferences between female field crickets and acoustically orienting parasitoid flies. Behav. Ecol. 7: 279–285.

Wahid, M. B. 1978. The biology and economic impact of the weta, *Hemiandrus* sp. (Orthoptera: Stenopelmatidae) in an apricot orchard, Horotane Valley. M.Sc. thesis, Canterbury University, Christchurch, New Zealand.

Wakeland, C. 1959. Mormon crickets in North America. Tech. Bull. U.S. Dept. Agric. 1202: 1–77.

Walker, T. J. 1969. Acoustic synchrony: Two mechanisms in the snowy tree cricket. Science 166: 891–894.

——1975. Stridulatory movements in eight species of *Neoconocephalus* (Tettigoniidae). J. Ins. Physiol. 21: 595–603.

Walker, T. J., and D. Dew. 1972. Wing movements of calling katydids: Fiddling finesse. Science 178: 174–176.

Walker, T. J., and M. D. Greenfield. 1983. Songs and systematics of Caribbean *Neoconocephalus* (Orthoptera: Tettigoniidae). Trans. Am. Entomol. Soc. 109: 357–389.

Wallace, A. R. 1891. Natural Selection and Tropical Nature. Macmillan, London.

Warchalowska-Sliwa, E. 1998. Karyotype characteristics of katydid Orthopterans (Ensifera, Tettigoniidae), and remarks on their evolution at different taxonomic levels. Fol. Biol.-Krakow 46: 143–176.

Warwick, S., D. Raubenheimer, and S. Simpson. 1998. Nutrition and the evolution of spermatophylax quality in *Gryllodes sigillatus*. Abstract presented at the seventh International Behavioral Ecology Congress, Asilomar, Cal.

Watson, P. J. 1998. Multi-male mating and female choice increase offspring growth in the spider *Neriene litigiosa* (Linyphiidae). Anim. Behav. 55: 387–403.

Watson, P. J., and R. Thornhill. 1994. Fluctuating asymmetry and sexual selection. Trends Ecol. Evol. 9: 21–25.

Wedell, N. 1991. Sperm competition selects for nuptial feeding in a bushcricket. Evolution 45: 1975–1978.

———1993a. Evolution of nuptial gifts in bushcrickets. Ph.D. thesis, zoology, University of Stockholm, Stockholm.

———1993b. Spermatophore size in bushcrickets: Comparative evidence for nuptial gifts as a sperm protection device. Evolution 47: 1203–1212.

———1994a. Dual function of the bushcricket spermatophore. Proc. R. Soc. Lond. [B] 258: 181–185.

———1994b. Variation in nuptial gift quality in bushcrickets (Orthoptera: Tettigoniidae). Behav. Ecol. 5: 418–425.

———1997. Ejaculate size in bush-crickets: The importance of being large. J. Evol. Biol. 10: 315–325.

———1998. Sperm protection and mate assessment in the bushcricket *Coptaspis* sp. 2. Anim. Behav. 56: 357–363.

Wedell, N., and A. Arak. 1989. The wartbiter spermatophore and its effect on female reproductive output (Orthoptera: Tettigoniidae, *Decticus verrucivorus*). Behav. Ecol. Sociobiol. 24: 117–125.

Wedell, N., and T. Sandberg. 1995. Female preference for large males in the bushcricket *Requena* sp. 5 (Orthoptera: Tettigoniidae). J. Ins. Behav. 8: 513–522.

Welch, A. M., R. D. Semlitsch, and H. C. Gerhardt. 1998. Call duration as an indicator of genetic quality in male gray tree frogs. Science 280: 1928–1930.

Werren, J. H., M. R. Gross, and R. Shine. 1980. Paternity and the evolution of male parental care. J. Theor. Biol. 82: 619–631.

West-Eberhard, M. J. 1983. Sexual selection, social competition, and speciation. Q. Rev. Biol. 58: 155–183.

Westneat, D. F., and P. W. Sherman. 1993. Parentage and the evolution of parental behavior. Behav. Ecol. 4: 66–77.

Wheeler, W. M. 1890. Note on the oviposition and embryonic development of *Xiphidium ensiferum* Scud. Ins. Life 2: 222–225.

———1893. A contribution to insect embryology. J. Morphol. 8: 1–160.

White, G. 1789. The Natural History of Selborne. Oxford University Press, London.

Whitesell, J. J. 1969. Biology of United States coneheaded katydids of the genus *Neoconocephalus* (Orthoptera: Tettigoniidae). M.Sc. thesis, entomology, University of Florida, Gainesville.

Whitesell, J. J., and T. J. Walker. 1978. Photoperiodically determined dimorphic calling songs in a katydid. Nature 274: 887–888.

Whitten, M. J. 1991. Australian insects in scientific research. Pages 236–251 in CSIRO, ed. Insects of Australia. Melbourne University Press, Melbourne.

Wickler, W. 1968. Mimicry in Plants and Animals. World University Library, London.

———1985. Stepfathers in insects and their pseudo-parental investment. Z. Tierpsychol. 69: 72–78.

Wiegmann, D. D. 1999. Search behaviour and mate choice by female field crickets, *Gryllus integer*. Anim. Behav. 58: 1293–1298.

Wiklund, C., A. Kaitala, V. Lindfors, and J. Abenius. 1993. Polyandry and its effect on female reproduction in the green-veined white butterfly (*Pieris napi* L.). Behav. Ecol. Sociobiol. 33: 25–33.

Will, M. W., and S. K. Sakaluk. 1994. Courtship feeding in decorated crickets: Is the spermatophylax a sham? Anim. Behav. 48: 1309–1315.

Williams, G. C. 1966. Adaptation and Natural Selection. Princeton University Press, Princeton, N.J.

——1975. Sex and Evolution. Monographs in Population Biology 8. Princeton University Press, Princeton, N.J.

Wirth, W. W., and J. L. Castner. 1990. New neotropical species of "stick-tick" (Diptera: Ceratopogonidae) from katydids. Fla. Entomol. 73: 157–160.

Withycombe, C. L. 1922. Notes on the biology of some British Neuroptera (Plannipenna). Trans. Entomol. Soc. (Lond.) 1922: 501–594.

Woodward, B. D. 1986. Paternal effects on juvenile growth in *Scaphiopus multiplicatus* (the New Mexico spadefoot toad). Am. Nat. 128: 58–65.

Young, G. R. 1985. Observations on the biology of *Segestes decoratus* Redtenbacher (Orthoptera: Tettigoniidae), a pest of coconut in Papua, New Guinea. Gen. Appl. Entomol. 17: 58–65.

——1987. Some parasites of *Segestes decoratus* Redtenbacher (Orthoptera: Tettigoniidae) and their possible use in the biological control of tettigoniid pests of coconuts in Papua New Guinea. Bull. Entomol. Res. 77: 515–524.

Zacharuk, R. Y. 1985. Antennae and sensillae. Pages 1–59 in G. A. Kerkut and L. I. Gilbert, eds. Comprehensive Insect Physiology, Biochemistry and Pharmacology, Vol. 6. Nervous System: Sensory. Permagon Press, New York.

Zeuner, F. E. 1939. Fossil Orthoptera Ensifera. British Museum of Natural History, London.

Zhiyun, J., J. Zhigang, and S. K. Sakaluk. 1998. Nutritional status influences investment by male katydids. Abstract presented at the seventh International Behavioral Ecology Congress, Asilomar, Cal.

Zuk, M. 1987. Variability in attractiveness of male field crickets (Orthoptera: Gryllidae) to females. Anim. Behav. 35: 1240–1248.

Zuk, M., and G. R. Kolluru. 1998. Exploitation of sexual signals by predators and parasitoids. Q. Rev. Biol. 73: 415–438.

Zuk, M., L. W. Simmons, and L. Cupp. 1993. Calling characteristics of parasitized and unparasitized populations of the field cricket *Teleogryllus oceanicus*. Behav. Ecol. Sociobiol. 33: 339–343.

Index

Acanthodis, 85, 261
Acanthoplus, 53, 57, 83, 117–118, 142, 195, 261
Acrididae, xii, 14, 37, 47, 49, 64, 66, 77, 130, 173
 oviposition, 49
Acripeza, 80, 135, 139, 261, Plate 18
Agamermis, 75, 261
Aganacris, 80, 261
Agraeciini, 29, 37, 39, 49, 53, 80–81, 98, 135
Alcock, J., 199, 268
Alexander, R. D., 17, 84, 100, 114, 116, 119, 130, 167, 173, 174, 192, 195, 199
Allard, H., 116
Allen, G., 71, 73, 175, 200–206, 214–215, 221
Amblycorypha, 61, 62, 69, 119, 172, 175, 261
Anabrus. See Mormon cricket
Anatomy and morphology, 28, 42, 62
Ancistrocercus, 60, 63, 261
Ancistrura, 135, 262
Ander, K., 114, 115
Andersson, M., 161, 183, 246, 252
Andrade, M., 131, 159
Anonconotus, 135, 261
Antaxius, 136, 261
Anthophiloptera, 52, 261
 origin of name, 59
Ants. *See* Hymenoptera, Formicidae
Anurans
 frogs and toads, 173, 176, 208
 poison-arrow frog, 224

 salamander, 176
Apteropedetes, 115, 261
Arachnida
 Araneae (spiders), 68, 76, 80, 131, 159, 176
 harvestman, 225
 mites (Acarina), 75, 254
 Scorpionida, 76
Arachnoscleis, 97, 262
Arak, A., 169, 191, 193, 194
Arethaea, 53, 262
Armored ground crickets. *See* Hetrodinae
Arnold, S., 237, 246
Arnqvist, G., 255
Atlanticus, 60, 61, 262
Auditory spiracles and tracheae, 75, 105–109
 evolutionary origins of, 114–115
 See also Ears and hearing
Austrodectes, 136, 262
Austrosaginae, 29, 30, 35, 36, 71, 73, 102–103, 113, 115, 117, 175, 200–206
Austrosalomona, 135, 262

Baier, L., 117, 118
Bailey, W. J., 34, 63–66, 75–77, 83, 86, 95–97, 102, 105, 106, 108–114, 117–118, 141, 172, 181–182, 187, 191–196, 204–205, 207, 210, 214, 223, 226–232, 234, 235, 253–255
Balboa, 135, 262
Balloon-winged katydids. *See* Tympanophorinae

309

Bancroft, H. H., 1, 4
Bands. *See* Gregarious behavior
Barbitistes, 52, 135, 262
Barendse, W., 45–47
Barrientos, L., 57, 63, 66, 131, 141, 190
Bartram, John, xi, 19
Basking behavior and thermoregulation, 7, 8, 78
Bateman, A., 237, 246–247
Bateman, P., 151, 158, 175, 207
Bateman slopes, 237, 246–247
Bats, 77, 83, 100, 102, 212–213, 226
  big brown, 208
  greater mouse-eared, 77
  lesser mouse-eared, 77, 208
  long-eared bats, 226
  Micronycteris, 77, 209, 212
  Tonatia silvicola, 207–208
  Trachops cirrhosus, 207
Beetles. *See* Coleoptera
Beier, M., 256, 257, 259
Belocephalus, 134, 135, 141, 262
Belwood, J., 17, 29, 59–60, 77, 80, 83, 100, 208, 210, 212, 221
Bennet-Clark, H., 109
Berglund, A., 237
Birds, 155
  Australian magpies, 83
  button quail, 222
  owls, 213, 215
  phalarope, 224
  sandpiper, 224
Birkhead, T., 16
Blattodea, 19, 21
Bliastes, 112, 124, 262
Boldyrev, B. T., 11, 50, 119, 123–127, 133, 137, 139–143, 158
Bradbury, J., 258
Bradyporinae, 29, 30, 39, 50, 57, 64–65, 83, 84, 87, 99–100, 109, 115, 117, 133, 135, 145, 147, 148, 150, 172, 173, 175, 207, 233, 244
Bradyporus, 39, 64, 80, 87, 262
Broad-winged katydids. *See* Phaneropterinae
Brockmann, H. J., 69–70, 210, 213
Brooks, D., 15, 140
Brown, W. D., 130–131, 138, 166–168, 171, 177–178
Brunellius, G., 42, 116–117, 125, 138–139
Burk, T., 16, 73, 166, 199, 200, 205, 207, 213, 217
Burmeister, H., 92
Bush-cricket
  origin of name, xi
  See also Katydids
Busnel, R.-G., 39, 100, 117, 125, 133, 136
Butlin, R., 169, 187

Cade, W., 71, 195, 199
Caedicia, 53, 135, 141, 262
  origin of common name, 53
Caelifera, 19–20
Camel cricket. *See* Rhaphidophoridae
Camouflage. *See* Defense against predators
Cannibalism, 4
Cherrill, A. J., 78
Chlorodectes, 136, 262
Choeroparnops, 100–101, 262
Clonia, 81, 262
Clutton-Brock, T., 15, 238, 240, 246
Cocconotus, 208, 262
Cockroaches. *See* Blattodea
Coleoptera, 236
  Bruchidae, 176
  Carabidae, 77
  Cerambycidae, 85
  Cicindelidae, 80
  Lampyridae, 174, 195, 199
  Meloidae, 130, 160
Color polymorphism, 64–66
Comparative method, 14–15, 28, 102, 114–116, 128, 139–148, 210–217
Competition for food, 61, 230, 235–236
Competition for mates, 7–8, 161–164, 191–197
  and fighting, 164, 196–197
  and mate "holding," 255–256
  and role of body size, 164, 254
  "satellite" males and alternative mating strategies, 58–59, 102, 164, 193–197
  and songs, 117–118, 164, 195–196
  territory and spacing behavior, 163–165, 192–194
Cone-head. *See* Copiphorini
Conocephalinae, 29–30, 37, 39, 45–47, 60–61, 73–74, 80, 81, 91, 97, 109, 117–118, 135, 145, 147, 171
Conocephalini, 29, 37, 60–61, 70, 73, 74, 84, 87–88, 91, 135, 161–165, 168–171, 178–181, 183–188, 190, 192–197, 208, 210, 211, Plates 21, 22
Conocephalus, 37, 49–54, 60, 66, 73, 83, 87, 91, 117–118, 131, 134, 135, 168–171, 178–181, 183, 187–188, 190, 194, 210, 213, 262
Controls on katydid populations, 68, 78–79
Cooloolidae, 22, 23
Coniungopterini, 29, 37
Copiphora, 59–60, 120, 135, 140, 263
Copiphorini, 29, 37, 45–47, 52, 54, 56–57, 64, 66, 69, 70, 77, 81, 83–84, 87, 102–103, 105, 117–118, 120, 135, 140, 141, 171, 173–174, 194, 200, 207, 208–209, 211–214

Coptaspis, 135, 263
Copulation duration, 141–142, 158
Coulee cricket, 4, 65
Courtship feeding, 9, 11–12, 14–16, 24, 63, 119–160, 168, 171–172, 176, 178–183, 187, 190, 223–225, 227, 231, 233, 235, 242–244, 247–250, 257, 260
  adaptive significance of, 133, 137–139
  diversity of, 130–133, 223–235
  effects on female fitness, 131, 150–153
  as investment in offspring, 159–160, 248–251
  protective device, 158–159
  See also Spermatophylax
Courtship role reversal, 7–9, 17, 160, 222–260
  and food availability, 225–232, 234–247
  and mate quality, 245
  and operational sex ratio, 236–243
  and parasites, 232
  and parental investment, 237–238, 248–251
  and rates of reproduction by the sexes, 237–238, 251
  in risk taking, 210, 211, 216, 234
  and time-out from mating, 237–238, 251
Cowan, F., 2–4, 12, 57, 64, 85
Crista acustica, 104–106, 114
Crypsis. *See* Defense against predators, crypsis and camouflage
Cumming, J., 139
Cyphoderris, 23, 24, 44, 84, 94, 130
Cyrtaspis, 135, 263

Dadour, I., 102, 118, 169, 193
Darwin, C., 7, 15–16, 18, 122, 128, 165, 199, 205, 221–223, 237, 247, 251, 252
Davies, N., 155
Decticita, 132, 263
Decticoides, 63, 263
Decticus, 39, 40, 52, 55, 64, 74, 120–121, 123, 131, 133, 136, 148–149, 161–162, 263. *See also* Wart Biter
Defense against predators, 78–88, 100, 203–206
  aposematism and mimicry, 65–66, 80–81, 85–87
  chemical defenses, 80, 85–88
  crypsis and camouflage, 28, 29, 30, 37, 65, 79–83, 86
  disturbance sounds, 83–85, 99
  protective spines, 40, 80–83
  startle displays, 80–81, 83–88
De Geer, Baron, 90, 116
De Luca, P., 171, 180–181
Diapause, 53–57
Dicranostomus, 256, 263

Diets and food preferences, 4, 59–63, 226–228, 235
Digger wasps. *See* Hymenoptera, Sphecidae
Diptera, 176
  Drosophila, 176, 237
  ears of, 72, 114
  Empididae, 130, 139, 223, 257–258
  ormiine parasitoids (Tachinidae), 70–73, 114, 200–207, 213–214, 219
  Sarcophagidae, 73
  stick ticks (Ceratopogonidae), 75
Diurnal predatory katydids. *See* Meconematinae
Docidocercus, 136, 208, 263
Ducetia, 29, 263
Dugatkin, L., 161, 183

Ears and hearing, 26–27, 72, 102–114, 204, 252–256
  katydid "pinnae," 109–110
  neuroethology of, 102–114
  and sexual selection, 106–107, 252–256
  use in detecting the direction of sound, 109–114
  See also Auditory spiracles and tracheae
Eberhard, W., 16, 130, 137–138, 166–167, 199, 255
Ecology. *See* Katydids, effects on ecological community
Economic importance. *See* Katydids, as pests
Eggs and hatching, 1, 12, 13, 49–55, 57, 162
Eisner, T., 130, 160
Elephantodeta, 31, 108, 112, 195–196, 263
Eluwa, M., 53, 57, 128
Emlen, S., 17, 133, 236
Ensifera, 19–27, 123–124, 129, 140–141
  phylogeny, 20–23, 115, 140
Enyaliopsis, 78, 263
Ephippiger, 39, 57, 65, 100, 117, 123, 125, 133, 135, 172, 173, 175, 233, 244, 263
Ephippigerida, 135, 263
Eremopedes, 40, 263
Eubliastes, 32, 263
Euconocephalus, 56, 57, 263
Eugaster, 88, 133, 264
Eupholidoptera, 60, 136, 264
Euthypoda, 57, 264
Evans, H. E., 5, 53, 68, 90
Evolutionary tree. *See* Phylogeny

Fabre, J. H., 14–15, 40, 50, 53, 60, 95, 116, 119–123, 127, 131, 133, 148–149, 161–162, 199–200, 222, 232
False leaf katydids. *See* Pseudophyllinae

Feaver, M., 66, 74, 76, 88, 116, 159, 161–166, 175, 181, 183, 185, 191–193, 196, 197, 210, 217–218, 220
Feilner, Captain John, 1, 3, 6, 7, 12, 57, 122–123
Female choice. *See* Mate choice
Field cricket. *See* Gryllidae; *Gryllodes*; *Gryllus*
Fishes, 176
  pipefish, 223–224
Fitzpatrick, S., 257
Flies. *See* Diptera
Flightless predatory katydids. *See* Austrosaginae
Flook, P., 19–21, 27
Fluctuating asymmetry, 175, 183, 205
Food. *See* Diets and food preferences
Fossil record, 20, 43–44, 95–96
Fullard, J., 168, 187–188, 226
Fungi, 75
Funk, D., 225, 257–258

Gampsocleis, 43–44, 104, 136, 194, 243, 264
Gangwere, S., 59–63
Gerhardt, H. C., 166, 177, 183, 187, 190
Gerhardt, U., 123, 126, 129, 131, 143
Giant leaf katydids. *See* Phyllophorinae
Gillette, C. P., 6, 12, 122
Gillott, C., 138
Gnathoclita, 255–256, 264, Plate 24
Goreau, M., 92, 99, 107
Gorochov, A. V., 27–30, 44–45, 59, 148
Grasshoppers and locusts. *See* Acrididae; Caelifera
Gravenreuthia, 108, 264
Great-green grasshopper, 40, 51, 52, 60, 75, 92, 116–117, 136, 169, 188–190, 194, 212
  anatomy, 42
  diets and food preferences, 60
  females attracted to calling males, 116–117, 188–190
  song, 18–19, 189
  See also Tettigonia
Greenfield, M., 109, 117, 167–169, 171–175, 180, 190, 194
Gregarine protozoan parasites, 75, 232, 245
Gregarious behavior, 63–67
Ground katydids. *See* Bradyporinae
Growth, 1, 57–59
Gryllacrididae, 22, 26, 50
Gryllidae (Grylloidea), 21–23, 26, 33, 45, 71, 90–92, 109, 114–116, 124, 130, 131, 139, 140, 142, 150–151, 171, 176, 177, 199, 205–206
  caged pets, 33
  song, 114–116, 171, 177–178

Gryllodes, 23, 131, 139, 143, 150–151, 154–156, 158
Gryllotalpidae, 22–23, 26, 124
Gryllus, 21, 71, 142, 166, 175, 205–206
Guilding, Rev. L., 95, 116, 118
Gwynne, D. T., 7–14, 20, 22–23, 27, 28, 44, 47, 54, 56, 64, 79, 94–97, 100, 114–115, 117, 119, 127–129, 131, 133–137, 139–147, 151–156, 158, 160, 168, 171, 172, 176, 178–183, 187–188, 191, 197, 198, 210, 213, 216, 221, 223, 225–232, 234–235, 239–257
Gymnoproctus, 83, 141, 264

Habitats of katydids, 18, 40–41, 49, 228
Haenschiella, 94, 264
Haglidae, 21–25, 27, 47, 50, 84, 94, 105, 114–115, 117–118, 130
  eggs and hatching, 50
  sound production, 44, 84
Hartley, C., 52, 54–57, 64–66, 99, 119, 181, 187, 233
Harvey, P., 142, 144
Hayashi, F., 130–132, 139–140, 143
Hearing. *See* Ears and hearing
Hedrick, A., 192
Helfert, B. V., 80–81
Heller, K.-G., 35, 53, 60, 76, 79, 82, 84, 92, 97, 117, 133, 136, 147, 152–156, 187, 206–207, 209, 212–214, 217–221
Helversen, D. von, 197
Hemiandrus, 74
Hemideina, 21, 25, 256, 257
Hemiptera and Homoptera
  Belostomatidae, 223, 239
  Cicadidae, 114
  as competitors with katydids, 61
  Reduviidae, 85
Hemisaga, 36, 102–103, 264
Hetrodinae, 29, 30, 39, 40, 57, 63, 70, 83, 87–88, 109, 133, 142, 195
Hexacentrus, 35, 57, 60, 112, 264
Holmes, Oliver Wendell, 18, 49, 89, 97
Homotrixa, 71, 73, 200–206
Horatosphaga, 108, 264
Horsehair worms. *See* Nematomorpha
Hoy, R., 105
Hubbell, S., 18
Human culture and katydids, 1–4, 6, 18–19, 29–30, 33, 35, 39–41, 43, 45, 49, 53, 64, 89–90, 92, 96
Hump-winged crickets. *See* Haglidae
Hunt, J., 205
Hybridization between species, 45–47, 183–190
Hymenoptera
  Chalcidae, 75
  Cynipidae, 52

Hymenoptera (*continued*)
  Encyrtidae, 75
  Eulophidae, 75
  Formicidae, 75
  Pompilidae (spider wasps), 80
  Scelionidae, 75
  Sphecidae (digger wasps), 5–6, 68–71, 210–217, 221
  Vespidae, 60

Idiarthron, 208, 264
Ingrisch, S., 33, 49, 52–60, 97, 180
Insara, 207, 264
Ischnomela, 208, 264
Isophya, 52, 124, 131, 264

Kalmring, K., 100, 104–105, 107
Kaltenbach, A., 59, 81, 85
Kathirithamby, J., 75–76
Katydids
  as caged singing pets, 29–30, 33, 35, 37, 39, 43–45, 96
  characters of the family, 27–28
  effects on ecological community, 60, 61, 77, 78–79
  as food for humans, 6, 77
  as hosts for parasites, 69–76, 166
  number of instars, 57–58
  number of species, xi, 18
  origin of name, xi, 18–19, 53
  as pests, 32–33, 61, 63–64, 55, 75, 78
  in poetry, 18, 49, 52, 89
  as predators, 59–61
  as prey, 4–6, 7, 68–80, 166
Kawanaphila, 50, 82, 94–95, 106, 117–118, 131, 136, 149, 152, 154–155, 157, 172, 181, 222–233, 235, 239–245, 247–251, 252–256, 265, Plate 23
  crypsis, 82
  diets and food preferences, 59
  origin of name, 59
  oviposition, 50
Kevan, D. K. McE., 19–20, 28–29, 43, 52, 53, 57, 63–64, 67, 74, 80, 142, 210
Key, K., 66, 86
Kindvall, O., 78–79
Kirby, W., and W. Spence's *Introduction to Entomology*, 19, 90
Koringkreiks. *See* Hetrodinae
Kvarnemo, C., 235, 245

Lacipoda, 108, 265
Lakes-Harlan, R., 72–73, 103–104, 114–116, 207, 214
Latimer, W., 100, 118, 171, 182, 212
Leaf katydids. *See* Phaneropterinae
Lehmann, G., 206–207, 213–214
Lepidoptera, 61, 66, 86

Leptophyes, 31, 51–53, 135, 149, 152, 265
Lewis, D., 110
Lipotactes, 34, 55, 58–59, 265
Lipotactinae, 29–30, 33–34, 55, 58–59
Listroscelidinae, 29–30, 35, 50, 60, 70, 81, 109, 112, 115, 117, 118, 124, 145, 147, 148, 172, 181–183, 207, 232, 235, 239, 243, Plates 5, 19
Lizards, 76, 207
Lloyd, J., 84, 99, 166, 174, 195, 199
Locust. *See* Acrididae; Caelifera
Loher, W., 138
Long-legged katydids. *See* Mecopodinae
Lorch, P., 259
Lymbery, A., 56, 82–83, 86

Macroxiphus, 80–81, 265, Plates 14–17
MacVean, C., 1–5, 75, 86
Mantids. *See* Mantodea
Mantodea, 19–20, 61, 79
Manuweta, 256
Markl, H., 100
Mason, A., 97–98, 110, 226
Mate choice, 8–10, 13, 14, 15–16, 119, 129–130, 160, 164–191, 228–230, 242, 244–245
  for choruses, 168–169
  cryptic choice, 129, 137–138, 167, 178, 198
  cued by diet, 234–235
  cued by mate-encounter rate, 234–235
  and fluctuating asymmetry, 175, 183–184
  and good genes, 16, 171–172, 175–178
  for goods and services, 16, 129–130, 160, 167, 171–172, 174–183, 187
  for large females, 15, 227, 245, 247
  for large males, 9, 171–172, 174–183
  for leading callers, 169–171
  by males, 8–10, 224, 225, 227–235, 242, 244–245, 247
  and mate quality, 16, 171–183, 190–191, 245
  and passive attraction, 167
  and runaway sexual selection, 167, 175
  and sensory exploitation, 167
  for species recognition, 16, 171–172, 183–190
  for young females, 155, 232–233
Mate feeding. *See* Courtship feeding
Mating behavior. *See* Courtship feeding; Mate Choice; Sexual selection; Songs and Sound production
Mating costs and risks of mating, 16, 71–73, 133–136, 199–221
Mating position in katydids, 119–120, 149, 197
Mbata, K. J., 39, 57, 63, 83, 117, 118, 142, 195
McIntyre, M., 273

Meadow katydids. *See* Conocephalini; *Conocephalus*; *Orchelimum*
Meconema, 52, 67, 99, 100, 115, 135, 217, 265
Meconematinae, 29–30, 34–35, 52, 67, 99, 100, 115, 135, 217
Mecopoda, 33, 141, 265
Mecopodinae, 29–30, 32–33, 50, 57, 59, 76, 84, 141
Mecoptera, 130–131, 159
Megaloptera, 131, 132, 139–140
Metaballus, 112, 136, 239, 246, 247, 260, 265
Metaplastes, 197, 265
Metholche, 265, Plate 6
Metrioptera, 40, 43, 52, 54, 67, 74, 78, 117–118, 136, 265
Metrioptera roeselii, 55, 57, 67, 136, 142
  adventive in North America, 67
  diapause, 55
  nematode parasites, 74
  number of instars, 57
  oviposition, 52
  as prey, 67
  wing-length polymorphism, 67
Michelsen, A., 109, 111, 114
Microcentrum, 51, 53, 265
Micro katydids. *See* Microtettigoniinae
Microsporidian protozoans, 75
Microtettigoniinae, 29–30, 39
Migration, 64–67, 181
Mimetica, 106, 110, 265
Mimicry, 79–82, 85. *See also* Defense against predators
Mites. *See* Arachnida, mites (Acarina)
Moffett, T., 90
Møller, A., 175, 178, 198
Molting, 57–58
Montealegre, F. Z., 27, 63, 97, 115
Mormon cricket, 1–14, 28, 53, 57, 58, 60, 64–65, 70, 71, 73–75, 78, 80, 85–87, 112, 119, 122–123, 134, 165, 194, 211, 213, 215–216, 225, 226, 231, 232, 236, 239–241, 244–248, 252, Plates 1, 9
  bands and gregarious behavior, 1–6, 64–65, 85–86
  defense against predators, 85–86
  diet, 4
  as food, 6
  natural enemies, 4–6, 70, 71, 73–75, 78, 211, 213, 215–216
  as a pest, 1–4, 64–65
  sexual behavior and role-reversal in mating, 6–14, 211
Morris, G. K., 17, 44, 47, 74, 87, 93–98, 100–101, 117, 118, 120, 162, 169–171, 177, 180–181, 183–185, 187–188, 190, 192, 194, 196–197, 208, 209, 213
Mygalopsis, 45–47, 56, 73, 81–84, 102–103, 106, 118, 193–194, 211, 214, 265
  defense against predators, 81–84
  hearing and acoustic behavior, 83–84
  as host of a parasitoid fly, 73
  life cycle, 56
  origin of name, 84
  speciation, 45–47
Myopophyllum, 100–101, 265

Naskrecki, P., 18, 28–30, 35
Nastonotus, 27, 265
Natural enemies, 4–6, 67–78
Natural History of Selborne, 92
Neduba, 93
Nematoda (mermithids), 74–75
Nematomorpha, 73–74, 166, 218
Neoconocephalus, 56, 60, 61, 64, 69, 73, 87, 105, 117–118, 133, 140–141, 171, 173–174, 194, 200, 208–209, 213, 266
Nickle, D., 29, 77, 80–82, 86, 99, 119
Nuptial gifts and meals. *See* Courtship feeding

Oecanthus, 130–132, 171, 177–178
Orchelimum, 50, 60, 73–74, 80, 87–88, 117, 159–166, 175, 178–179, 183–187, 192, 194, 196–197, 211, 217–218, 266, Plates 21, 22
  songs, 185
Ormia, 73, 200, 205, 213
Orophus, 57, 266
Orthoptera, 14, 19–27
  phylogeny, 21–22
Otte, D., 18, 20, 114, 116, 118, 130, 173
Oviposition. *See* Eggs and hatching
Owens, I., 251, 257
Oxyaspis, 112, 266
Oxycous, 112, 266

Pachysaga, 36, 73, 113, 117, 266
Palmodes, 5–6, 7, 68, 70, 213, 215–216
Panacanthus, 83–84, 266
Panoploscleis, 99, 259, 266
Pantecphylus, 84, 266
Parasites of katydids. *See* Natural enemies
Parental investment, 17, 128, 129, 143–160, 237–238, 242, 248–250
Parker, G. A., 137, 167, 198, 236, 238, 242, 251
Parthenogenesis, 35, 141
Peranabrus. *See* Coulee cricket
Phaneroptera, 52, 53, 54, 73, 135, 267, Plate 3
Phaneropterinae, 29–31, 51–55, 57–59, 61–62, 65–66, 69–70, 72–73, 79–82, 84–85, 93, 99, 109, 112, 117, 124, 131, 132, 136, 140, 141, 145, 147, 149–158, 172, 175, 195, 197, 213, Plates 2, 3, 8, 18, 20
Phasmatodea (Phasmida), 66, 79, 86

Phasmodes, 34, 136, 267
Phasmodinae, 29–30, 33–34, 79, 115, 140–142
Phisis, 106, 108, 267
Pholidoptera, 52, 117–118, 208, 267, Plate 7
Phonotaxis by females to the male's call, 116–119, 168–191
Phyllophora, 99
Phyllophorinae, 29–32, 99, 109, 115
Phylogeny
  of Ensifera, 20–23, 115, 140
  of orthopteroid insects, 19–21
  of Tettigoniidae, 27–30
Platycleis, 52, 54, 64, 75, 136, 267
Poecilimon, 53, 72–73, 117, 131, 135, 149–158, 178, 206–207, 213–215, 217–220, 267
Polichne, 53, 135, 267
Pollen and nectar katydids. *See* Zaprochilinae
Polyancistrus, 75, 267
Polysarcus, 105, 107, 136, 267
Population regulation of katydids, 78–79
Powell, John Wesley, 6
Predators of katydids. *See* Natural enemies
Predatory katydids. *See* Saginae
Pristonotus, 136, 267
Proctor, H., 124, 167
Promeca, Plates 10, 11
Prophalangopsidae. *See* Haglidae
Protohermes, 125
Pseudophyllinae, 29–30, 52, 57, 59–60, 63, 66, 70, 75, 77, 79–82, 84–85, 97, 100–101, 109, 112, 115, 124, 131, 140, 141, 207–210, 212, 255–259, Plates 4, 10, 13, 24
Pseudosubria, 49, 267
Pseudotettigonia, 95–96, 267
Pterochroza, 85, 267
Pterophylla, 18, 19, 21, 30, 52, 57, 63, 66, 131, 141, 190, 268. *See also* True katydid

Ragge, D., 20, 22, 57, 114, 116, 212
Receptivity in females, loss after mating (refractory period), 127–128, 135, 137, 144–145, 241, 251
Reinhold, K., 48, 57, 131, 152, 177
Rentz, D. C., 19, 20, 22, 28–30, 34–35, 37, 40, 44, 52, 59, 60, 63, 85, 94, 115, 119, 132–133, 141, 200, 209, 256
Reproductive isolation, behavioral, 56. *See also* Mate Choice, for species recognition
Requena, 35, 111, 117, 124, 127–128, 135, 148–158, 172, 176, 181–183, 207, 232, 235, 239–240, 243, 250, 254, 268
Rhachidorus, 41, 268

Rhaphidophoridae, 21–23, 26, 104
Rheinlaender, J., 103–104
Riley, C. V., 18, 49, 51–53, 75, 89, 96
Ritchie, M., 172, 175–176, 244
Robert, D., 71, 114
Robinson, D., 65, 119
Robinson, M., 57, 82
Römer, H., 102–104, 106, 182, 194, 253–254
Ruspolia, 37, 38, 52, 54, 56, 64, 66, 77, 117, 124, 135, 141, 158
Rust, J., 95–96
Ryan, M., 167, 187

Saga, 52, 55, 57, 137, 141
Sagebrush crickets. *See Cyphoderris*
Saginae, 29–30, 35–36, 52, 55, 57, 81, 85, 137
Sakaluk, S., 126, 131, 137, 139, 141, 143, 147, 149–158, 199–200, 205, 243, 250
Schatral, A., 100, 118, 169, 176, 182, 191, 232
Schizodactylidae, 21–22
Schul, J., 171, 188–190
Sciarasaga, 71, 73, 175, 200–206, 214–215, 268
Scopiorinus, 208, 268
Scorpionfly. *See* Mecoptera
Scudderia, 53, 61, 74, 171, 177, 268, Plate 8
Segestes, 75, 78, 268
Segestidea, 57, 75, 76, 78, 268
Sex-determination and sex-determining mechanisms, 177
Sex role reversal. *See* Courtship role reversal
Sexual conflict, 126, 129, 133, 137–139, 241
Sexual dimorphism, 37–38, 106–107, 112, 251–260
Sexual selection, 15–17, 245–246
  and the evolution of courtship gifts, 126–160
  factors controlling, 222–251
  and genitalia, 255–256
  for large body size, 174–183, 252, 256
  and mate quality, 245
  measuring, 237, 245–248
  and sex difference in risk taking, 7, 199–221
  and speciation, 47–48
  and weaponry, 255–257
Shapiro, L., 45, 183–187, 207, 211, 218
Shaw, K., 172–173, 175, 191, 194
Shelly, T., 235
Shield-backed katydids. *See* Tettigoniinae
Signaling using substrate vibrations, 100–102, 168, 180–181, 191
  as a cue for female choice, 180–181, 191
  drumming, 100–101
  as a response to bat predation, 209
  tremulation, 100–101, 168, 180–181, 209

Simmons, L. W., 63, 118, 131, 136, 137, 139, 142, 154–156, 158, 178, 181, 183–184, 198, 223, 225–227, 230–233, 235–236, 238–251, 253
Smith, R., 126, 225
Snedden, A., 171–174
Snodgrass, R., 4, 64, 122, 137
Songs and sound production, 7, 8, 13, 15, 71, 83–85, 89–99, 102–103, 107, 114–119, 163–197, 203–215, 226–230, 233, 234, 239, 240, 254–255
 and attraction of natural enemies, 16–17, 203–215
 and competition for mates, 102–103, 117–118, 191–197
 for courtship, 118
 energetic costs of, 207
 evolutionary origins of, 114–116
 by female katydids, 97, 99, 118–119, 195–196, 258–260
 and male coyness, 209–210, 229–230, 233–234, 241
 species specificity, 183–190
 synchronous singing, 168–174
 variation in duty cycle, 207–210, 229–230, 233–234
Speciation, 45–48
Species diversity, 43–48, 53, 184–186
Species recognition. See Mate choice, for species recognition
Spelaeala, 68, 268
Spermatadesm, 126
Spermatheca, 11, 123–127, 129, 137
Spermatodose, 42, 126
Spermatophore, 9–12, 31, 32–34, 38, 41, 150, 151
 diversity among animals, 123–124
 See also Spermatophylax
Spermatophylax, 9–12, 14, 15, 31, 32–34, 38, 41, 63, 119–161, 168, 171–172, 176, 178–183, 187, 190, 203, 206, 209, 223, 225, 227, 233, 235, 243, 247–248
 adaptively tailoring to mate quality and mating frequency, 155–156, 243–344
 cost to males, 133–136, 164, 203, 206, 209
 function of, 122–160
 as investment in offspring, 128, 129, 143–159
 origin and evolution, 139–148
 as a protective device, 125–129, 143–159
 protein content, 135–136, 145–148, 183–184
 variation in size, 134–136
Sperm competition and confidence of paternity, 126, 129, 137–138, 140–160
Sperm transfer, 11, 123–127
Sphex wasps, 67–70, 210, 213, 215

Sphyrometopa, 173, 268
Spiders. See Arachnida, Araneae (spiders)
Spiny predatory katydids. See Listroscelidinae
Spooner, J., 94, 119, 191
Steiroxys, 74, 268
Stenopelmatidae, 21–23, 25, 27, 50, 58–59, 74, 105, 114–115, 256–257
Stetharasa, 268, Plate 4
Stick insects. See Phasmatodea (Phasmida)
Stick katydids. See Phasmodinae
Stictophaula, 52, 269
Stilpnochlora, 124, 269
Strepsiptera, 75–76, 269
Stridulation mechanisms, 83–85, 89–99
Suga, N., 104–105
Sword, G., 66

Tachytes wasps, 70, 213, 218, 225
Tamarins, 77
Taxonomy, 14, 21–22, 29–30
Tettigonia, 40–42, 51–52, 55, 60, 74–75, 78, 92, 116–118, 136, 169, 171, 188–190, 194, 225, 269. See also Great-green grasshopper
Tettigoniinae, 29–30, 39–44, 50–55, 57–58, 60–61, 63–67, 69, 70, 78, 84, 93, 108, 109, 112, 115, 117–118, 136, 140, 145–147, 194, 225, 233, Plates 1, 7, 9
Therobia, 72–73, 206, 214–215
Thliboscelus, 165, 269
Thornhill, R., 128, 130–131, 145, 159, 167, 175, 178, 182–183, 199, 221, 238, 252
Tinkham, E., 18, 40
Tinzeda, 136, 269
Torbia, 269, Plate 2
Tree cricket. See Oecanthus
Tremulation. See Signaling using substrate vibrations
Trivers, R. L., 7, 15, 17, 128, 133, 200, 238, 249
True katydid, xi, 18, 19, 21, 30, 52, 89, 97, 208
Tuckerman, J., 171, 177
Tympana. See Ears and hearing
Tympanophora, 37, 38, 140, 269
Tympanophorinae, 29–30, 35, 37–38, 50, 59, 140
Typophyllum, 269, Plates 12, 13

Ultrasonic communication, 96–97, 104–105, 226–227
Uromenus, 135, 148–158, 207, 233, 269
Uvarovites, 43, 269

Vahed, K., 129, 132, 133, 136, 141, 143–147, 149–159
Veria, 39, 269

Vestria, 86, 269
Vetralla, 52, 269
Vibration signals. *See* Signaling using substrate vibrations
Vickery, V., 40, 67, 115, 210
Vincent, A., 251

Wagner, W., 205
Wakeland, C., 2, 4, 6, 64
Walker, T. J., 93, 173, 213
Wallace, A. R., 79
Wart biter, 39, 52, 78, 108, 131, 136, 148–149, 154, 157–158
  origin of name, 39
  See also Decticus
Watson, P., 177
Wedell, N., 126–129, 131, 136–139, 141, 143–148, 151–158, 176, 181–182, 198, 211
Weta. *See* Stenopelmatidae

Wheeler, W. M., 49, 53–54
White-faced decticus. *See Decticus*
White, Gilbert, 92
Wickler, W., 57, 80, 151
Williams, G. C., 222, 238
Wing polymorphism and wing length, 66–67
  and mate choice, 181

Xestoptera, 208, 209, 212, 269

Yersinella, 136, 269

Zabalius, 53, 57
Zaprochilinae, 28–29, 33, 50, 59, 79, 82, 94–95, 117–118, 136, 140, 149, 172, 223, 226–232
Zeuneria, 108, 269
Zoraptera, 130
Zuk, M., 176, 199, 203, 207